Osiris

A RESEARCH JOURNAL DEVOTED
TO THE HISTORY OF SCIENCE
AND ITS CULTURAL INFLUENCES

EDITORIAL OFFICE
UNIVERSITY OF PENNSYLVANIA
3401 WALNUT STREET, SUITE 460B
PHILADELPHIA, PENNSYLVANIA, 19104-6228, USA

SUGGESTIONS FOR CONTRIBUTORS TO OSIRIS

OSIRIS is now devoted to thematic issues, conceived and compiled by guest editors.

1. Manuscripts should be **typewritten** or processed on a **letter-quality** printer and **double-spaced** throughout, including quotations and notes, on paper of standard size or weight. Margins should be wider than usual to allow space for instructions to the typesetter. The right-hand margin should be left ragged (not justified) to maintain even spacing and readability.

2. Bibliographic information should be given in **footnotes** (not parenthetically in the text), typed separately from the main body of the manuscript, **double-** or even **triple-spaced,** numbered consecutively throughout the article, and keyed to reference numbers typed above the line in the text.

a. References to **books** should include author's full name; complete title of the book, underlined (italics); place of publication and publisher's name for books published after 1900; date of publication, including the original date when a reprint is being cited; page numbers cited. *Example:*

[1]Joseph Needham, *Science and Civilisation in China,* 5 vols., Vol. I: *Introductory Orientations* (Cambridge: Cambridge Univ. Press, 1954), p. 7.

b. References to articles in **periodicals** should include author's name; title of article, in quotes; title of periodical, underlined; year; volume number, Arabic and underlined; number of issue if pagination requires it; page numbers of article; number of particular page cited. Journal titles are spelled out in full on first citation and abbreviated subsequently. *Example:*

[2]John C. Greene, "Reflections on the Progress of Darwin Studies," *Journal of the History of Biology,* 1975, *8*:243–272, on p. 270; and Dov Ospovat, "God and Natural Selection: The Darwinian Idea of Design," *J. Hist. Biol.,* 1980, *13*:169–174, on p. 171.

c. When first citing a reference, please give the title in full. For succeeding citations, please use an abbreviated version of the title with the author's last name. *Example:*

[3]Greene, "Reflections" (cit. n. 2), p. 250.

3. Please mark clearly for the typesetter all unusual alphabets, special characters, mathematics, and chemical formulae, and include all diacritical marks.

4. A small number of **figures** may be used to illustrate an article. Line drawings should be directly reproducible; glossy prints should be furnished for all halftone illustrations.

5. Manuscripts should be submitted to OSIRIS with the understanding that upon publication **copyright** will be transferred to the History of Science Society. That understanding precludes OSIRIS from considering material that has been submitted or accepted for publication elsewhere.

OSIRIS (SSN 0369-7827) is published once a year.

Subscriptions are $39 (hardcover) and $25 (paperback).

Address subscriptions, single issue orders, claims for missing issues, and advertising inquiries to *Osiris,* The University of Chicago Press, Journals Division, P.O. Box 37005, Chicago, Illinois 60637.

Postmaster: Send address changes to *Osiris,* The University of Chicago Press, Journals Division, P.O. Box 37005, Chicago, Illinois 60637.

Osiris is indexed in major scientific and historical indexing services, including *Biological Abstracts, Current Contexts, Historical Abstracts,* and *America: History and Life.*

Hardcover edition, ISBN 0-226-84883-3
Paperback edition, ISBN 0-226-84884-1

Instruments

Edited by
Albert Van Helden and Thomas L. Hankins

Osiris

A RESEARCH JOURNAL
DEVOTED TO THE HISTORY OF SCIENCE
AND ITS CULTURAL INFLUENCES
SECOND SERIES VOLUME 9 1994

OSIRIS 1993 SECOND SERIES VOLUME 9

INSTRUMENTS

Cover: Observational instruments in the seventeenth and twentieth centuries: NASA's wide field–planetary camera and Johannes Hevelius with his telescope (inset). See pages 102, 8.

Introduction: Instruments in the History of Science

By Albert Van Helden and Thomas L. Hankins

IN THE 1943 VOLUME of *Isis* there appeared a two-page article entitled "Traduttore-Traditore: A Propos de Copernic et de Galilée," written by Alexandre Koyré. Besides the normal problem of accuracy in translating scientific texts from another era, Koyré argued that there is the additional danger of "involuntarily substituting *our* conceptions and *our* habits of thought for those, entirely different, of the author." Among the examples he cited was a passage at the beginning of the third day of Galileo's *Dialogues concerning Two New Sciences,* translated by Henry Crew and Alfonso de Salvio, in which Galileo's *comperio* was rendered in English as "I have discovered by experiment." Koyré argued that the word "experiment" was simply added by the translator who, "obviously committed to empiricist epistemology, could not imagine that one could demonstrate or discover something other than *by experiment.*" It was no wonder, according to Koyré, that the legend of Galileo the empiricist and experimenter was so firmly established in America, for even the best American historians cited the translation by Crew and de Salvio.[1]

Koyré set out to change all that. In his *Etudes galiléennes* of 1939 he drew a different portrait of Galileo, a Platonist, who found things in his head and used instruments only by way of illustration. In the same year that his little article appeared in *Isis,* there appeared in *The Philosophical Review* his article "Galileo and the Scientific Revolution of the Seventeenth Century," in which he made the following statement:

> Good physics is made *a priori.* Theory precedes fact. Experience is useless because before any experience we are already in possession of the knowledge we are seeking for. Fundamental laws of motion (and of rest), laws that determine the spatio-temporal behaviour of material bodies, are laws of a mathematical nature. Of the same nature as those which govern relations and laws of figures and numbers. We find and discover them not in Nature, but in ourselves, in our mind, in our memory, as Plato long ago has taught us.[2]

For Koyré a scientific instrument served only to illustrate a conclusion reached first by logical reasoning. He asserted that the experiments with inclined planes that

[1] Alexandre Koyré, "Traduttore-Traditore: A Propos de Copernic et de Galilée," *Isis,* 1943, *34:*209–210.

[2] Alexandre Koyré, "Galileo and the Scientific Revolution of the Seventeenth Century," *The Philosophical Review,* 1943, *52:*333–348; repr. in Koyré, *Metaphysics and Measurement: Essays in Scientific Revolution,* ed. M. A. Hoskin (London: Chapman & Hall; Cambridge, Mass.: Harvard Univ. Press, 1968), pp. 1–15, quoting from p. 13.

Galileo described in his *Discourses* were merely rhetorical devices. In an article published in 1953, Koyré quoted Galileo's description of his instruments and experimental procedure and then commented:

> A bronze ball rolling in a "smooth and polished" wooden groove! A vessel of water with a small hole through which it runs out and which one collects in a small glass in order to weigh it afterwards and thus measure the times of descent (the Roman water-clock, that of Ctesibus, had been already a much better instrument): what an accumulation of sources of error and inexactitude.
>
> It is obvious that the Galilean experiments are completely worthless: the very perfection of their results is a rigorous proof of their incorrection.[3]

For the empiricist epistemology of Crew and de Salvio, Koyré substituted an extreme idealist epistemology that became very influential in the history of science. In 1965 Norwood Russell Hanson wrote in an important article in *Science:* "Centuries of scholarship to the contrary notwithstanding, Galileo was not a great experimental scientist. He was no experimental scientist at all, not as we would know one."[4] In this approach, the history of science was the history of theory. Experiment and measurement took place after the fact and were not of prime interest; instruments were considered at best "reified theories," in Bachelard's term.[5] The debate was in large part ideological and focused more on acceptable method than on historical accuracy. Whether Galileo actually built his instruments or only imagined them was to be determined from his philosophical goals. Because he obviously *did* build telescopes and proportional compasses, one would think that he might equally well have built inclined planes, but because, from Koyré's point of view, they served no purpose, it seemed clear to Koyré that Galileo never built them.[6]

Such an extreme idealist position invited counterargument. In 1961 Thomas B. Settle, then a graduate student at Cornell University, published the results of his attempt to repeat as faithfully as possible the Galilean experiment criticized by Koyré. Settle showed that with the tools available to him Galileo could easily have achieved the accuracy he claimed in his *Discourses on Two New Sciences.* Settle's dissertation, finished in 1966, gave a more exhaustive account of Galileo's experiments and, though never published, became the *locus classicus* of the study of Galileo's experiments.[7]

Over the past several decades, Settle's pioneering study has been followed up by several scholars, including Stillman Drake and Ronald Naylor, and we may now

[3] Alexandre Koyré, "An Experiment in Measurement," *Proceedings of the American Philosophical Society,* 1953, *97:* 222–237; repr. in Koyré, *Metaphysics and Measurement,* pp. 89–117, quoting from p. 94.

[4] Norwood Russell Hanson, "Galileo's Discoveries in Dynamics," *Science,* 1965, *147:*471.

[5] Gaston Bachelard, *Les intuitions atomistiques* (Paris: Boivin, 1933), p. 140. See also Bachelard, *L'activité rationaliste de la physique contemporaine* (Paris: Presses Universitaires de France, 1951).

[6] In *From the Closed World to the Infinite Universe* (Baltimore: Johns Hopkins Press, 1957), Koyré devotes part of a chapter to Galileo's discoveries with the telescope, but he says little about the instrument besides calling it "the first scientific instrument" (p. 90). Further, according to Koyré, the instrument had little bearing on the question. In *The Astronomical Revolution: Copernicus, Kepler, Borelli,* trans. R. E. W. Maddison (Ithaca, N.Y.: Cornell Univ. Press, 1973), Koyré mentions the instrument only in passing in connection with Giovanni Borelli's work on Jupiter's satellites (p. 468).

[7] Thomas B. Settle, "An Experiment in the History of Science," *Science,* 1961, *133:*19–23; and Settle, "Galilean Science: Essays in the Mechanics and Dynamics of the *Discorsi*" (Ph.D. diss., Cornell Univ., 1966).

safely conclude that Galileo's instruments were not abstractions but actual apparatus that both suggested and tested theory.[8] If Galileo's argument in his *Discourses* was mathematical, that was because mathematical arguments could claim a certainty that purely qualitative arguments could not; it was not because mathematics made instruments unnecessary.

Even with Settle's correction of Koyré's idealism, most historians of science continued to regard instruments as unproblematic. They assumed that instruments were employed then as they are now, and that their interest for the historian was largely antiquarian. Key to this assumption was the notion that scientific principles resided in theory and perhaps in experimental method, but not in instruments. Instruments helped to quantify concepts, but did not contain or initiate them. More recently, however, historians have recognized that the role of instruments in experimental science has been much more complex. In the seventeenth century it was unclear how instruments like the telescope, microscope, and air pump should be used to obtain natural knowledge. Less problematical, but still unusual, was apparatus like the inclined plane and the pendulum. In either case, the instruments were new and there was no established convention for using them or for validating their results.[9]

Albert Van Helden has argued elsewhere that the modern scientific instrument was born during the period between 1550 and 1700.[10] Earlier devices that we would recognize as scientific instruments were designed for measurement (if we except musical instruments, surgical instruments, and the like), while the new instruments of the Scientific Revolution seldom measured anything, at least not at first. Also the new instruments came from the tradition of natural magic, where the emphasis was on deception and entertainment, not on careful and dispassionate observation. Giambattista della Porta gets close to the ideas of the barometer and the telescope in his *Magia naturalis,* but then uses air pressure to produce a "vessel wherewith as you drink, the liquor shall be sprinkled about your face" and lenses and mirrors to perform "other merry sports."[11] It is not surprising, then, that instruments were not always accepted enthusiastically and that their value and their trustworthiness had to be demonstrated. Instruments like the telescope, which we accept as quintessentially "scientific," were suspect when they first appeared.

Deborah Warner, one of the contributors to this volume, has warned us recently that it is dangerous even to talk about "scientific instruments" in the seventeenth century, because that term did not become common until the nineteenth. In England after 1650 such instruments were called "philosophical" in contrast to the older

[8] See e.g., Stillman Drake, "Renaissance Music and Experimental Science," *Journal for the History of Ideas,* 1970, *31:*483–500; Drake, "The Role of Music in Galileo's Experiments," *Scientific American,* June 1975, *232:*98–104; Drake, *Galileo at Work: His Scientific Biography* (Chicago: Univ. Chicago Press, 1978); R. H. Naylor, "Galileo and the Problem of Free Fall," *British Journal for the History of Science,* 1974, *7:*105–134; and Naylor, "Galileo: Real Experiment and Didactic Demonstration," *Isis,* 1976, *67:*398–419. See also James MacLachlan, "A Test of an 'Imaginary' Experiment of Galileo," *Isis,* 1973, *64:*374–379.

[9] The most notable exploration into the status of instruments in the seventeenth century is Steven Shapin and Simon Schaffer, *Leviathan and the Air-Pump: Hobbes, Boyle, and the Experimental Life* (Princeton: Princeton Univ. Press, 1985).

[10] Albert Van Helden, "The Birth of the Modern Scientific Instrument, 1550–1700," in *The Uses of Science in the Age of Newton,* ed. John G. Burke (Berkeley/Los Angeles: Univ. California Press, 1983), pp. 49–84.

[11] Giambattista della Porta, *Natural Magick* (trans. of 2nd ed., 1589), ed. Derek J. Price (New York: Basic Books, 1957), pp. 357, 393.

measuring instruments, which were called "mathematical." Nehemiah Grew's 1681 catalogue of instruments belonging to the Royal Society distinguished between "instruments relating to natural philosophy" and "things relating to the Mathematicks," and the distinction was picked up by the instrument makers in the eighteenth century.[12] Thus the process of determining what was acceptable practice in natural philosophy also required a decision about what were acceptable instruments. If we agree that an important part of the Scientific Revolution was the creation of an experimental method, then the creation of conventions for the proper use of instruments—that is, deciding what kinds of instruments should be admitted into natural philosophy and what constituted their proper use—was crucial to the entire scientific enterprise. The role of instruments has changed, of course, as science has changed since the seventeenth century, both in its methods and in its social organization. By studying instruments we can better understand how the changes have taken place.

In raising instruments from the subordinate position where Koyré placed them to a more elevated status, we assume that they are not merely tools for testing theory or exploring ideas. Because instruments determine what can be done, they also determine to some extent what can be thought. Often the instrument provides a possibility; it is an initiator of investigation. The scientist asks not only: "I have an idea. How can I build an instrument that will confirm it?" but also: "I have a new instrument. What will it allow me to do? What question can I now ask that it was pointless to ask before?"

There is, of course, an ambiguity in the word *instrument.* Francis Bacon, in his second aphorism wrote: "Neither the naked hand nor the understanding left to itself can effect much. It is by instruments and helps that the work is done, which are as much wanted for the understanding as for the hand. And as the instruments of the hand either give motion or guide it, so the instruments of the mind supply either suggestions for the understanding or cautions."[13] Bacon considered the instrument to be both a method and a tool. The word has always had a broad meaning, and if it had any special meaning in classical Latin it was in law, not in natural philosophy.[14]

The word also has multiple meanings within natural philosophy. Scientific instruments can be the means for producing natural wonders for the edification of man, as in natural magic; they can be models or analogies to nature, as in the case of orreries or ether models; they can be extensions of the senses, such as the telescope and microscope; they can be measuring devices, as in the case of meters, micrometers, or gauges; they can be the means for creating extreme conditions that do not occur naturally on the earth, as in the case of the air pump and the particle accelerator; they can be apparatus for controlling and analyzing phenomena, as in the case of the pendulum or chemical apparatus; and they can be the means of visual or graphic display, as in the case of recording devices. Undoubtedly many more kinds

[12] Deborah Jean Warner, "What Is a Scientific Instrument, When Did It Become One, and Why?" *Brit. J. Hist. Sci.,* 1990, *23:*83–93. Warner, who is curator of scientific instruments at the Smithsonian Institution, claims not to know what a scientific instrument is. It is a practical problem. What should she collect?

[13] Francis Bacon, *Novum Organum,* aphorism II, in *Francis Bacon: A Selection of His Works,* ed. Sidney Warhaft (Toronto: Macmillan, 1965), p. 331.

[14] *Oxford Latin Dictionary,* ed. P. G.W. Glare (Oxford: Clarendon Press, 1968–1982), s.v. *instrumentum.* As we would expect, the *Oxford English Dictionary* gives special meanings for instruments of law and music.

of scientific instruments could be added to this list, but any attempt to be exhaustive would not do justice to the inventiveness of those who have used instruments to investigate nature. Sometimes ambiguity is a virtue, and until we have a better understanding of the role of instruments in natural science, we are better off leaving to the term "scientific instrument" its traditional vagueness. Perhaps it is best to say that instruments are the technology of science, a technology that has expanded greatly since the seventeenth century.

In addition to the variety in the *kinds* of scientific instruments there is also variety in their *use*. Instruments serve different purposes, and in this volume we explore four different ways in which they are used.

In the first place, *instruments confer authority*. This would be obvious if the instrument provided a clear-cut way of determining who is right in a scientific dispute, but frequently a scientist will claim more authority than the instrument reasonably provides. Albert Van Helden describes the authority claimed by competing telescope makers in the seventeenth century, and Jan Golinski shows how the English chemists resisted Lavoisier's claim to authority based on his superior instruments. Bruce Hunt describes the pressure coming from engineers, not scientists, for an authoritative standard by which to measure resistance, leading to the units of electricity that we use today. While it has been customary to say that the modern scientific community is a republic where ideas circulate freely without resort to authority, the truth appears to be otherwise.

Second, *instruments are created for audiences*. These audiences consist not only of the scientists who use the instruments, but also the patrons who pay for them. Bruce Hevly shows how patronage (or "grant support" in modern parlance) shapes the design and intended use of instruments. Robert Smith and Joseph Tartarewicz show how an extremely complex instrument like the Hubble Space Telescope reflects an equally complex network of conflicting interests and ambitions among its builders and supporters. Deborah Warner, on the other hand, uses the case of magnetic instruments to show how two traditionally *different* audiences—natural philosophers and mathematical practitioners—in fact used similar instruments for similar purposes.

Third, *instruments can act as bridges between natural science and popular culture*. Instruments provide metaphors for writers and poets, they have an important pedagogical role in illustrating and confirming theory, and they define for the public what is acceptable science. Thomas Hankins studies the position of the ocular harpsichord on the margin between science and art and shows how different approaches to nature suggest different instruments. Simon Schaffer shows how a pedagogical instrument like Atwood's machine was not used as intended or as we would have expected. Instead it served the social and political purposes of the Cambridge scientific establishment. And Thatcher Deane argues that the astronomical instruments of Ming dynasty China were more useful as symbols of the emperor's heavenly mandate than as tools for observation.

And fourth, even in their traditional site, the laboratory, *the role of instruments changes when they are used to study living organisms*. Robert Frank shows that in neurophysiology there is a very close connection between the instrument and the organism under investigation such that the instrument almost becomes an extension of the organism. And Timothy Lenoir shows how in carrying out research on the perception of color and tone, Hermann Helmholtz borrowed from the telegraph industry, not only its detectors and recording instruments, but also the general properties of a system of communication.

With the exception of Thatcher Deane's chapter on Ming dynasty astronomy, all of the chapters in this book are on Western science since 1600, which alerts us to the likelihood that the instruments we study are highly culture bound. Koyré's denigration of instruments and the empiricists' praise of them are reflections of the constant and sometimes hostile confrontation between theory and experiment characteristic of Western scientific culture. Looking at instruments in a non-Western society teaches us that their use and intended purpose is not obvious, and warns us, by reflection, that the role of instruments in Western science is sure to be even more complex than it was in China. The important question to ask is not whether Koyré was right or wrong about the importance of instruments. Subsequent historians and philosophers have proved him wrong. What we need to ask is, rather, *how* instruments have worked to determine and, perhaps, even to define the methods and content of science.

The philosophical debate over whether theory drives experiment or experiment drives theory has tended to obscure the independent role of instruments in science. Instruments come and go, but not necessarily in phase with the vicissitudes of experiment and theory. The traditional mix of experiment and theory needs a new ingredient—instruments. It is not just a matter of getting the quantities right; we need an entirely new recipe.[15] The following essays are intended as a step in that direction.

[15] Peter Galison, "History, Philosophy, and the Central Metaphor," *Science in Context,* 1988, *2:*197–212.

INSTRUMENTS & AUTHORITY

Figure 1. *Johannes Hevelius with one of his telescopes. From Hevelius,* Selenographia, sive Lunae descriptio *(Gdansk, 1647), pp. 42–43.*

Telescopes and Authority from Galileo to Cassini

*By Albert Van Helden**

WHEN IT WAS UNVEILED TO THE WORLD in the dramatic discoveries of Galileo, the telescope was an entirely new sort of instrument. Mathematical instruments, such as the dioptra and the astrolabe—and yes, even Galileo's proportional compass—were part of a long tradition, certified by usage and custom in the restricted realm of the mathematical subjects. The dioptra was based on the well-known principles of Euclidean and Ptolemaic optics; the astrolabe embodied the technique of mathematical projection of a sphere on a plane; and the proportional compass was based on the theory of proportions going back to Eudoxus. The telescope, however, was a different sort of instrument. The optical principles involved were not understood, and the results it produced cut across traditional disciplinary boundaries because they affected cosmology. Why and how were its results accepted, and what strategies did its users employ to legitimate their discoveries?

The problem was that although the observatory could be made into a public space, the actual telescopic observation remained, with some exceptions, a private act. Seventeenth-century practitioners employed a mix of strategies for convincing their public of the truth, or at least trustworthiness of their observations. These included demonstration when possible; witnessing; virtual witnessing by means of pictorial representations; and, most important, appeals to the superiority of one's telescopes. Although they were on the whole successful in this endeavor, no completely satisfactory method for assuring the reliability of an observation was found until the twentieth century, with the advent of space astronomy.[1]

I. PERSONAL AUTHORITY OF THE OBSERVER

Although there were perhaps some isolated antecedents in the Western tradition, the observatory emerged as a public space between the end of the sixteenth and the middle of the eighteenth century. Beginning in the 1570s, Tycho Brahe revolution-

* Department of History, Rice University, Houston, Texas 77251–1892.

[1] The problem of authority was in this case exacerbated by the lack at the beginning of the seventeenth century of an established visual language with which to communicate one's observations. See Mary G. Winkler and Albert Van Helden, "Representing the Heavens: Galileo and Visual Astronomy," *Isis,* 1992, *83:*195–217; and Winkler and Van Helden, "Johannes Hevelius and the Visual Language of Astronomy," in *Renaissance and Revolution: Humanists, Scholars, Craftsmen and Natural Philosophers in Early Modern Europe,* ed. J. V. Field and Frank A. J. L. James (Cambridge: Cambridge Univ. Press, 1993), pp. 95–114. See also Steven Shapin and Simon Schaffer, *Leviathan and the Air-Pump: Hobbes, Boyle, and the Experimental Life* (Princeton: Princeton Univ. Press, 1985).

ized the construction of naked-eye measuring instruments, exhausting the limits that the technology of his day placed on their construction. The new devices, constructed under his supervision, were installed in his observatory, and before they were put into use they were carefully calibrated. The measurements were made by a staff of assistants, often under the supervision of one of Tycho's trusted senior adjutants and sometimes of Tycho himself, and the results were entered in notebooks that Tycho regarded as his personal property.[2] Although specific measurements were published in support of various arguments, Tycho never shared the complete record of his measurements with the world at large. Only slowly, in the course of the seventeenth century, did his actual observations become available to European astronomers.[3] The reasonably prompt publication of complete series of observations began only in the seventeenth and did not become routine until the nineteenth century.

What Tycho did share with the world was a full description of his instruments and observatory, complete with pictures, in his *Astronomiae instauratae mechanica* (1598).[4] Those who had never seen his observatory and instruments could, therefore, nonetheless see the marvelous instruments that were central to the restoration of astronomy and the accuracy of which certified the whole endeavor. Tycho's authority, then, rested squarely on his instruments, and this became the model for authority in telescopic astronomy in the seventeenth century.

Whereas the principles underlying Tycho's instruments were well understood, those that governed the operation of the telescope were, at least initially, a mystery, and the discoveries made with it were sufficiently unprecedented and important to throw doubt on the instrument itself. Beginning in March 1610, when *Sidereus nuncius* was published, Galileo led a vigorous campaign for the acceptance of his discoveries and the instrument as a new tool with which to observe the heavens. Central to this campaign was the question of the reality of the "Medicean Stars," the four satellites of Jupiter.

In *Sidereus nuncius* Galileo presented all his observations of Jupiter's moons, sixty-four in all, made between 7 January and 2 March 1610. He did so not only in words, but also in diagrams.[5] One could cast one's eye over the sequence of diagrams and see the motions of the moons. Obvious as it may appear to us in retrospect, presenting a series of actual observations by means of a sequence of diagrams was new. We find antecedents only in the presentation of the paths of comets.[6] Galileo's shrewd approach was convincing to some, notably Johannes Kepler, but others needed much more to be convinced. It is in Galileo's efforts to convince others and in Kepler's efforts to verify and certify the discovery that we first see the problems of authority and verification clearly outlined.

[2] Victor E. Thoren, *The Lord of Uraniborg: A Biography of Tycho Brahe* (Cambridge/New York: Cambridge Univ. Press, 1990), passim.

[3] J. L. E. Dreyer, *Tycho Brahe: A Picture of Scientific Life and Work in the Sixteenth Century* (London, 1890) (New York: Dover, 1963; Gloucester, Mass: Peter Smith, 1977), pp. 370–375.

[4] See *Tycho Brahe's Description of His Instruments and Scientific Work as Given in* Astronomiae instauratae mechanica, trans. and ed. Hans Ræder, Elis Strömgren, and Bengt Strömgren (Copenhagen: Munksgaard, 1946).

[5] See Galileo Galilei, *Sidereus nuncius, or the Sidereal Messenger*, trans. with introduction, conclusion, and notes by Albert Van Helden (Chicago: Univ. Chicago Press, 1989), pp. 64–83. Pp. 87–113 provide a convenient summary.

[6] See, e.g., the title page of Peter Apian's *Ein kurtzer bericht d'Observations und urtels des Jüngst erschinnen Cometen* (Ingolstadt, 1532). For the earlier efforts of Paolo Toscanelli (which remained

During the Easter holidays of 1610, Galileo stopped in Bologna on his way from Padua to Florence to stay at the house of his colleague Giovanni Antonio Magini, the professor of astronomy at the university. Many in Bologna, including Magini, were skeptical of Galileo's discoveries, and therefore an opportunity presented itself for a demonstration. In the presence of a number of learned men, Galileo showed his telescope and let others observe earthly and celestial things through it. They agreed that for earthly objects the instrument performed as promised but that in the heavens it was not reliable. Although Galileo's notes show that on the first night two and on the second night all four of the satellites were visible,[7] none of the gentlemen present were able to see satellites around Jupiter. Just how serious a failure this first recorded effort at a public demonstration of telescopic discoveries was can be gauged from an account written by one very biased participant:

> Galileo Galilei, the mathematician of Padua, came to us in Bologna and he brought with him that spyglass through which he sees four fictitious planets. On the twenty-fourth and twenty-fifth of April I never slept, day and night, but tested that instrument of Galileo's in innumerable ways, in these lower [earthly] as well as the higher [realms]. On Earth it works miracles; in the heavens it deceives, for other fixed stars appear double. Thus, the second evening I observed with Galileo's spyglass the little star that is seen above the middle one of the three in the tail of the Great Bear, and I saw four very small stars nearby, just as Galileo observed about Jupiter. I have as witnesses most excellent men and most noble doctors, Antonio Roffeni, the most learned mathematician of the University of Bologna, and many others who with me in a house observed the heavens on the same night of 25 April, with Galileo himself present. But all acknowledged that the instrument deceived. And Galileo became silent, and on the twenty-sixth, a Monday, dejected, he took his leave from Mr. Magini very early in the morning. And he gave no thanks for the favors and the hospitality, because, full of himself, he hawked a fable. . . . Thus the wretched Galileo left Bologna with his spyglass on the twenty-sixth.[8]

Now it may very well be that the participants were in this case militantly skeptical and even hostile, but skepticism is to be expected in science. The problem went much deeper. First, even today those who look through a telescope for the first time have a difficult time seeing what they are supposed to see. The eye and the brain have to become accustomed to the instrument—to figure out how and where to look for the image.[9] Some learn this more quickly than others, but at a public occasion, when each individual could try his luck with the instrument for only a few minutes, there could not have been much of an opportunity for training oneself. Second, Galileo's instrument had a field of view of perhaps 15 arc minutes, and it was difficult (especially for the uninitiated) to find Jupiter, let alone see its moons. Galileo could give advice and help, but he could not actually show the moons in the heavens. Third, the telescope is not exactly an equal-opportunity research instrument. Besides the vagaries of the perceptual apparatus from brain to brain, there is the inescapable

---

in manuscript) see Jane L. Jervis, *Cometary Theory in Fifteenth-Century Europe* (Studia Copernicana, 26) (Wroclaw: Ossolineum; Dordrecht/Boston: D. Reidel, 1985), pp. 77–84.

[7] *Le opere di Galileo Galilei,* edizione nazionale, ed. Antonio Favaro, 20 vols. (Florence: Barbèra, 1899–1909; repr. 1929–1939, 1964–1966) (hereafter ***Opere***), Vol. III, p. 436.

[8] Martin Horky to Johannes Kepler, 27 April 1610, *ibid.,* Vol. X, p. 343.

[9] Vasco Ronchi, *Optics: The Science of Vision,* trans. Edward Rosen (New York: New York Univ. Press, 1957), pp. 188–189.

fact that some eyes are better than others. On several occasions I have taken a group of my students to look for Jupiter's satellites through a replica of one of Galileo's telescopes—students who were convinced the moons were really there—and the results were always mixed. Some saw all that were visible, some saw one or two, and some saw none at all. No matter how public the occasion, the actual observing remains an individual, private act.

It appears that Galileo had more luck in Tuscany, for the Grand Duke Cosimo de' Medici and his relative Giulio were able to see the satellites through Galileo's instrument under his guidance.[10] Cosimo honored Galileo by offering him a position but made no public pronouncement on the reality of the Medicean Stars.[11]

In the meantime Kepler was becoming increasingly worried. Upon reading *Sidereus nuncius,* he had, without being able to see Jupiter's moons for himself, penned a very favorable reaction to Galileo's discoveries and published it under the title *Conversation with the Sidereal Messenger.* As reports reached his ears that many denied Galileo's discoveries and scoffed at them, still unable personally to verify the existence of the moons, he begged Galileo to send him reports by witnesses. Six months after the publication of *Sidereus nuncius*, however, the only witnesses Galileo could mention were himself, the grand duke, and Giulio de' Medici. When Kepler asked Galileo to send him a telescope that was good enough to show the satellites, Galileo pleaded that he did not have any to give.[12] His strategy was to send instruments to important patrons, not to mathematicians.

One of these patrons, the elector of Cologne, who was in Prague in August and September 1610, gave Kepler the use for a few days of a telescope Galileo had sent. Kepler made the most of the opportunity. On the night of 1 September 1610, when he first had a good view of some of Jupiter's moons, he observed with a young astronomer named Benjamin Ursinus: "We followed the procedure whereby what one observed he secretly drew on the wall with chalk, without its being seen by the other. Afterwards we passed together from one picture to the other to see if we agreed." They were certain of three of the moons. About the fourth they were in doubt, Ursinus more so than Kepler. At the end of the description of the observation, Kepler noted again: "On this day the above-mentioned Ursinus was an observer [*spectator*] and an eye-witness [*testis oculatus*] to me."[13]

On 4 September Kepler barely saw two of the moons but his investigation was impeded by thick clouds. The next night the sky was clear but the light of the Moon made observing difficult. That night a visiting Englishman, Thomas Seget, observed with him. On the seventh he again observed with Ursinus (*Testis Ursinus*), and they were joined by Tycho Brahe's son-in-law, Franz Ganzneb Tengnagel, who saw one of the moons. But Kepler states that he had been advised (*admonitus*) of its presence. Finally, on the ninth, Kepler and Seget saw three satellites disposed in the same way,

[10] Francesco Sizzi, *Dianoia* (1611), in Galileo, *Opere,* Vol. III, p. 207. See also Galileo to Kepler, 19 Aug. 1610, *ibid.,* Vol. X, p. 422.

[11] Vincenzo Giugni to Galileo, 5 June 1610, p. 368; and Galileo to Giugni, 25 June 1610, p. 380; *ibid.,* Vol. X.

[12] Kepler to Galileo, 9 Aug. 1610, pp. 413–417; and Galileo to Kepler, 19 Aug. 1610, pp. 421–422; *ibid.*

[13] Johannes Kepler, *Narratio de observatis a se quatuor Iovis satellitibus erronibus* (1611), in Galileo, *Opere,* Vol. III, pp. 185–186.

but Abraham Schultetus (presumably Schultz), although advised (*sed admonitus*), saw only one.[14]

Kepler, then, not only verified the existence of Jupiter's satellites, he provided witnesses to certify his observations and his account. These witnesses were men of good credentials. Ursinus was described as "a student of astronomy who, because he loves the art and has determined to cultivate philosophy, would never consider undermining, right at the beginning, the credibility necessary to a future astronomer by false evidence." Seget was characterized as "a man already known in the books and letters of celebrated men, to whom therefore the reputation of his name is dear to his heart"; Tengnagel was Archduke Leopold's "Privy Council"; and Schultetus was "the Imperial Treasurer for Silesia."[15]

Here was a recipe for certifying one's observations: make the observatory a public space by enrolling fellow observers of high social status or other excellent credentials, have them draw what they see independently, and then compare results, thus confirming the observations by means of witnesses. Kepler even went so far as to tell the reader of his published tract on the subject that Prague was his witness that these observations were not sent to Galileo and that during this period he had not written to Galileo although he owed him a reply. He could therefore not have had any direct help from the Florentine.[16] This legalistic approach stands in stark contrast with Galileo's didactic approach.

Important as Kepler's independent verification was (it was among the first), the decisive victory came not in Prague but in Rome. There, the mathematicians of the Collegio Romano had equipped themselves with increasingly good telescopes, and by late 1610 they were able to see the satellites of Jupiter and verify Galileo's other discoveries as well.[17] When in the spring of 1611 Galileo went to Rome, Cardinal Bellarmine, the head of the Collegio Romano, asked the college's mathematicians for a judgment of Galileo's discoveries. Their reply, which stated that they had verified the discoveries, was the certificate of legitimacy for the telescope and the discoveries made with it. There was a catch, however: Father Clavius and his colleagues were careful to specify that the appearances were really as Galileo described them, but that their interpretation was another matter.[18] The telescope was thus certified as a "mathematical," not a "philosophical" instrument (if in 1611 such a thing could exist).

Galileo, his instrument, and his discoveries were celebrated in Rome, and in this receptive environment his didactic approach bore many fruits. At a feast arranged by Federico Cesi, at which Galileo was inducted into the Accademia dei Lincei, he was able to guide a number of men to seeing the satellites of Jupiter for themselves.

[14] *Ibid.,* pp. 186–187.

[15] *Ibid.,* pp. 184, 186, 187.

[16] *Ibid.,* p. 184. Kepler had written to Galileo on 9 Aug. 1610 that whether by his zeal Galileo would deceive the world was not so much a philosophical problem as a legal question (*ibid.,* Vol. X, p. 415). The analogy between witnesses to an experiment and witnesses in a legal case was made explicitly by Robert Boyle and the Royal Society half a century later: see Steven Shapin, "Pump and Circumstance: Robert Boyle's Literary Technology," *Social Studies of Science,* 1984, *14:*481–520, at pp. 487–490.

[17] Christopher Clavius to Galileo, 17 Dec. 1610, *Opere,* Vol. X, pp. 484–485.

[18] Roberto Bellarmino to the mathematicians of the Collegio Romano, 19 April 1611, pp. 87–88; and the mathematicians to Bellarmino, 24 April 1611, pp. 92–93; *ibid.,* Vol. XI.

It was at this time, too, that the instrument was given the name by which we know it.[19] Especially in the Italian setting, then, we must see the events in Rome in the spring of 1611 as highly symbolic. In the world of patronage and religious authority in which Galileo moved, he could not have hoped for more ringing endorsements. The modern equivalents of the mathematicians' judgment and the induction into the Accademia dei Lincei might be something like having one's discoveries certified by the National Academy of Sciences and being awarded a Nobel Prize.

At the very beginning of the life of the telescope, then, we can see the problems of verification and certification of knowledge clearly outlined. One solution, which can be applied in some special cases, was developed by Galileo shortly afterwards, in the next controversy. Here the issue was the nature of sunspots. Direct observation of the Sun, especially through a telescope, is a foolhardy and dangerous business. The earliest observations by Thomas Harriot, in 1610, made when the Sun was behind a thin veil of clouds and close to the horizon, were sufficiently problematic for Harriot to check his own observations by switching eyes.[20] Johannes and David Fabricius, after discovering the spots, observed them through a camera obscura without a lens.[21] Convenient as this arrangement was, it could only show very large spots. In 1612 Galileo's pupil Benedetto Castelli invented the method that became standard, projecting the Sun's image through a telescope.[22] In this arrangement sunspots could be studied by several observers at once, and thus the observation itself could become a public act. But although others tried to adapt this method to observations of the Moon and planets, the light-gathering power of the instruments was not great enough. Projection techniques could only be used for observing sunspots and solar eclipses.

Even among those who agreed with Galileo on the reliability of the telescope and the general nature of the discoveries made with it, there were quibbles over minor matters. One interesting case that clearly illustrates the authority structure in matters telescopic was the shape of Saturn. In July 1610 Galileo saw Saturn as three separate bodies: a larger central globe flanked by two smaller globes that almost touched it. Were these lateral bodies satellites like those of Jupiter? Apparently not, for they did not move with respect to the central body. (They did, however, disappear in 1612, only to reappear shortly afterwards in the same guise.)[23] During this same period others did not see Saturn in this way. In their report to Bellarmine, the mathematicians of the Collegio Romano stated: "We have observed that Saturn is not round, as Jupiter and Mars are to be seen, but of an ovate and oblong figure, in this fashion oOo; we have not seen the two little stars on the sides clearly enough

[19] Edward Rosen, *The Naming of the Telescope* (New York: Henry Schuman, 1947).

[20] See, e.g., Harriot's observations of 8 Dec. 1610 and 19 Jan. 1611 (old style), Harriot Papers, West Sussex Record Office, HMC 241/8, p. 1; reproduced in *Thomas Harriot: Renaissance Scientist,* ed. John W. Shirley (Oxford: Clarendon Press, 1974), Plate IV, pp. 118–119; texts transcribed in John North, "Thomas Harriot and the First Telescopic Observations of Sunspots," *ibid.,* pp. 129–165, at p. 132.

[21] Johannes Fabricius, *De maculis in sole observatis, et apparente earum cum sole conversione, narratio* (Wittenberg, 1611), C4r-C4v.

[22] Galileo, *Istoria e dimostrazioni intorno alle macchie solari e loro accidenti* (1613), *Opere,* Vol. V, pp. 136–137, also trans. Stillman Drake, in *Discoveries and Opinions of Galileo* (Garden City, N.Y.: Doubleday, 1957), pp. 115–116.

[23] Galileo to Belisario Vinta, 30 July 1610, *Opere,* Vol. X, p. 410. For 1612 see Galileo, *Istoria e dimostrazioni,* pp. 237–238, trans. Drake, *Discoveries and Opinions,* pp. 143–144 (both cit. n. 22).

separated from the one in the middle to be able to say that they are separate stars."[24] It appears that the mathematicians were influenced by what Galileo had seen: they described Saturn as being oval but drew it as three separate bodies.

They were not the only ones to see Saturn this way. In *Tres epistolae de maculis solaribus* of 1612, Christoph Scheiner, perhaps already influenced by what Galileo had reported, described Saturn as sometimes oval and sometimes triple-bodied.[25] Galileo's reply is instructive. In his first letter on sunspots he wrote:

> But as to the supposition of Apelles that Saturn is sometimes oblong and sometimes accompanied by two stars on its flanks, Your Excellency may rest assured that this results either from the imperfection of the telescope or the eye of the observer, for the shape of Saturn is thus: , as shown by perfect vision and perfect instruments, but appears thus: where perfection is lacking, the shape and distinction of the three stars being imperfectly seen. I, who have observed it a thousand times at different periods with an excellent instrument, can assure you that no change is to be seen in it.[26]

Not only, Galileo argued, were his telescopes better than those of his rivals, but his eyes were better as well. And he made his claim stick. Henceforth others deferred to him when it came to such observations. In fact, the triple-bodied guise that Galileo claimed for Saturn became the dominant way of conceptualizing the problem of its appearances until the 1640s. In an observation made in 1633, for instance, Pierre Gassendi described Saturn as shaped like a cocoon but drew it as triple-bodied.[27] A few years later Gassendi asked Galileo for a telescope because he thought those available to him were not very good. Galileo obligingly sent him one of his.[28]

Galileo, then, claimed authority for himself as an observer and for his instruments. The fact that he had made virtually all the initial discoveries with the telescope led others to accept it. During and even after his lifetime this authority was virtually unchallenged. Christopher Wren, writing in 1658, expressed Galileo's authority as well as any:

> The incomparable Galileo, who was the first to direct a telescope to the sky—although the telescope had then only recently been invented and was not yet in all respects perfected—so overcame yielding nature, that all celestial mysteries were at once disclosed to him. And with the crystal sceptre he almost overcame not only the lonely multitude of the Milky Way, the crowd of nebulae, the earth-like Moon, horned Venus and the spotted Sun, but even triple-bodied Saturn. His successors are envious because they believe that only to succeeding Lyncei is it granted to add to the discoveries of Galileo.[29]

After the certification of 1611 Galileo had little trouble fighting off challenges to his observations and the quality of his instruments until he became entirely blind in

[24] Mathematicians to Bellarmino (cit. n. 18), p. 93.

[25] Christoph Scheiner, *Tres epistolae de maculis solaribus* (1612), in Galileo, *Opere,* Vol. V, p. 31.

[26] Galileo, *Istoria e dimostrazioni* (cit. n. 22), trans. Drake, *Discoveries and Opinions,* pp. 101–102.

[27] Pierre Gassendi, *Opera omnia,* 6 vols. ed. H. L. Habert de Montmor (Lyons, 1658; repr. Stuttgart–Bad Cannstatt: F. Frommann, 1964), Vol. IV, p. 142.

[28] Nicolas Claude Fabri de Peiresc to Galileo, 17 April 1635, *Opere,* Vol. XVI, pp. 259–262; and references to the telescope in Gassendi, *Opera omnia,* Vol. IV, passim, e.g., pp. 474, 479.

[29] Albert Van Helden, "Christopher Wren's *De corpore Saturni,*" *Notes and Records of the Royal Society of London,* 1968, *23:*213–229, on p. 219.

the late 1630s. His discoveries were accepted on his personal authority, which was raised, as the example of Wren shows, to almost divine proportions. The underlying problem remained, however, and became pressing as others began to carve out reputations in telescope making and observing. The first to come forward was Francesco Fontana of Naples, a man with no formal education who made his living as an instrument maker. Beginning in about 1637, Fontana's telescopes made a stir outside his native Naples, and some of his observations began to circulate in pictorial form. Galileo only grudgingly admitted that Fontana's telescopes magnified more than his own, but he dismissed the observations as crude and without merit. Nothing new was seen through these new instruments.[30]

Whatever the merit of Fontana's skills as a lens grinder at this time, his telescopes were superior to those of Galileo in one important respect: because they were astronomical telescopes (using a convex instead of a concave ocular), their field of view was much greater than that of the Galilean instrument. Therefore, they supported higher magnifications.[31] In fact, Fontana launched a telescope race that lasted for half a century. But he did something else as well: he circulated his observations by means of pictures, thus making the distant reader a "virtual witness" of, indeed a partner in, his observations. In 1646 he published the first picture book of telescopic astronomy. Unfortunately the pictures were not well drawn and engraved, and in his observations Fontana was often unable to distinguish between the effects of imperfect optics and heavenly phenomena. For these reasons many scorned them. Evangelista Torricelli wrote to a friend that he had received the book "of the foolishnesses observed, or rather dreamed, by Fontana. If you want to see insane things, that is, absurdities, fictions, effronteries, and a thousand similar outrages, I will send you the book."[32]

A year after its appearance, Fontana's book was eclipsed by Johannes Hevelius's *Selenographia,* a book devoted to the study of the Moon but containing observations of the planets, sunspots, and satellites. Hevelius's claim for authority was multifaceted. First, he claimed an excellent memory and imagination and great patience. Second, when he undertook the task of studying the Moon, he searched everywhere for good telescopes but could not procure them. Therefore he made his own: "Thus I had to act as an artisan before an observer of the heavens."[33] His book gives a full description of telescope making and the various configurations used in different observations—all this with beautiful illustrations (Figures 1 and 2). He therefore claimed that his telescopes were the best that could be obtained. Commenting on a representation of Saturn (Figure 3, about which more below), he wrote: "And with a long tube of exquisite manufacture, I have been able to observe accurately that this is the true figure of Saturn and to consider everything properly, so that everyone

[30] Galileo to Elia Diodati, 15 Jan. 1639, *Opere,* Vol. XVIII, p. 18. On Fontana see Gino Arrighi, "Gli 'occhiali' di Francesco Fontana in un carteggio inedito di Antonio Santini nella Collezione Galileiana della Biblioteca Nazionale di Firenze," *Physis,* 1964, *6:*432–448.

[31] Albert Van Helden, "The 'Astronomical Telescope,' 1611–1650," *Annali dell'Istituto e Museo di Storia della Scienza di Firenze,* 1974, *1:*13–36, at pp. 27–28.

[32] Francesco Fontana, *Novae coelestium terrestriumque rerum observationes* (Naples, 1646); and Torricelli to Vincenzo Renieri, 25 May 1647, *Le opere dei discepoli di Galileo Galileo,* Vol. I: *Carteggio 1642–1648,* ed. Paolo Galluzzi and Maurizio Torrini (Florence: Giunti-Barbèra, 1975), p. 366.

[33] Johannes Hevelius, *Selenographia, sive Lunae descriptio* (Gdansk, 1647; New York: Johnson Reprint, 1967), "Ad Lectorem," pp. [ii] (cf. p. 209), [iv].

Figure 2. *Hevelius's lathe for grinding lenses. From* Selenographia, *pp. 6–7.*

who is guided by a desire to discover the truth can trust in this indefatigable observation." He then repeated the argument that Galileo had made, that with an inferior telescope the observer would see something quite different.[34] Galileo himself was not immune from Hevelius's criticism. Commenting on the pictures of the Moon in *Sidereus nuncius,* in which the large central spot is exaggerated, Hevelius argued that "Galileo did not have a sufficiently good telescope or could not devote sufficient care to those observations of his or, what is most likely, was ignorant of the art of picturing and drawing that serves this work very well, and no less than acute vision,

[34] *Ibid.,* pp. 43–44.

patience, and toil."[35] Hevelius, then, claimed not only that he was a better telescope maker and more careful observer than Galileo, but that he was also a better artist.

But making accurate observations and drawing them on paper was not enough. In order for all inhabitants of the world to become readers or at least spectators of Hevelius's observations, they had to be published. Here lay his third claim to authority. Ordinarily an artist and engraver replicated the astronomer's drawings under his supervision. But how could the astronomer guarantee that the artist and then the engraver had truly represented what the astronomer had observed? Hevelius solved this problem by dispensing with these artisans altogether, drawing and engraving all the astronomical illustrations for *Selenographia* himself. And to assure the meticulous reader who noticed that his lunar pictures were slightly elliptical, he gave their original diameter in fractions of the Paris, Rhineland, and Gdansk feet, and explained that the paper shrunk ten parts in width to six parts in length as it dried.[36]

Hevelius thus produced a seamless path by which information went from his keen eye through his mind, equipped with excellent imagination and memory, to his skilled hand, and from there directly to the eye of the reader, without the intervention of anyone else whose interpretation or error might cast doubt on the reliability of the picture. It was as though the reader of *Selenographia* was observing the heavens through Hevelius's eye. Moreover, Hevelius's pictures (if sometimes wrong) were models of clarity in the northern artistic tradition of naturalistic representation.[37] This method of presenting observations to the public was the best that could be hoped for before photography. Others, not as universally skilled as Hevelius, had to put up with the interposition of the engraver between their eye and that of the reader, in microscopy as well as telescopic astronomy. Two decades later, Robert Hooke pointed to precisely this problem in his *Micrographia.*[38]

Because of his illustrations and his meticulous method, Hevelius became an international authority figure in telescopic astronomy overnight. His instruments were considered by many to be the best in Europe, and his sharpsightedness, skill, and patience as an observer were celebrated.

From an early point, then, the authority of instruments was intertwined with personal authority. A strong argument can be made that after about 1612 Galileo's lead in telescope making had disappeared and that others had instruments of comparable quality. Yet Galileo ruled until his death as the undisputed master of telescopic astronomy. Likewise, it is to be doubted that in 1647 Hevelius made the best telescopes in Europe. Those of Fontana in Naples, Johannes Wiesel in Augsburg, and, shortly afterwards, Eustachio Divini in Rome were in all likelihood better. Yet for the next

[35] *Ibid.,* p. 205. See, e.g., Galileo, *Sidereus nuncius* (cit. n. 5), pp. 44–46.

[36] Hevelius, *Selenographia* (cit. n. 33), pp. 210–211, 214. With one exception, all the plates showing figures of the heavens are marked *Aut. Sculps.,* or *autor sculpsit.* The exception is plate Q (pp. 226–227), in which the full moon is rendered in the manner of a terrestrial map and in which the legend held by the putti only indicates Hevelius as *autore.* In contrast, the legends of the two other views of the full moon, plates P (pp. 222–223) and R (pp. 262–263) read, respectively, "accurata observata adumbrata aerique incisa per Johannem Hevelium," and "summa diligentia observata delineata aerique incisa a Johanne Hevelio." The engravings of equipment are not signed but are in Hevelius's style.

[37] Svetlana Alpers, *The Art of Describing: Dutch Art in the Seventeenth Century* (Chicago: Univ. Chicago Press, 1983).

[38] Michael Aaron Dennis, "Graphic Understanding: Instruments and Interpretation in Robert Hooke's *Micrographia,*" *Science in Context,* 1989, *3:*309–364, on pp. 314–315.

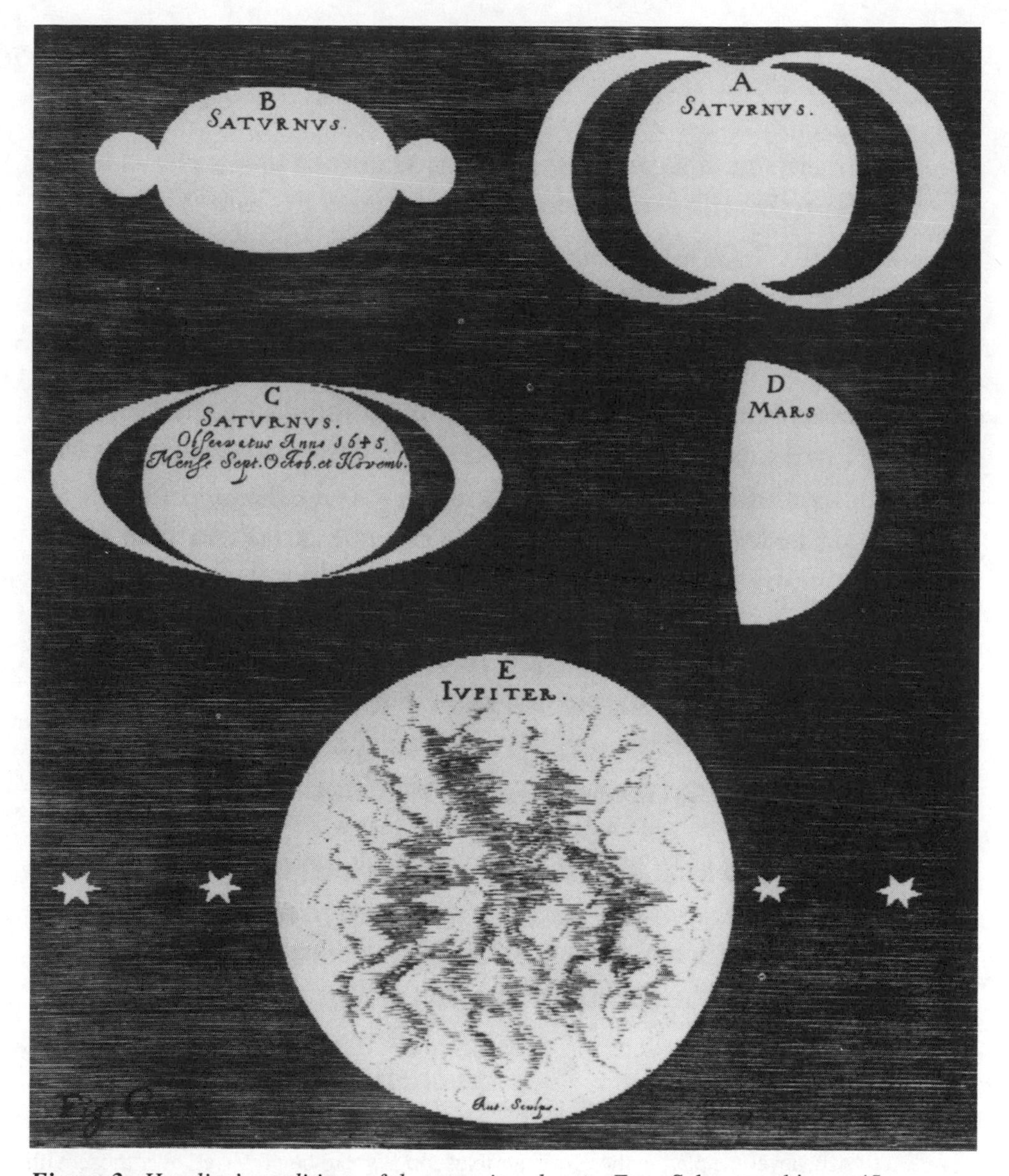

Figure 3. *Hevelius's renditions of the superior planets. From* Selenographia, *p. 45.*

fifteen years or so Hevelius was considered Europe's premier telescopic observer, and his reputation lingered on until after 1673, when he became embroiled in a controversy with John Flamsteed and Robert Hooke over his continued use of naked-eye measuring instruments.[39] We shall shortly take up the separation of authority of the observer from the authority of the instrument maker, beginning about 1650. But first we must deal with the question of how disputes about discoveries were settled.

[39] Eugene Fairfield MacPike, *Hevelius, Flamsteed, and Halley: Three Contemporary Astronomers and Their Mutual Relations* (London: Taylor & Francis, 1937), pp. 1–16, 75–102.

II. SETTLING DISPUTES

In 1643, a Capuchin friar named Antonius Maria Schyrlaeus de Rheita published a little book entitled *Novem stellae circa Jovem, circa Saturnum sex, circa Martem nonnullae* ("Nine stars around Jupiter, six around Saturn, and several around Mars"), announcing the discovery of a number of new satellites of the superior planets. The book contained essays by the polymath Juan Caramuel Lobkowitz, vicar-general (in absentia) of the Cistercian order in Great Britain, and Pierre Gassendi, one of Europe's foremost astronomers. Lobkowitz upheld Rheita's claim; Gassendi dismissed it. Rheita claimed that the new satellites could only be seen with a new sort of telescope of his invention, a claim similar to the one made by Galileo when he discovered the Medicean Stars in 1610.[40] It was perhaps for this reason that in his book of astronomy of 1646 Fontana claimed to have verified these new satellites: to show that his telescopes were just as good as Rheita's (though he did not mention Rheita).[41] Gassendi argued that Rheita had mistaken fixed stars for satellites, and others generally followed him in that judgment and dismissed Rheita's claim. Yet Gassendi could not claim instruments good enough to compete with Rheita's. The definitive refutation therefore did not come until 1647, when in his *Selenographia* Hevelius, who claimed even better telescopes, showed by means of detailed maps of the fixed stars in the regions where Rheita had observed Jupiter that the claimed satellites were indeed fixed stars (Figure 4).[42] This was the end of Rheita's claim.

Another controversy, much more difficult to settle, arose over the problem of Saturn's apparent shape. Over the years many astronomers had printed diagrams or pictures of Saturn's appearances. These varied enormously, and no single theory could account for them all. In 1656 Christiaan Huygens announced the discovery of a satellite of Saturn (now called Titan)—a claim that was verified quickly by others—and then in 1659 he published *Systema Saturnium,* in which he announced his hypothesis that the planet was circled by a ring to explain its strange appearances. But many of the published observations, including some by Hevelius (see Figure 3), could not be accounted for by the ring hypothesis, and Huygens therefore explained them away as due to either inferior telescopes or faulty observations. He thus cast himself in the role of the judge of the observations of others. By what authority did he do this? Huygens's argument was quite simple:

> In this investigation we ask that it be conceded to us that, because with our telescopes we have discovered the satellite of Saturn for the first time and because we see it distinctly whenever we please, therefore our telescopes are to be preferred over those with which others have not been able to reach that satellite, even though they were attentively observing Saturn every day; and that consequently the results of our observations concerning the shape of the planet are also to be considered true in each case when we and they saw different figures simultaneously.[43]

[40] *Novem stellae circa lovem, circa Saturnum sex, circa Martem nonnullae, a P. Antonio Rheita detectae & Satellitibus adiudicatae; De primis (& si mavelis de universis) D. Petri Gassendi iudicium; D. Ioannis Caramuel Lobkowitz eiusdem ludicij censura* (Louvain, 1643), pp. 9–10.

[41] Fontana, *Novae observationes* (cit. n. 32), pp. 109–110.

[42] Hevelius, *Selenographia* (cit. n. 33), pp. 49–66.

[43] Huygens, *Systema Saturnium* (1659), *Oeuvres complètes de Christiaan Huygens,* 22 vols. (The Hague: Nijhoff, 1888–1950) (hereafter ***Oeuvres***), Vol XVI, pp. 21–353, esp. pp. 271 (quotation)–287. See also Huygens, *De Saturni luna observatio nova* (1656), Vol. XV, pp. 173–177; and Albert Van Helden, "Saturn and His Anses," *Journal for the History of Astronomy,* 1974, *5:*105–121.

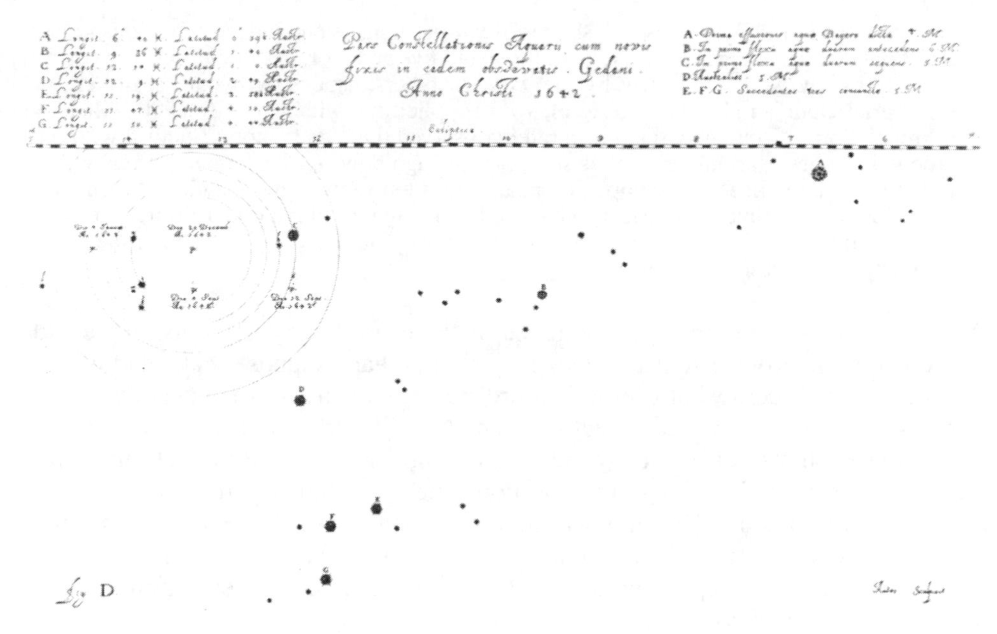

Figure 4. *Hevelius's visual refutation of Rheita's supposed discovery of new satellites of Jupiter. From* Selenographia, *pp. 34–35.*

Now in fact Huygens's solution of this long-standing puzzle was a mental one, in which his Cartesianism was in all likelihood instrumental; it was surely not a product of better telescopes.[44] It is not surprising, therefore, that this relative newcomer's audacious claim for the superiority of his telescopes met with anger on the part of those whose reputations were established.

In his tract *Dissertatio de nativa Saturni facie* (1656), published three years before Huygens's *Systema Saturnium,* Hevelius had put forward his own theory of Saturn's appearances. In *Selenographia* he had shown Saturn with its appendages fully deployed as an egg-shaped body with two attached arcs (the observation about whose veracity he had assured the reader; see above and Figure 3), while when the handles were invisible the planet appeared round. In *De nativa Saturni facie* he argued that the egg-shaped planet with its two appendages rotated on an axis perpendicular to the length of the configuration. Huygens now claimed that this egg-shaped central body was an error.[45] Hevelius expressed his frank opinion of Huygens's argument in a letter to his friend the astronomer Ismael Boulliau:

> Impetuously, Huygens attaches more faith to his own observations than to all others . . . [made] with his most splendid tubes with which, whenever he pleases, he clearly sees the Saturnian [satellite], which others are not able to see. For this reason he asserts that the rest of his observations are more correct than the rest of the observations of others.

[44] Albert Van Helden, "'Annulo Cingitur': The Solution to the Problem of Saturn," *J. Hist. Astron.*, 1974, *5:*155–174, on pp. 160–161.

[45] Hevelius, *Selenographia* (cit. n. 33), pp. 43–45; Hevelius, *Dissertatio de nativa Saturni facie* (Gdansk, 1656), pp. 1–34; and Huygens, *Systema Saturnium* (cit. n. 43), pp. 277–279.

> I reply that I likewise possess telescopes of various lengths and indeed of various types, not only with two lenses but also with three and five and more. These lenses were made partly by my work and partly by the work of others, especially by the most skillful Imperial Optician [Johannes] Wiesel, who supplied me with a tube for 500 Polish or French florins. You know that I accurately observed that companion of Saturn with the benefit of these various telescopes several years ago now. . . . Yes, indeed, I remember that I already noticed that companion near Saturn ten years or more ago. But being too careless at that time, I held it to be a fixed star. Thus on this point I do not concede anything to him. And therefore I see no reason why our observations do not merit the same faith as those of Huygens.

When it came to the elliptical appearance of the central body, Hevelius expressed himself in even stronger terms: "Does Huygens perhaps suppose that I and others are not able to discern what is elliptical or spherical, or that it was invented by my mind as he writes . . . , or rather that I dreamed it? No, by Hercules!"[46]

To Huygens himself Hevelius wrote in much milder tones, stating that future observations when the handles were in a more open position would show who was right.[47] As it turned out, Hevelius was simply wrong on this issue, and as the ring hypothesis was confirmed (with some modifications, however) by others over the next several years, his reputation as a maker of excellent telescopes suffered in the measure by which Huygens's increased.

Hevelius was not the only one to dispute Huygens's claim for the superiority of his instruments. In Rome, Eustachio Divini, who had eclipsed Francesco Fontana as a telescope maker and was thought by many to be the foremost practitioner of the art in Europe, took issue with Huygens for the same reason. Huygens criticized not his *observation* of Saturn but rather his *representation,* published in a broad sheet, and Divini felt obliged to respond.[48] In 1660, the year after Huygens's *Systema Saturnium* came out, there appeared in Rome a tract entitled *Brevis annotatio in Systema Saturnium,* under the authorship of Divini. Although Divini was perfectly capable of writing a book (he went on to write several), he was not a Latinist: much of the tract was in fact written by the French Jesuit Honoré Fabri. Their combined attack took much the same approach as Hevelius's. Divini disputed Huygens's claim for the superiority of his telescopes. In fact, he reported that from 1653 to 1658 he had often seen Saturn flanked by two little globes in the manner in which Galileo had first seen the planet, and this with telescopes of higher magnification and larger aperture than those of Huygens. Once the foundation of Huygens's entire argument was demolished, the complete range of reported observations was open again, and Fabri had little trouble making up a different hypothesis to explain Saturn's appearances. He postulated two large light-reflecting satellites and two smaller light-absorbing satellites that revolved in orbits about points *behind* the planet.[49]

[46] Hevelius to Ismael Boulliau, 9 Dec. 1659, Paris, Bibliothèque Nationale, MSS Collection Boulliau XXV (FF 13042), fols. 89v-90r, 90v.

[47] Hevelius to Huygens, 13 July 1660, *Oeuvres,* Vol. III, p. 91.

[48] Huygens, *Systema Saturnium,* p. 279. Divini had the outline of the formation of planet and ring correct, but he or the engraver added shading so that the figure looked like a ball superimposed on an elliptical ring. For the broadsheet see G. Govi, "Della invenzione del micrometro per gli strumenti astronomici," *Bullettino di Bibliografia e di Storia delle Scienze Matematiche e Fisiche,* 1887, *20*:614–615.

[49] Eustachio Divini, *Brevis annotatio in Systema Saturnium Christiani Hugenii* (1660), in Huygens, *Oeuvres,* Vol. XV, pp. 403–437, on pp. 407–415, 423–437.

Contrived as Fabri's hypothesis was, Huygens could not easily dismiss the attack on his instruments and his hypothesis because Divini and Fabri had shrewdly cast their tract in the form of a letter to Prince Leopold de' Medici, the very patron to whom Huygens had dedicated his *Systema Saturnium.* Divini, or rather Fabri, wrote: "Everything written by me I want to be subject to Your most impartial opinion and honest judgment, and rightly so in my opinion, since all men of letters honor You as a Maecenas, and the makers of more refined works as a singular patron."[50]

Prince Leopold, who was to be the judge in this dispute, assigned the problem to his academy, the Accademia del Cimento. The scientists of this body, dedicated to the idea that no hypothesis should be accepted until it had been "tested and retested" (the motto was *provando e riprovando*), did a thorough and imaginative job. They built models of both hypotheses (Figure 5) and observed them from various distances with telescopes of differing magnification and optical quality. Since, however, they knew what the models looked like, they went so far as to call in others who did not know what they were supposed to see, among them illiterate persons, and carefully recorded their responses as well. Their judgment was that Fabri's model failed to reproduce the appearance of Saturn, which they observed at the same time through the best telescopes available in Florence—instruments actually made by Divini. Huygens did not escape entirely unscathed, however. In *Systema Saturnium* he had proposed a ring of substantial thickness whose outer edge absorbed all light and was therefore invisible except on the planet itself, where it appeared as a dark band. The academicians reported that no matter what they did to the ring model, they could not make the edge invisible. In their opinion, therefore, the ring had to be so thin as to be a mere surface in order for it to be invisible when seen edge on.[51]

Although Huygens was given to know that in Florence his hypothesis had been judged far superior to Fabri's, he was asked not to boast publicly about his victory. Such an action would have embarrassed Prince Leopold, whose position was delicate: Huygens was a Protestant and an avowed Copernican, while Fabri was an influential Catholic cleric and, true to the ruling of 1616, believed the Earth to be the center of the universe. The appeal to an impartial authority of high social standing in this dispute was thus only partially successful. In fact, the controversy continued with a published reply by Huygens and a rejoinder by Divini and Fabri, by which time Fabri's position had become sufficiently untenable for him to backtrack a bit and admit that he could hardly help seeing a ring around the planet.[52] Scientists in Paris and London, in the meantime, had little difficulty in deciding which hypothesis was preferable. Naturally, Divini's reputation suffered accordingly.

How could the observers in Florence know whose telescopes were better, and was that not what this argument had been all about? Since so much of Huygens's argument depended on his claim that his telescopes were superior, would a direct comparison of telescopes perhaps be more fruitful? At the beginning of the controversy Prince Leopold requested that Huygens report to Florence the distance at which a person of good vision could distinguish a letter or character through his best telescope, and include a sample of the unit of length he used. Huygens complied with

[50] *Ibid.,* p. 407.

[51] See Albert Van Helden, "The Accademia del Cimento and Saturn's Ring," *Physis,* 1973, *15:* 237–259.

[52] *Ibid.,* pp. 249–250, 253–259.

Figure 5. *The model that the Accademia del Cimento used to test Christiaan Huygens's ring hypothesis. From* Oeuvres complètes de Christiaan Huygens, *(The Hague: Nijhoff, 1888–1950), Vol. III, pp. 154–155.*

this request.[53] But apparently his report did not settle the relative merits of the instruments of the two protagonists. In the meantime, it was reported to Huygens from Rome, Divini and his allies were contemplating another way to decide the issue: they were "thinking of naming a place and a time at which, the author of this work [i.e., Huygens] and the corrector [i.e., Divini] being gathered with some excellent mathematicians, the tubes will be compared and the matter will be examined more thoroughly, and a judgment will be made of the phenomena of Saturn that have been

[53] Carlo Dati to Nicolaas Heinsius, 16 March 1660, Huygens, *Oeuvres,* Vol. III, pp. 42–43.

published; on the condition that he who is convicted of errors leaves the arena and pays the expenses."[54] Divini never issued such a challenge directly to Huygens. Presumably the distance separating them was too great. In Rome, however, the idea of comparing telescopes directly in this manner was an accepted practice and somewhat of a spectator sport.

III. PERSONAL AUTHORITY OF THE INSTRUMENT MAKER

As long as Galileo was alive, his personal authority ensured that Florence and the Medici Court would be the center of telescopic astronomy. Upon his death, however, it appeared that the torch would pass to Naples, where Francesco Fontana was making the best telescopes in Italy, and that henceforth authority would come simply from the power and quality of one's telescopes, especially since no spectacular new discoveries were forthcoming. In order to rescue the honor of Florence and the Medici, no less a scientist than Evangelista Torricelli took up the art of telescope making, putting himself in direct competition with Fontana. This was no easy task, however, and it took Torricelli several years before he could match the Neapolitan's instruments in magnification and optical quality.[55] But how could one prove whose telescopes were better?

In 1645 Torricelli was notified by a correspondent that in Rome an excellent telescope by Fontana had been compared directly with one of Torricelli's instruments. In this *paragone,* the Neapolitan instrument was judged to excel the Florentine one by a wide margin, but there was a question about the contest's fairness. Was the Torricelli telescope one of his best, or just a mediocre one? Who was in control of the event?[56] Problematic as such *paragoni* were, they became common in Rome. In the late 1640s Eustachio Divini bested Fontana in them, and he ruled Italy until he was bested by a newcomer in the 1660s.

In 1664 the Accademia del Cimento in Florence became involved in a celebrated controversy in Rome between Divini and Giuseppe Campani, a young clockmaker who had recently switched his efforts to optical instruments. In a series of confrontations, supervised as much as possible by the Roman agents of Prince Leopold and his academy, the *paragone* was refined in successive steps. First, the instruments had to be of the same length (the seventeenth-century measure of power), and the test had to be conducted under circumstances that did not favor either party. Second, an objective standard for testing the instruments had to be developed. If the first requirement presented the greatest human problem—Campani and his seconds were very shrewd in their efforts to gain an advantage—the second presented the greater scientific problem.[57]

It had become fairly common to use a piece of writing as the standard in rating the power of a telescope. The Accademia del Cimento began with this method, specially printing standard sheets filled with lines of poetry set in successively smaller type.

[54] Gillis François de Gottigniez to Gregorius à St. Vincento, [1660], *ibid.,* pp. 59–60.

[55] Maria Luisa Righini Bonelli and Albert Van Helden, "Divini and Campani: A Forgotten Chapter in the History of the Accademia del Cimento," *Annali dell'Istituto e Museo di Storia della Scienza di Firenze,* 1981, *6*(1):3–176, on pp. 6–7.

[56] Michel Angelo Ricci to Evangelista Torricelli, 20 Aug. 1645, *Opere dei discepoli,* Vol. I (cit. n. 32), pp. 263–264.

[57] Righini Bonelli and Van Helden, "Divini and Campani" (cit. n. 55), pp. 15–25.

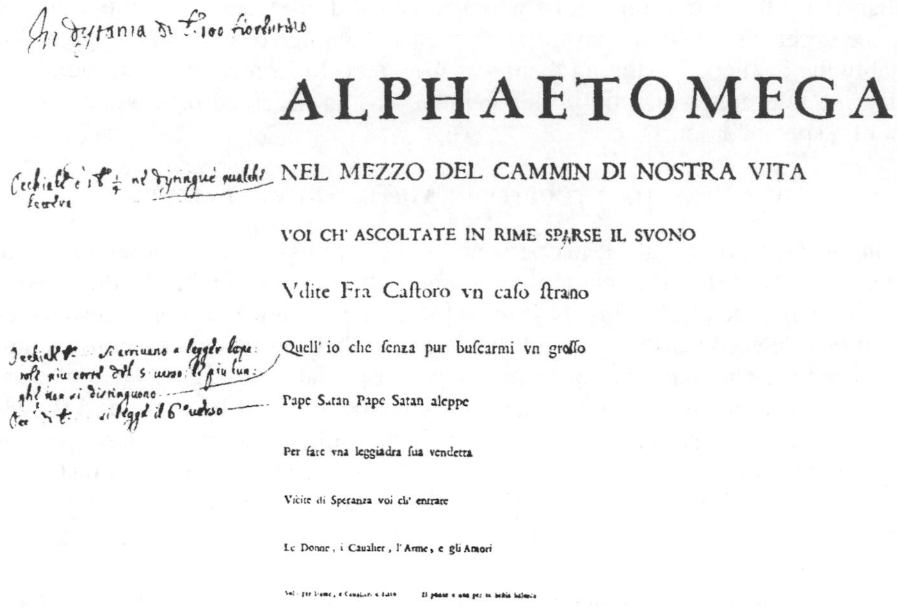

ALPHAETOMEGA

NEL MEZZO DEL CAMMIN DI NOSTRA VITA

VOI CH' ASCOLTATE IN RIME SPARSE IL SVONO

Vdite Fra Caſtoro vn caſo ſtrano

Quell' io che ſenza pur buſcarmi vn groſſo

Pape Satan Pape Satan aleppe

Per fare vna leggiadra ſua vendetta

Viuite di Speranza voi ch' entrate

Le Donne, i Caualier, l'Arme, e gli Amori

Figure 6. *The Accademia del Cimento's first test sheets for comparing telescopes. The annotations refer to telescopes with which the lines indicated were read at a distance of 100 Florentine braccia (58 meters). Courtesy of the Biblioteca Nazionale, Florence.*

It was found, however, that this standard was inadequate. Often reading only one word could give away an entire line. Who in Italy did not know, for instance, that the second line of the sheet (Figure 6), *Nel mezzo del cammin di nostra vita,* is the first line of Dante's *Inferno?* For this reason a standard of nonsense words, mere groupings of letters, was made up next. But now that one had to concentrate on individual letters, it was found that some were easier to recognize than others, especially those with risers and hangers. The next test sheet therefore employed only capital letters. Finally, it was found that the imprint caused by the type created a shadow that could vary with the obliqueness of the illumination, thus making the letters sometimes easier and sometimes harder to recognize. For this reason, after printing the sheets were carefully beaten flat with a wooden mallet, so that no imprints remained. The result of this effort (Figure 7) was something very much like the eye charts developed two centuries later to test refractive errors of the eye.[58]

The Accademia del Cimento was thus successful in at least developing a standard target on which to test telescopes. But the method was not published. When by 1665 the results were in, the academy's report on experiments, *Saggi di naturale esperienze* (1666), was already in press, and there was no effort to add this new and interesting material. A few years later the Accademia del Cimento ceased to exist. Perhaps more important, Campani's telescopes were certified in another way at the

[58] *Ibid.,* pp. 31–36.

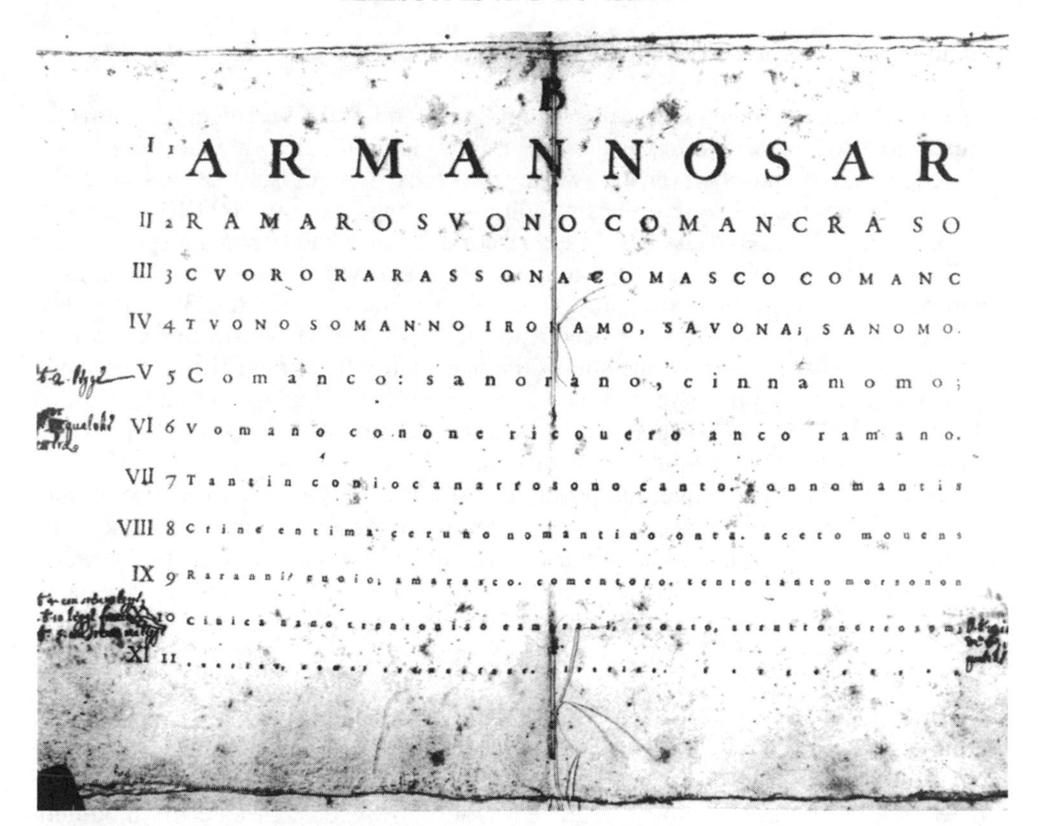

Figure 7. *The Accademia del Cimento's final test sheet for comparing telescopes. Courtesy of the Biblioteca Nazionale, Florence.*

same time. The astronomer Giovanni Domenico Cassini preferred Campani's telescopes to those of all others. With these instruments Cassini discovered a number of celestial novelties at this time: the shadows of Jupiter's satellites on its disk; subtle, more or less permanent features in Jupiter's belts and zones that allowed him to determine the planet's rotation period; and permanent surface features on Mars by which he also determined the rotation period of that planet.[59] Cassini's observations were the wonder of the age, and they made Campani's telescopes famous all over Europe.

When he went to Paris a few years later, Cassini took his Campani telescope with him, and the munificence of the Sun King allowed him to order a number more over the next fifteen years. With these Cassini discovered four satellites of Saturn, the division in Saturn's ring named after him, and several other phenomena.[60] If

[59] Giovanni Domenico Cassini, *Lettera astronomica al sign. Abb. Ottavio Falconieri, sopra l'ombra de pianetini medicei in Giove, con le tavole dell'osservazioni ne due mesi seguenti* (Rome, 1665); Cassini, *Lettere astronomiche al sign. Abb. Ott. Falconieri, sopra le varietà delle macchie osservato in Giove, e loro diurne revoluzioni* (Rome, 1665); and Cassini, *Martis circa proprium axem revolubilis, observationes Bononienses* (Bologna, 1666).

[60] A. F. O'D. Alexander, *The Planet Saturn: A History of Observation, Theory, and Discovery* (London: Faber & Faber, 1962), pp. 111–116.

Campani was Europe's premier telescope maker, Cassini was its finest telescopic astronomer.

Cassini's position in telescopic astronomy at the end of the seventeenth century is comparable to that of Galileo at the beginning. Galileo had combined the role of telescope maker and observer, but we do not know how closely he was actually involved in the manual work of lens grinding—perhaps more in 1609/10 and much less later in his life. By Cassini's time telescope making had become a specialized craft, and the practitioner had emerged, often through writing his own books, as a member of the scientific community. Cassini therefore availed himself of the best product the craft could offer and used it to make a series of discoveries almost as dazzling as Galileo's. And if the site of the astronomy had now shifted to France, Cassini added to the prestige of Louis XIV as much as Galileo had added to the glamor of Cosimo II. Their authorities were strengthened by their positions in the bureaucracies of these absolute heads of state.

Cassini's personal prestige, his position at the French Observatoire Royale and Académie Royale, and his access to the best telescopes in Europe made his discoveries accepted by everyone. Some of these were almost impossible to verify. The second pair of satellites of Saturn he discovered in 1684 were never seen by Christiaan Huygens, who had discovered Saturn's largest satellite (Titan) in 1655 and was in the 1680s again actively grinding lenses for very powerful telescopes.[61] In fact, early in the eighteenth century English astronomers were beginning to doubt the existence of these bodies—never seen in England—until astronomers at the Paris observatory mounted a very long telescope (no doubt a Campani instrument) and made some measurements of them. It was not until 1718 that James Pound managed to observe these satellites with a 123-foot telescope (made by Huygens!), especially mounted for that observation.[62]

The mere possession of a Campani telescope could not, however, guarantee the veracity of observations. Shortly afterwards, the Roman astronomer Francesco Bianchini published his *Hesperi et Phosphori,* in which he described surface features of Venus and determined its rotation period. His observations, made with Campani instruments, could not be verified by others, and indeed they never were. But many accepted them nevertheless. In 1805 William Herschel, whose large reflecting telescopes were the wonder of his age, and who had discovered two more satellites of Saturn, announced that Saturn's body was not even, but protruded a little halfway between its equator and poles—the so-called square-shouldered appearance of Saturn. Again, these discoveries could not be verified by his contemporaries, who did not have instruments that matched his in power. Later the satellites were found to be real enough, but the square-shouldered appearance turned out to be an illusion.[63]

Thus, while often astronomical discoveries could be verified by other observers, in some cases they could not, and there was no easy solution to this problem.

[61] Christiaan Huygens, *Cosmologia, Oeuvres,* Vol. XXI, pp. 193–194, 302, 778–779.

[62] Jacques Cassini, "Nouvelles découvertes sur les mouvements des satellites de Saturne," *Mémoires de l'Académie Royale des Sciences: Mathématique et Physique,* 1714 (pub. 1717), pp. 361–378; and "A Rectification of the Motions of the five Satellites of Saturn; with some accurate Observations of them, made and Communicated by the Reverend Mr. James Pound, R. S. Soc.," *Philosophical Transactions of the Royal Society,* 1718, *30* (355):768–774 (see p. 769 for the earlier failures).

[63] Francesco Bianchini, *Hesperi et Phosphori nova phaenomena, sive observationes circa planetam Veneris* (Rome, 1728); and Alexander, *Planet Saturn* (cit. n. 60), 147–148.

Whether or not claims were accepted depended for the most part on the prestige of the observer and the reputed quality of his instruments. The situation finally came to a head in the notorious episode of the canals of Mars.[64] The final undoing of Giovanni Schiaparelli and Percival Lowell led professional astronomers (who were becoming astrophysicists) to turn to photographic evidence as the only objective method of certifying discoveries, even though, because of atmospheric trembling, this method was far inferior to actual observation.

The issue is now more or less resolved. The light gathered from distant objects is sensed not by eyes or photographic plates, but by infinitely more sensitive charged-couple devices; it can be stored digitally and displayed at will to many observers.[65] These may argue about the interpretation, but the images are there for all to see. The act of observing has thus become an infinitely repeatable public activity, and the observatory has become a public space in the full sense of the word.

[64] William Sheehan, *Planets and Perception: Telescopic Views and Interpretations, 1609–1909* (Tucson: Univ. Arizona Press, 1988), esp. Ch. 16.

[65] See the article by Robert W. Smith and Joseph N. Tatarewicz in this volume.

Precision Instruments and the Demonstrative Order of Proof in Lavoisier's Chemistry

By *Jan Golinski**

> If it is true that a controversy approaches its conclusion by the accumulation of facts that impinge upon it, it is only so provided these "facts" are without ambiguity in their implications. For otherwise, twisted by the rival hypotheses, and sometimes with so many more words that they convey less sense, these "facts" so multiply the extraneous questions that controversies become endless. Thus prejudice and imagination freely hold sway and logic is replaced by fashion.
>
> —Jean-André Deluc (1790)[1]

THAT, AFTER MORE THAN A DECADE OF DEBATE about the fundamentals of chemical theory, Jean-André Deluc should express frustration is not really surprising. As an upholder of the traditional theory of phlogiston and an opponent of the new theory of Antoine-Laurent Lavoisier, Deluc feared that the controversy would never end. Each new fact could be interpreted in different ways by the two sides and, rather than resolving the debate, seemed to bring ever more subjects into doubt. Reason, supposedly the secure path to scientific truth, seemed incapable of deciding the issue. The same sentiment was voiced nearly simultaneously by the English chemist James Keir, who tried to curb the hopes of his fellow phlogistonist Joseph Priestley, who looked forward to an imminent compromise. Keir cautioned that "there are wonderful resources in the dispute about phlogiston, by which either party can evade, so that I am less sanguine than you are in my hopes of seeing it terminated."[2]

Historians are interested in controversies for much the same reasons that historical participants like Deluc and Keir found them so frustrating. As "facts" accumulate on each side, less and less appears to be certain. Instead, debate ramifies across an ever wider range of questions. Phenomena, methods, apparatus, personal competence, assumptions and principles—all may become issues in dispute. Hence if controversies become prolonged, more and more background assumptions and practices

* Department of History, University of New Hampshire, Durham, New Hampshire 03824-3586.

[1] "Lettre de M. DeLuc à M. De La Métherie, sur la nature de l'eau, du phlogistique, des acides & des airs," *Observations et Mémoires sur la Physique, sur l'Histoire Naturelle, et sur les Arts et Métiers,* 1790, 36:144–154, on p. 153.

[2] James Keir to Joseph Priestley, [n.d., 1789?], in *A Scientific Autobiography of Joseph Priestley (1733–1804),* ed. Robert E. Schofield (Cambridge, Mass.: MIT Press, 1966), pp. 252–253, on p. 253.

are exposed to view. Much recent historical work has shown the value of disputes as sites for examining scientific practice as a social activity. In controversies it is particularly clear how many elements of historical context shape rival interpretations of nature and how many "wonderful resources" are available to those trying to close the issue.[3]

Lavoisier's "chemical revolution" presents itself as an underexploited field for such study. There has been relatively little work on the dynamics of the controversy, which ebbed and flowed throughout the 1780s and into the following decade. Perhaps historians have been too concerned with trying to grasp in essentialist terms the real nature of Lavoisier's achievement or assessing whether it deserves the label "revolution." The process of persuasion undertaken by Lavoisier and his allies in the 1780s tends to be regarded as an aftermath to the main events. And yet what Carl Perrin called the "triumph of the antiphlogistians" was no walkover, but a lengthy process that deserves detailed investigation. Controversy ranged over numerous issues of fact and swelled to embrace methodological, linguistic, and social questions. Lavoisier's system as a whole was articulated in the context of this debate. The mapping of the struggle in its temporal, geographical, and social dimensions is a large-scale task, but one that promises considerable rewards in historical understanding of the processes of science.[4]

Such a mapping cannot be attempted here, though a step can be taken towards it by surveying the role of instruments in the controversy. Trevor Levere has recently reminded us of the importance of Lavoisier's novel instruments, including the calorimeter and the balance, and of their role in the campaign against phlogiston. Frederic L. Holmes has pointed out how radical a break this apparatus marked with the "longue durée" of the eighteenth-century chemical laboratory. And Arthur Donovan has argued that Lavoisier's instrumentation signals his transfer into chemistry of the methods of the more mathematized physical sciences.[5] In this article I build upon this work to place Lavoisier's apparatus against the background of the controversy surrounding his new chemistry. My aim is to use the circumstances of dispute to expose the assumptions and practices governing his deployment of this particular technology.

[3] For sociological work on controversies, see H. M. Collins, *Changing Order: Replication and Induction in Scientific Practice* (London/Beverly Hills: Sage, 1985); and Collins, ed., *Knowledge and Controversy: Studies of Modern Natural Science,* special issue of *Social Studies of Science,* 1981, *11*(1). Historical studies include Steven Shapin and Simon Schaffer, *Leviathan and the Air-Pump: Hobbes, Boyle and the Experimental Life* (Princeton: Princeton Univ. Press, 1985); Martin J. S. Rudwick, *The Great Devonian Controversy: The Shaping of Scientific Knowledge among Gentlemanly Specialists* (Chicago: Univ. Chicago Press, 1985); and James A. Secord, *Controversy in Victorian Geology: The Cambrian-Silurian Dispute* (Princeton: Princeton Univ. Press, 1986).

[4] Carleton Perrin, "The Triumph of the Antiphlogistians," in *The Analytic Spirit: Essays in the History of Science in Honor of Henry Guerlac,* ed. Harry Woolf (Ithaca, N.Y.: Cornell Univ. Press, 1981), pp. 40–63. Other work on the controversy includes Karl Hufbauer, *The Formation of the German Chemical Community (1720–1795)* (Berkeley/Los Angeles: Univ. California Press, 1982); John G. McEvoy, "The Enlightenment and the Chemical Revolution" in *Metaphysics and Philosophy of Science in the Seventeenth and Eighteenth Centuries: Essays in Honour of Gerd Buchdahl,* ed. R. S. Woolhouse (Dordrecht: Kluwer Academic, 1988), pp. 307–325; and some of the essays in Arthur Donovan, ed., *The Chemical Revolution: Essays in Reinterpretation, Osiris,* 2nd ser., 1988, *4.*

[5] Trevor H. Levere, "Lavoisier: Language, Instruments, and the Chemical Revolution" in *Nature, Experiment and the Sciences,* ed. Levere and W. R. Shea (Dordrecht: Kluwer Academic 1990), pp. 207–233; Frederic Lawrence Holmes, *Eighteenth-Century Chemistry as an Investigative Enterprise* (Berkeley: Office for History of Science and Technology, Univ. California, 1989), esp. Ch. 5; and Arthur Donovan, "Lavoisier and the Origins of Modern Chemistry," *Osiris,* 1988, *4*:214–231.

Controversy enables us to see how material apparatus is embedded in specific settings of practice that enable it to function as a tool of investigation and persuasion. We shall see that Lavoisier had to mobilize particular personnel and their skills to craft and use his instruments. He forged links with practitioners of the exact sciences, trained in the French mathematical engineering tradition, and with skilled instrument makers.[6] He expended substantial financial resources on the construction of his apparatus. He mastered difficult techniques of measurement and calculation—in calibration, for example. He also constructed social and literary "technologies," managing the audiences at set-piece experimental demonstrations and conveying the results in a written form that stressed the accuracy of the procedures and the high standard of proof thereby achieved.[7] In the ongoing controversy, many aspects of this form of practice were made explicit in the course of challenges to, and defenses of, Lavoisier's claims.

Outside Lavoisier's own setting, his instruments did not always convey their hoped-for persuasive potential. Many resources enabled opponents such as Priestley and Keir to evade the purported implications of his experiments. Priestley articulated a radically different model of scientific practice and condemned Lavoisier's supposed accuracy as the spurious result of excessively complex experimental contrivances. For Priestley, his own inability to replicate the French experiments was a reason not to trust them. Thus discussion of instruments was implicated in wider debates about the way science should be practiced. Arguably, the controversy was not just about the facts of chemical phenomena but about how science should be carried on. In the face of such radical disagreement, Lavoisier's instruments simply could not convey their meaning unequivocally.

Nonetheless, the controversy was eventually brought to a close, albeit in a prolonged and confused way that deserves further investigation. It seems clear that the extension of Lavoisier's practices of instrumental use played a part in this process. His victory (to a certain extent a posthumous one) relied upon transmitting a culture of experimental practice to support diffusion of his instruments and replication of the phenomena they produced.

I. LAVOISIER'S INSTRUMENTAL STRATEGY

By the late 1770s, Lavoisier had convinced himself of the need to reform chemical theory fundamentally and to dispense with the notion of phlogiston. In his "crucial year" of 1772 he had studied the calcination of metals and the combustion of sulphur and phosphorus and had established the fixation of air in these processes. He then followed up Priestley's isolation of "dephlogisticated air," repeating the operation

[6] The classic studies of Lavoisier's instruments are Maurice Daumas, *Lavoisier: Théoricien et expérimentateur* (Paris: Presses Universitaires de France, 1955), esp. Ch. 6; Daumas, "Les appareils d'experimentation de Lavoisier," *Chymia* 1950, 3:45–62; and Daumas, "Precision of Measurement and Physical and Chemical Research in the Eighteenth Century," in *Scientific Change: Historical Studies,* ed. A. C. Crombie (London: Heinemann, 1963), pp. 418–430. On the French mathematical engineering tradition see C. Stewart Gillmor, *Coulomb and the Evolution of Physics and Engineering in Eighteenth-Century France* (Princeton: Princeton Univ. Press, 1971), esp. Ch. 1; and Charles Coulston Gillispie, *Science and Polity in France at the End of the Old Regime* (Princeton: Princeton Univ. Press, 1980), pp. 506–552.

[7] For this use of the term *technologies* see Steven Shapin, "Pump and Circumstance: Robert Boyle's Literary Technology," *Soc. Stud. Sci.,* 1984, *14*:481–520.

for producing it by heating red mercury calx. In 1776–1777 he determined that this "purest part of the air" was what combined with solid substances in the course of their calcination or combustion. He disclosed its role as the portion of the atmosphere consumed in respiration and ascribed to it the power to make substances acidic that was to give it its name, *oxygen* (the acid generator).[8]

Lavoisier's readiness to deploy new apparatus, borrowing it from disciplines usually considered beyond the bounds of chemistry, had been characteristic of his work since his early researches in mineralogy and geology. Already in the 1760s he had been using thermometric and barometric measurements in geological surveys and developing hygrometric methods for analyzing mineral water. It was in the 1780s, however, that he began to exploit physical instrumentation with greater consistency and to deploy it in his campaign against traditional chemical theory. His collaboration with the mathematical physicist Pierre-Simon de Laplace in 1782–1783 has been illuminated by a classic study by Henry Guerlac and in a stimulating recent paper by Lissa Roberts.[9] The two collaborators designed and used a new instrument, which they introduced in a jointly written "Mémoire sur la chaleur" in 1783. As Roberts points out, the (initially unnamed) machine was presented as a purportedly unproblematic measuring device for heat exchanges in reactions, the authors professing that it had no particular implications as to the nature of heat. The naming of the device as a *calorimeter* occurred subsequently in the context of Lavoisier's systematic reconstruction of the disciplinary profile of chemistry in his *Traité élémentaire de chimie* (1789). Roberts also shows that other experimenters, such as Josiah Wedgwood and Adair Crawford, experienced difficulties in replicating Lavoisier and Laplace's experiments and disputed the working of their machine. An anonymous writer appears to have reflected the general appraisal, when he wrote in 1797 that, "little reliance . . . can be placed on the accuracy of this much-boasted process of the French chemists," although their results had been presented "with all the precision of the new school."[10]

In the case of the calorimeter, an initial attempt to build a consensus around the supposedly theory-neutral use of a measuring machine was succeeded by a more explicitly theoretical deployment of the instrument. Lavoisier was also, by the mid 1780s, making use of other new apparatus to try to secure acceptance of his theories of combustion, acidity, and the composition of water. In this period Lavoisier's instruments were just as much at issue as his substantive theoretical claims. At the beginning of the decade he had no allies among leading chemists. Most remained convinced that phlogiston was a material entity released from burning bodies. Indeed, phlogiston acquired a new lease on life, in the view of many, when it was identified by the Irish chemist Richard Kirwan as the basis of "inflammable air" (the

[8] Henry Guerlac, *Lavoisier—The Crucial Year: The Background and Origin of His First Experiments on Combustion in 1772* (Ithaca, N.Y.: Cornell Univ. Press, 1961); Guerlac, *Antoine-Laurent Lavoisier: Chemist and Revolutionary* (New York: Scribners, 1975); and Frederic L. Holmes, *Lavoisier and the Chemistry of Life: An Exploration of Scientific Creativity* (Madison: Univ. Wisconsin Press, 1985).

[9] Henry Guerlac, "Chemistry as a Branch of Physics: Laplace's Collaboration with Lavoisier," *Historical Studies in the Physical Sciences,* 1976, *7*:193–276; and Lissa Roberts, "A Word and the World: The Significance of Naming the Calorimeter," *Isis,* 1991, *82*:198–222.

[10] T. H. Lodwig and W. A. Smeaton, "The Ice Calorimeter of Lavoisier and Laplace and Some of Its Critics," *Annals of Science,* 1974, *31*:1–18; and *Critical Examination of the First Part of Lavoisier's Elements of Chemistry* (London, 1797), pp. 20–21.

gas Lavoisier was to call "hydrogen"). Kirwan articulated a theory of combustion that won considerable support. In his view the phlogiston released by a burning body combined with dephlogisticated air to form fixed air, which then united chemically with the residue of the solid to form a calx or acid. Kirwan's account had the appeal of accommodating the weight gain that was agreed to occur in instances of combustion and calcination while maintaining the existence of phlogiston.[11]

In the face of this alternative to his theory of combustion, Lavoisier's fortunes turned on a new issue introduced into the debate in the early 1780s: the composition of water. In 1781 Henry Cavendish produced water from a mixture of inflammable and dephlogisticated airs ignited by an electric spark. The experiment emerged from the tradition of eudiometry, in which the "goodness" of a sample air was measured by phlogistication (in this case by sparking with inflammable air) and measuring the diminution in volume. The production of water in the reaction was a quite unexpected result. Cavendish canvassed a couple of explanations, suggesting that the more likely one was that water was part of the composition of both airs and was released on their combination by a kind of condensation reaction.[12]

Lavoisier seized on this result, repeating Cavendish's experiment before its long-delayed publication. In June 1783, with the assistance of Laplace and in the presence of Charles Blagden (Cavendish's assistant) and witnesses from the Académie des Sciences, Lavoisier ignited jets of the two airs over mercury in a sealed glass vessel. The experiment made use of two pneumatic chests that he had recently had constructed for storing the gases. Lavoisier immediately announced a new interpretation of the reaction, stating that water was the sole product of combination of the two gases and hence was not an element, as Cavendish and all other chemists had maintained, but a compound.[13] This interpretation explained two classes of phenomena that had previously constituted troublesome anomalies for his theory. The inflammable air generated by metals when they dissolved in acids could now be explained as a product of the decomposition of water, while the reduction of lead calx and other calxes by inflammable air could be understood in terms of the combination of the gas with oxygen from the calx (or oxide) to synthesize water.

Lavoisier's interpretation of the reaction was not, however, accepted by other chemists. Cavendish, when he finally published the account of his own experiment in 1784, referred to Lavoisier's antiphlogistic explanation but professed himself unconvinced: "As the commonly received principle of phlogiston explains all phenomena, at least as well as Mr. LAVOISIER'S, I have adhered to that."[14] James Watt, the Birmingham steam-engine manufacturer and a friend of Priestley, made a similar distinction between the facts reported by the French experimenters and the interpretive gloss they had laid over them. Watt wrote to Deluc that he had no reason to doubt the credibility of the factual report that water was the sole product of the reaction and its weight equal to that of the two gases: "From the character you give

[11] Richard Kirwan, "Remarks on Mr. Cavendish's Experiments on Air," *Philosophical Transactions of the Royal Society,* 1784, *74*:154–169; and Michael Donovan, "Biographical Account of the Late Richard Kirwan, Esq.," *Proceedings of the Royal Irish Academy,* 1850, *4*: lxxxi–cxviii.

[12] Henry Cavendish, "Experiments on Air," *Phil. Trans.,* 1784, *74:*119–153.

[13] A. L. Lavoisier, "Mémoire dans lequel on a pour objet de prouver que l'eau n'est point une substance simple," *Oeuvres de Lavoisier,* ed. J. B. Dumas and Edouard Grimaux, 6 vols. (Paris: Imprimerie Nationale, 1864–1893), Vol. II, pp. 334–359.

[14] Cavendish, "Experiments on Air" (cit. n. 12), p. 152.

me of the gentlemen who made it, there is no reason to doubt of its being made with all necessary precautions and accuracy." He was, however, not convinced that the experiment was demonstrative of the compound nature of water or the nonexistence of phlogiston. Alternative ways of "solving the phenomena," which were "as plausible as any other conjectures which have been formed on the subject," remained open.[15] Watt's conjectured explanation was very like Cavendish's. Dephlogisticated air was water deprived of phlogiston and with its latent heat bound; inflammable air was phlogiston plus a little water and latent heat. When the two airs united, water was released along with heat.

Lavoisier was thus made aware that a more persuasive proof than the June demonstration was needed if his contention of the compound nature of water were to be accepted. In the autumn and winter of 1783/84 he labored to provide such a proof. His approach was to attempt more accurate measurement of the quantities of reactants and products, following the lead of Gaspard Monge, instructor in experimental physics at the military engineering academy, the Ecole Royale du Génie, at Mézières. In June and July 1783 Monge had conducted his own experiments on the synthesis of water independently of Lavoisier's, measuring the volumes and specific weights of the reactant gases and thus establishing their (almost exact) equality to the weight of water produced. Such a quantitative approach appealed to Lavoisier because it seemed to offer the rigor of a geometrical standard of proof, since "it is no less true in physics than in geometry that the whole is equal to its parts."[16]

To repeat the quantified synthesis experiment, Lavoisier would require new vessels capable of measuring the volumes of the gases used; the pneumatic chests he had employed with Laplace were not adequate for this purpose. He therefore enlisted the help of Jean-Baptiste Meusnier, a former pupil of Monge's at Mézières, who set to designing appropriate vessels and having them constructed by the instrument maker Pierre Mégnié. Meanwhile Lavoisier and Meusnier worked on an experiment to demonstrate the compound nature of water by decomposing it into its constituent gases. They passed steam through a red-hot iron gun barrel. The steam was taken to be decomposed, its oxygen uniting with the iron to form an oxide and its inflammable air emerging from the pipe to be collected along with undecomposed water. Lavoisier reported the success to the Académie in April 1784 and subsequently published the account. Although he admitted that the proportions of the constituents of water could not yet be calculated with "mathematical precision," since the gun barrel had also undergone oxidation on the external surface while being heated, he nonetheless proposed the experiment as a "demonstrative proof" that water was a compound.[17]

Again, however, dissension continued. Kirwan and Priestley denied to Lavoisier's experiments the implication their author sought to give them. Both insisted that the proposed analysis of water was no such thing. What had happened was that phlogiston (inflammable air) had been displaced from iron by combination of water with the metal. In February 1785 Priestley described to the Royal Society his own

[15] James Watt, "Thoughts on the Constituent Parts of Water," *Phil. Trans.*, 1784, *74*:329–353, esp. pp. 329, 333.

[16] Lavoisier, "Mémoire dans lequel on a pour objet" (cit. n. 13), p. 339.

[17] A. L. Lavoisier and J. B. Meusnier, "Mémoire où l'on prouve, par la décomposition de l'eau, que ce fluide n'est point une substance simple," *Oeuvres de Lavoisier* (cit. n. 13), Vol. II, pp. 360–373, esp. p. 371.

replication of Lavoisier and Meusnier's experiment, in which he used measurements of weights of reactants to show that the source of the inflammable air was the iron, not the water. As had Cavendish and Watt, Priestley charged Lavoisier with transgressing the convention that experimental philosophers should simply describe what they observed and not go beyond that to impose hypothetical rationalizations on the phenomena: "Whilst philosophers are faithful narrators of what they observe, no person can justly complain of being misled by them; for to *reason* from the facts with which they are supplied is no more the province of the person who discovers them, than of him to whom they are discovered."[18]

Facing this persistent opposition to his claim that water was a compound of two gases, Lavoisier continued to seek a more stringent and compelling proof, to push back the boundary that his critics had erected between the "facts" of experiment and what they insisted could only be an interpretation or "hypothesis." He worked to make the compound nature of water into a fact—a direct, unmediated inference from experiment, permitting no possibility of doubt. This was to be done by employing new, more refined apparatus to yield quantitative weight measurements of an unprecedented accuracy. His efforts culminated in a large-scale set-piece demonstration of the analysis and synthesis of water in the Paris Arsenal on 27 and 28 February 1785. On that occasion Lavoisier assembled all the elements of his form of experimental practice to convey a demonstrative proof of his claims; his apparatus was deployed in the full setting designed to maximize its persuasive efficacy.[19]

The analytic part of the 1785 experiment was relatively little changed from what Lavoisier and Meusnier had accomplished the previous year. But the operation to synthesize water from its component gases was performed with unprecedented care and very sophisticated new apparatus. Preparing for the experiment, Lavoisier further exploited his links with personnel trained in the mathematical engineering tradition and with the skilled instrument makers who served it. He continued to work with Laplace and Meusnier and recruited Monge to help with the synthesis. The instrument maker Mégnié produced the new pneumatic vessels designed by Meusnier towards the end of 1783 and was paid 338 livres for them. He also built two new balances for Lavoisier, using novel techniques to suspend the beams and damp their oscillations. The larger of these two was estimated to be capable of weighing one pound with an accuracy of about 1 in 100,000. For all his work for Lavoisier, Mégnié was paid 1,814 livres during the years 1783–1785. No more than 400 livres of this came from the Académie, the remainder from Lavoisier's personal wealth. In straightforward financial terms, Lavoisier was investing substantial resources in apparatus that would serve his purposes.[20]

The new pneumatic vessels were described in a paper published by Meusnier. Like the calorimeter, the instrument Lavoisier was subsequently to name the "gasometer" was introduced initially as an anonymous *appareil* or *machine*. It was, said

[18] Joseph Priestley, "Experiments and Observations Relating to Air and Water," *Phil. Trans.*, 1785, *75*:279–309, esp. p. 280.

[19] Maurice Daumas and Denis Duveen, "Lavoisier's Relatively Unknown Large-Scale Decomposition and Synthesis of Water, February 27 and 28, 1785," *Chymia*, 1959, *5*:113–129; and Holmes, *Lavoisier and the Chemistry of Life* (cit. n. 8), pp. 237–238.

[20] Daumas, *Lavoisier* (cit. n. 6), p. 149. Compare the 600 livres that Lavoisier paid the tinplate worker Naudin for the two calorimeters he constructed. Roberts, "Word and World" (cit. n. 9), makes the telling comparison with the average daily wages of a skilled worker: 1½–2½ livres.

Meusnier, a "universal instrument" to manipulate volumes of gases, "by a perfectly uniform flow, variable at will, and giving, at each instant, the measure of the quantity of air used with all the precision that one can desire.[21] The device was based on the principle of the pneumatic trough, used by numerous eighteenth-century chemists to store gases over water (see Figure 1). An upper tank, open at the bottom, was suspended from a counterweighted beam so that it could move up and down within a lower tank containing water. Pipes entered the tank to introduce the gas as it was prepared and to let it out as required. The crossbeam had arcs of circles mounted on each end to equalize frictional resistance to motion at all positions of the upper tank, which was suspended by a chain designed not to elongate under tension.

Meusnier's major design innovation was a means of ensuring a constant flow of gas out of the apparatus. To achieve this, a constant pressure had to be maintained. But as the upper tank descended it would displace water and thereby reduce the pressure on the gas. To compensate for this Archimedean thrust, Meusnier designed an ingenious modification to the beam arm that suspended the counterweight. The end of the arm was displaced parallel to the remainder of it, so that the moment of the counterweight around the fulcrum would be different for different positions of the beam. The linear variation in effective counterweight compensated for the linear variation in pressure of the gas due to Archimedes' principle. A long screw connected the displaced part of the arm to the rest of the beam, so that the degree of displacement could be adjusted to set a particular pressure. A scale mounted beside the screw enabled this to be measured. Another scale, on the arc of the arm suspending the tank, was read against a pointer mounted in a fixed position on the fulcrum pillar. This scale could be calibrated to give the volume of gas contained in the vessel. And the pressure of the gas could be read from a water manometer mounted on the outside of the lower tank and connected with the interior.

Two gasometers were required for the synthesis experiment: one each for the oxygen and the hydrogen (see Figure 2). Towards the end of December 1784 Lavoisier and his collaborators began operations to calibrate them. For each instrument this was a two-stage process, requiring several days' work. First, the screw of the counterweight arm was adjusted while the pressure of gas in the vessel was observed on the water manometer. Appropriate positions on the scale were noted to maintain different set pressures: one inch (of water, above atmospheric pressure), two inches, three inches, and so on. Each setting would produce a different, but constant, speed of gas flow. In the second stage the volumes of gas in the vessel, corresponding to different positions of the pointer against the other scale, were determined. This was done by filling bottles of known capacity with air drawn from the gasometer and noting the change in the pointer position. This was done several times, with bottles of different sizes, to overcome inequalities in the width of the tank at different heights. When the scale had been calibrated in terms of volume, the conversion could be made to weight by consulting prepared tables of the densities of the two gases under various conditions of temperature and pressure.[22]

[21] J. B. Meusnier, "Description d'un appareil propre à manoeuvrer différentes espèces d'air," in *Oeuvres de Lavoisier* (cit. n. 13), Vol. II, pp. 432–440.

[22] This account is drawn from Daumas and Duveen, "Lavoisier's Large-Scale Decomposition" (cit. n. 19); Meusnier, "Description" (cit. n. 21); and A. L. Lavoisier, *Traité élémentaire de chimie, présenté dans un ordre nouveau, et d'après les découvertes modernes,* 2 vols. (Paris, 1789), Vol. II, pp. 346–360.

The success of the calibrations depended, of course, on the various skills of the experimenters, not least their tacit knowledge of the apparatus they were handling. Accuracy was striven for in all the measurements taken. The position of the pointer against the limb scale was read to two places of decimals of a degree of arc, by use of an attached vernier. Verniers were apparently also fitted to the thermometer, which was read to one decimal place of a degree Réaumur, and to the barometer, which was read to one decimal place of a line of mercury. (One line is 1/12 of an inch, approximately 0.225 cm.) Lavoisier had been interested in the accuracy of thermometers and barometers since his early mineralogical work, and he now put to use the most advanced precision versions of these instruments.[23]

The calibrations completed, Lavoisier invited about thirty savants, including a dozen witnesses nominated by the Académie, to attend the demonstration. The course of the experiments has been well described by Daumas and Duveen.[24] The first analysis was begun on the morning of 27 February, and more than ten bell jars were filled with hydrogen gas and measured. This hydrogen was used in the first synthesis experiment later the same day. One of the two gasometers was loaded with the gas, the other having been filled with oxygen obtained from heating red mercury calx. A jet of hydrogen was led into the reaction vessel, which was filled with oxygen and connected with the oxygen-holding gasometer. After a few failures, the jet was ignited by a spark from an electrical machine owned by Lavoisier. While the constant flow of gases from the gasometers continued, the combustion was maintained for about three hours. The following morning, both gasometers were refilled and the combustion resumed, while a second analysis was performed to produce more hydrogen. This had in turn been consumed in the synthesis experiment by the end of the evening of 28 February. Further work was done on the following days, though it was less carefully witnessed and recorded. The water produced by the synthesis reaction was carefully weighed and analyzed, as was the residual gas in the reaction vessel. Finally, the weights of reactants and products in the two analyses and the one (discontinuous) synthesis were calculated.

The circumstances and results of the demonstration made it convincing to many of those who took part. Lavoisier had not so much mounted a show as provided an opportunity for his colleagues to participate in a well-organized team effort. The academicians inspected the apparatus, took measurements, and signed their names to the records of the results. The procedures of witnessing, recording, and certifying both guaranteed the authenticity of the results and gave the participants a stake in their validity. Monge and many of the mathematicians and physicists in the Académie confirmed their support for Lavoisier's doctrine. The crucial conversion was that of the chemist Claude-Louis Berthollet, who wrote enthusiastically to Blagden about the demonstration and, in a paper read to the Académie on 6 April 1785, announced that he had been convinced by "the beautiful experiment" that water

[23] Estimates of accuracy of measurement are drawn from examination of notes from the experiment surviving in the Lavoisier MSS, History of Science Collection, Cornell University Library, Ithaca, N.Y., MSS 3.04, 9.02a–d, 9.03a–d, 9.04a–c, 9.05, 9.06, 9.07a–b, 9.08, 9.09, 9.10, and 9.11a–c. On the accuracy of thermometers and barometers see Theodore S. Feldman, "Late Enlightenment Meteorology," in *The Quantifying Spirit in the Eighteenth Century,* ed. Tore Frängsmyr, J. L. Heilbron, and Robin E. Rider (Berkeley/Los Angeles: Univ. California Press, 1990), pp. 143–177, on pp. 156–157, 166.

[24] Daumas and Duveen, "Lavoisier's Large-Scale Decomposition" (cit. n. 19), pp. 123–126.

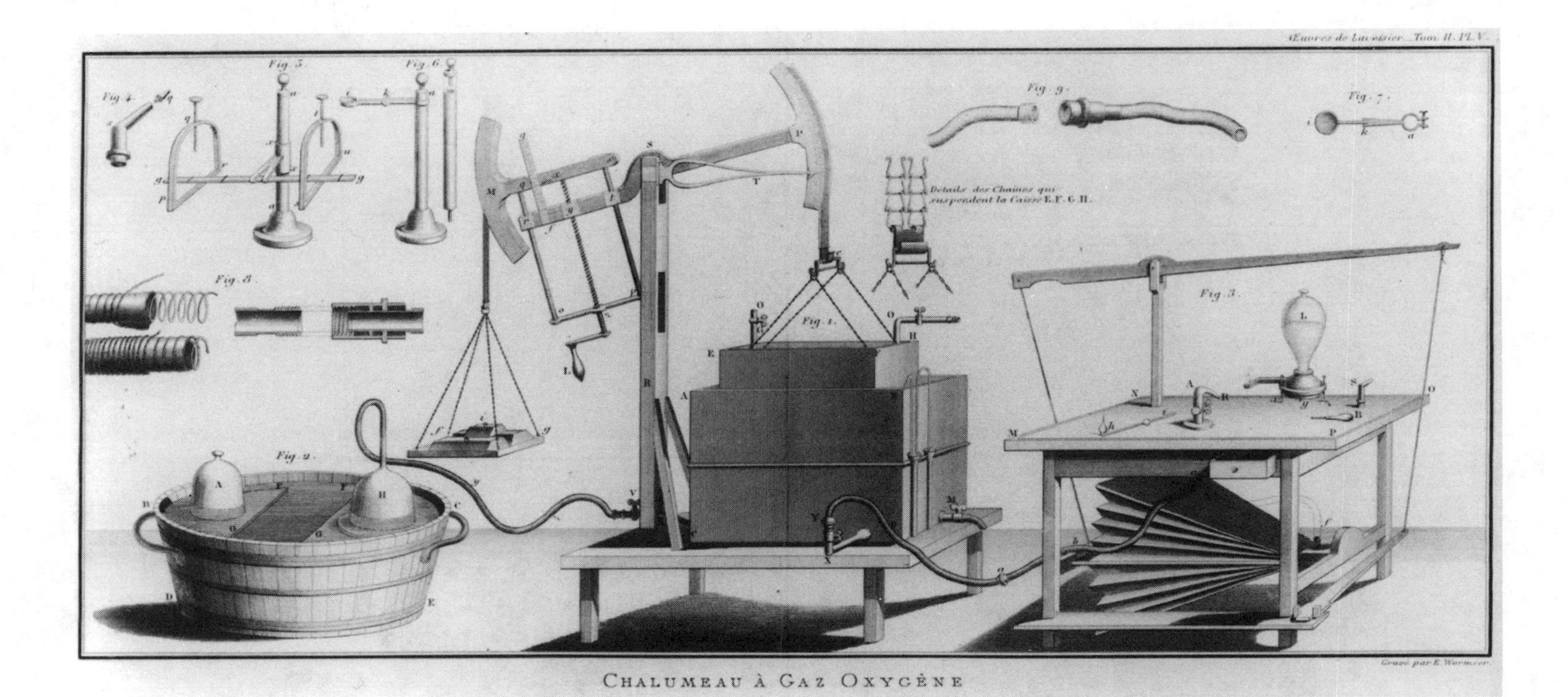

Figure 1. *Jean-Baptiste Meusnier's gasometer design, built by Pierre Mégnié. From* Journal Polytype des Sciences et des Arts, *1786*, 1, *plate* 2, *facing p. 44. Courtesy of the History of Science Collections, Cornell University Library.*

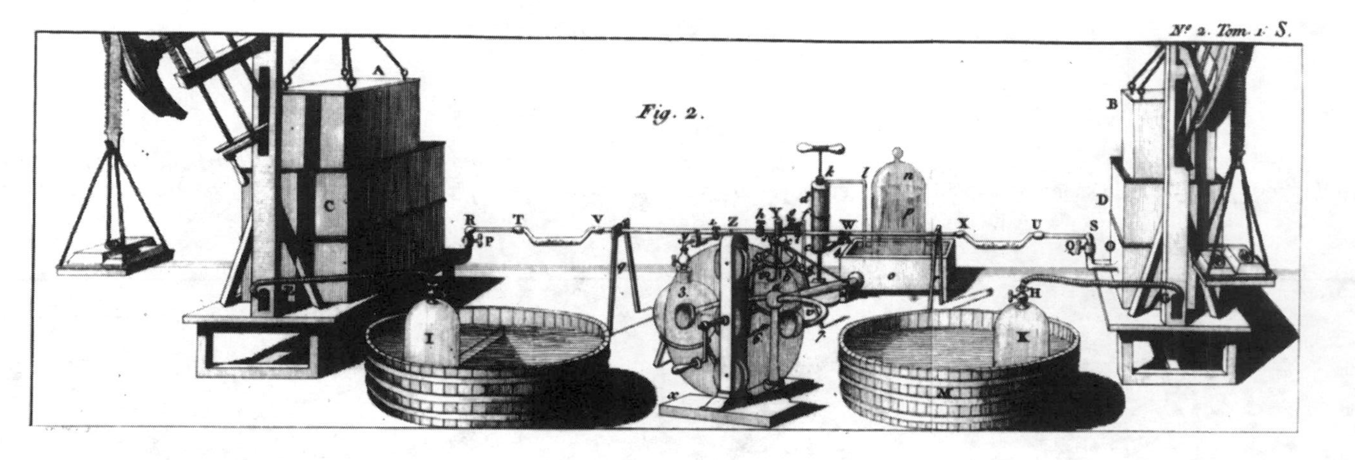

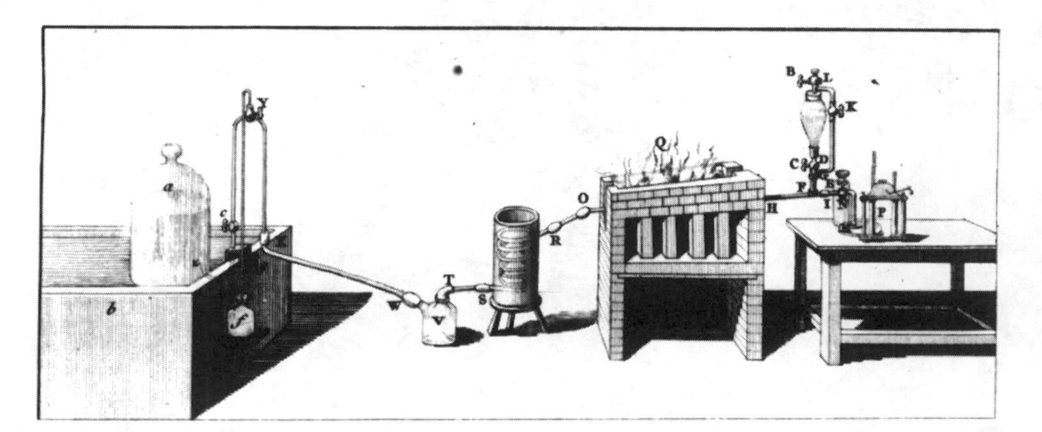

Figure 2. Top: *Lavoisier's apparatus for synthesizing water, used in the experiments of February 1785.* Bottom: *The apparatus for analyzing water. From* Oeuvres de Lavoisier *(Paris, 1864–1893), Vol. II, plate 5. By permission of the Syndics of the Cambridge University Library.*

was indeed a compound.[25] Not everyone was so convinced: the chemists Balthazar-Georges Sage and Antoine Baumé, who were present at the Arsenal, remained opposed to Lavoisier's theory.[26] But in general it appears that Lavoisier had managed the social technology of convincing his audience at the demonstration almost as successfully as he had manipulated the material technology of his apparatus.

Persuading a wider audience was however a different matter. On this occasion, Lavoisier's literary technology untypically let him down. A brief account of the experiment, probably based on drafts by Lavoisier, appeared under Meusnier's name in the *Journal Polytype des Sciences et des Arts* in February 1786.[27] Meusnier was apparently also charged with producing a comprehensive report, but this never emerged. In 1789 Berthollet was still apologizing for the fact that Meusnier's absence from Paris on military duties had prevented him from completing the work.[28] The account that did appear gave only an incomplete impression of the sophistication of the apparatus and the precautions taken in its use. No description was given of the calibration procedures for the gasometers or the meticulous care taken to ensure the accuracy of the measurements. Somewhat out of the blue, Meusnier announced a rounded-off figure for the proportions of the components of water: 85 percent oxygen and 15 percent hydrogen, by weight; a figure that was within the range of the results obtained but not unambiguously proven. The rather thin account was buttressed by a somewhat dogmatic and aggressive rhetoric. Readers were bluntly told that the description was "more than sufficient to lay hold of the certainty of the proposition" that water had just this composition. And the paper concluded by promising that methods of precision measurement offered the prospect of uniting chemistry with the other physical sciences and advancing to make discoveries of unprecedented certainty.[29]

II. THE CRITICS OF THE STRATEGY

Perhaps unsurprisingly, many chemists, particularly outside France, were not convinced. In Britain, Kirwan and Priestley continued to voice objections to the composition of water and gained significant support. In France, Jean-Claude de Lamétherie kept up a steady barrage of criticism for the remainder of the 1780s. In the *Observations et Mémoires sur la Physique,* of which he assumed the editorship in 1785, Lamétherie coordinated the anti-Lavoisian forces, publishing the views of such critics as Sage and Deluc, translating papers by Priestley and Keir, and launching regular attacks in his "Discours préliminaire" at the front of each annual volume.[30]

Various alternative interpretations of Lavoisier's prized analysis and synthesis reactions were advanced. In his *Essay on Phlogiston* (1787) Kirwan proposed that

[25] C.-L. Berthollet, "Mémoire sur l'acide marin déphlogistiqué," *Observations sur la Physique,* 1785, *26*:321–325, esp. p. 324; and H. E. LeGrand, "The 'Conversion' of C.-L. Berthollet to Lavoisier's Chemistry," *Ambix,* 1975, *22*:58–70, esp. pp. 67–68.

[26] Perrin, "Triumph of Antiphlogistians" (cit. n. 4), pp. 49, 55–56, 62.

[27] A. L. Lavoisier and J. B. Meusnier, "Développement des dernièrs expériences sur la décomposition et la recomposition de l'eau," in *Oeuvres de Lavoisier* (cit. n. 13), Vol. V, pp. 320–334.

[28] C.-L. Berthollet, "Considérations sur les expériences de M. Priestley," *Annales de Chimie,* 1789, *3*:63–114, on p. 70.

[29] Lavoisier and Meusnier, "Développement" (cit. n. 27), pp. 205–209; cf. Holmes, *Lavoisier* (cit. n. 8), p. 237.

[30] See, e.g., J. C. De Lamétherie, "Discours préliminaire," *Obs. Phys.,* 1787, *30*:3–45, esp. pp. 29–45; and Lamétherie, "Discours préliminaire," *Obs. Phys.,* 1789, *34*:3–55, esp. pp. 26–29.

inflammable and dephlogisticated airs would combine to form water only in the conditions of extreme heat that Lavoisier had used. At lower temperatures they would form fixed air, as in normal processes of combustion. Of the analysis experiment, Kirwan reiterated the traditional view that the inflammable air emerging from the gun barrel was phlogiston displaced from the iron by water.[31] Priestley concurred with this, but his view of the supposed synthesis was different from Kirwan's. When he had repeated Cavendish's original experiment, he had already concluded that the water produced by the reaction of the two gases was present in their composition in the gaseous state. The experiment showed only that all "airs" contained some proportion of water. There was, however, another product, namely "nitrous acid" (later called nitric acid), which was produced by the combination of the ponderable bases of the gases. Until the early 1790s Priestley continued to perform the experiment with what he still called "inflammable" and "dephlogisticated" airs and to record both water and nitrous acid as the products. Although Berthollet and other Lavoisians tried to cast doubt on Priestley's methods, asserting that the acid was the result of contamination of his oxygen by atmospheric nitrogen, Priestley consistently denied such contamination. His position was endorsed by Keir, Deluc, Lamétherie, and Watt, among others.[32]

As well as insisting upon these alternative experimental facts, Lavoisier's critics teased apart the rhetoric by which he constructed his claims. Lavoisier had sought to have his claims accepted as facts by asserting that they followed directly from precise quantitative measurements. In relation to the 1785 demonstration he wrote, according to Kirwan's translation:

> This double experiment . . . may be regarded as a demonstration, . . . if in any case the word Demonstration may be employed in natural philosophy and chemistry. . . . The proofs which we have given of the decomposition and recomposition of water being of the demonstrative order, it is by experiments of the same order, that is to say by demonstrative experiments, which they ought to be attacked.[33]

Rather than accepting the challenge to confront him on his own ground, however, Lavoisier's opponents used various strategies to disconnect what were agreed to be the facts from their purported implications.

Kirwan and William Nicholson, the editor of the second edition of Kirwan's *Essay on Phlogiston* (1789), focused on the assumed connection between precision and demonstration. Kirwan praised Lavoisier as "the first that introduced an almost mathematical precision into experimental philosophy," but denied that one or a few accurate experiments could overturn decades of chemical experimentation supporting the phlogiston theory. There was no possibility that a single experiment could

[31] Richard Kirwan, *An Essay on Phlogiston and the Composition of Acids,* 2nd ed., ed. William Nicholson (London, 1789), pp. 42–44.

[32] Priestley, "Experiments and Observations" (cit. n. 18), pp. 282, 294–295; Joseph Priestley, "Further Experiments Relating to the Decomposition of Dephlogisticated and Inflammable Air," *Phil. Trans.,* 1791, *81*:213–222; J[ames] K[eir], *The First Part of a Dictionary of Chemistry* (Birmingham, 1789), pp. 118–119; Deluc, "Lettre à M. De La Métherie" (cit. n. 1), pp. 145–146; J. C. de Lamétherie, "Mémoire sur l'air phlogistiqué (ou impur) obtenue par la combustion de l'air inflammable & de l'air pur," *Obs. Phys.,* 1789, *34*:227–228; and James Watt to Joseph Black, 8 June 1788, in *Partners in Science: Letters of James Watt and Joseph Black,* ed. Eric Robinson and Douglas McKie (London: Constable, 1970), pp. 166–167.

[33] Lavoisier, as quoted in Kirwan, *Essay on Phlogiston* (cit. n. 31), pp. 59–61.

be "demonstrative" of a conclusion as revolutionary as the notion that water was a compound. Kirwan insisted that "the book of nature should be interpreted like other books, the sense of which must be collected . . . from an attentive consideration of the whole." When this was done, contradictions between Lavoisier's claims and other well-established experimental results were not hard to find. Kirwan pointed out that the decomposition of water was supposedly effected by iron but not by charcoal, while other findings showed charcoal had a greater affinity for oxygen than did iron. The doctrine of affinities yielded several other problems for the antiphlogistic theory, as Lavoisier was obliged to acknowledge.[34]

Nicholson took a slightly different tack, in a sophisticated critique of Lavoisier for making excessive claims for accuracy of measurement. After discussing the likely experimental errors in measurements of weights and volumes, Nicholson concluded that Lavoisier's figures, with their long strings of decimals, "exhibit an unwarrantable pretension to accuracy." That being so, the rhetorical link from precision of measurement to certainty of conclusion was broken. Measurements that showed a quite spurious precision could not be taken as proof of what was asserted. Taking up Lavoisier's terminology of a "demonstrative order" of proof, Nicholson wrote: "When the real degree of accuracy in experiments is thus hidden from our contemplation, we are somewhat disposed to doubt whether the *exactitude scrupuleuse* of the experiments be indeed such as to render the proofs *de l'ordre demonstratif.*"[35]

In an earlier work Nicholson had already discussed and dismissed the possibility that experimental knowledge could attain the certainty of demonstration. Drawing upon the resources of British empiricist philosophy, Nicholson had argued in his *Introduction to Natural Philosophy* (1782) that experiments could not convey facts to the mind with sufficient immediacy to give them the intuitive certainty of mathematical truths. "The great perspicuity and certainty of mathematical knowledge," he wrote, "arises from the simplicity of the ideas employed, and their not depending on any external being." In experimental operations, however, it was not possible to present all the relevant aspects of the situation to the mind directly and simultaneously. Hence, "in general, we must be contented with less proof than demonstration."[36] The results of experiments could not, therefore, be treated like the axioms of geometry—demonstrative certainty did not inhere in the results of experiments, as Lavoisier maintained. This appears to have been the general view among British natural philosophers, who shared the empiricist position that the senses could not immediately perceive all the elements of a complex experimental situation. The English chemist Thomas Beddoes used the same terms in his discussion of the demonstrative status of the water composition experiments, in his *Observations on the Nature of Demonstrative Evidence* (1793):

> What for instance is it, that prevents me from being as certain, that water consists of hydrogene and oxygene airs, as of any proposition in Euclid?—nothing surely but the incompetency of my senses. . . . Now if I could perceive the small quantity of azotic air present separately uniting with a certain portion of the oxygene air to form acid, while the hydrogene air unites with the rest to form water; if I could see that the airs

[34] Kirwan, *Essay on Phlogiston* (cit. n. 31), pp. 7, 304, 317.

[35] *Ibid.*, pp. viii, xi.

[36] William Nicholson, *An Introduction to Natural Philosophy,* 2 vols. (London, 1782), Vol. I, pp. 1–6, quotations on p. 4.

> previously contain only a little or no water beforehand, and if there was no heat and light, I should have demonstrative evidence.[37]

Following a related line of critique, Priestley also took aim at the means by which Lavoisier had sought to demonstrate the factuality of his claims. Priestley identified and brought into question several elements of Lavoisier's material, social, and rhetorical techniques, subjecting to scrutiny the validity and reliability of instruments, and the form of practice within which they were put to use. Priestley's own epistemology stressed the autonomy of individual judgment and the equality of all observers, and in his view Lavoisier threatened to use his privileged access to instrumental resources to impose his own authority.[38] Priestley refused to submit his interpretive judgment to what he took to be such a naked assertion of power. He charged Lavoisier with using apparatus so complex that his experiments were liable to error and so expensive that they were impossible to replicate. The apparatus used by the French academicians in February 1785 was, he wrote, "extremely complex, as a view of their plates will shew, and mine was perfectly simple, so that nothing can be imagined to be less liable to be a source of error." For Lavoisier to use private resources to develop specially refined instrumentation seemed to Priestley to indicate his refusal to submit to the social validation of widespread replication. The synthesis experiment required, Priestley noted, "so difficult and expensive an apparatus, and so many precautions in the use of it, that the frequent repetition of the experiment cannot be expected; and in these circumstances the practised experimenter cannot help suspecting the certainty of the conclusion." In 1796 he was still insisting that the critical experiment on composition of water had "not been sufficiently repeated." Summarizing his position in 1800, he maintained: "Till the French chemists can make their experiments in a manner less operose and expensive . . . I shall continue to think my results more to be depended upon than theirs."[39]

Priestley and the other critics showed that to dispute Lavoisier's claims required analysis of the means by which those claims had been rendered as facts. Instruments that Lavoisier proposed as accurate and refined were portrayed by his opponents as unnecessarily complex and liable to error. His mobilization of specialist skills and investment of substantial financial resources in his apparatus were said to point towards an illegitimate concentration of instrumental power. This would deny other investigators the right to contribute their own observations or to replicate his experiments. His use of methods of precision measurement was also denounced as claiming an unjustified degree of accuracy, and its purported connection with a "demonstrative order" of proof was denied. In these respects Lavoisier's opponents pointed to the fragility of his experimental practice; they denied that the phenomena he had produced could be reproduced in other settings.

[37] Thomas Beddoes, *Observations on the Nature of Demonstrative Evidence* (London, 1793), pp. 108–109.

[38] Cf. McEvoy, "Enlightenment and Chemical Revolution" (cit. n. 4), pp. 205–209.

[39] Joseph Priestley, *Considerations on the Doctrine of Phlogiston and the Decomposition of Water* (Philadelphia, 1796), ed. William Foster (Princeton: Princeton Univ. Press, 1929), pp. 17, 34, 41; and Priestley, *The Doctrine of Phlogiston Established and that of the Composition of Water Refuted* (Northumberland, Pa., 1800), pp. xi, 48, 50, 76–77.

III. REPRODUCING LAVOISIER'S TECHNOLOGY

Although the controversy continued for several years, Lavoisier's instruments eventually proved themselves relatively robust tools for extending acceptance of his claims. He showed that it *was* possible to replicate his crucial experiments, including those on the composition of water. This required reproducing certain features of the setting of the original experiments at other sites. A certain amount of redesign of material technology and its supporting practices was also called for. To the extent that this succeeded and the same phenomena were taken to be reproduced elsewhere, the controversy was steadily closed in Lavoisier's favor.

One relatively well documented episode in this process is that concerning the Dutch chemist and experimental philosopher Martinus van Marum. Van Marum traveled to Paris in July 1785, a few months after the dramatic analysis-and-synthesis demonstration. Although unable to witness that event, he had a brief meeting with Lavoisier and several conversations with Monge regarding the new antiphlogistic theory. Returning to the Netherlands, he worked to replicate Lavoisier's experiments and, in February 1787, wrote to Lavoisier and Monge declaring his allegiance to the new chemistry.[40] He then devoted himself to converting other Dutch chemists. The critical necessity, as he realized, was for convincing and widespread replications of the relevant experiments, of which that on the synthesis of water seemed most important. This experiment, he noted, "had not previously been performed outside Paris." Dutch chemists had remained skeptical because "they had had no opportunity to see or to repeat experiments, the results of which formed the basic principles of the new chemical theory. Indeed, the necessary apparatus as made by the generous Lavoisier at his own expense could hardly be obtained, owing to its expensiveness and to the difficulty of constructing it with the precision required." In particular, the gasometers required for the synthesis experiment were prohibitively costly.[41]

Using the resources of Teyler's Museum in Haarlem, van Marum solved the problem by significantly simplifying Meusnier's design. Instead of a moving tank suspended by a counterweighted beam, van Marum's apparatus used two vessels: one containing the gas over water, the other a constant head of water (maintained by an adjustable tap on a feeder vessel) to keep up a steady pressure and regulate the gas flow. Linear scales attached to the side of the gas-containing vessel enabled the water level, and hence the gas volume, to be measured. With this simplified apparatus, van Marum first performed the synthesis experiment in Haarlem in 1791, "before all the

[40] Martinus van Marum to Lavoisier, 26 Feb. 1787; and van Marum to Monge, 26 Feb. 1787, in *Martinus van Marum: Life and Work,* ed. R. J. Forbes, E. LeFebvre, and J. G. Bruijn, 6 vols. (Haarlem: Tjeenk, Willink & Zoon, 1969–1976), Vol. I, pp. 193–194, 255–256. See also T. H. Levere, "Martinus van Marum and the Introduction of Lavoisier's Chemistry in the Netherlands," *ibid.,* pp. 158–286; and H. A. M. Snelders, "The New Chemistry in the Netherlands," *Osiris,* 1988, *4*:121–145, esp. pp. 127–130.

[41] Martinus van Marum, "Lettre à M. Berthollet, contenant la description d'un gazomètre construit d'une manière différente de celui de MM. Lavoisier & Meusnier," *Ann. Chimie,* 1792, *12*:113–140, trans. in *Martinus van Marum,* ed. Forbes, LeFebvre, and Bruijn (cit. n. 40), Vol. V, pp. 245–259, 241–242 (quotations). A parallel may be the slow replication outside France of Coulomb's determination of the law of electrostatic force, which was also embedded in the local practices of French engineering physics. See J. L. Heilbron, *Electricity in the Seventeenth and Eighteenth Centuries: A Study of Early Modern Physics* (Berkeley: Univ. California Press, 1979), pp. 475–476.

devotees of Physics or Chemistry who desired to be present." The result was a significant success for the Lavoisian theory. Van Marum recorded: "The simple and less costly gasometers I had used for this experiment were ordered here and imitated elsewhere, in order to repeat it in several places."[42]

Notwithstanding van Marum's claim that his apparatus allowed the synthesis experiment to be performed "with all the accuracy that can be desired," it seems clear that replicability was bought at the cost of some degree of precision. Van Marum declined to give any figures for the results of his experiments, saying only that they were "perfectly in agreement" with those of Lavoisier and his allies.[43] Nor did he give an account of the calibration of his gasometers or the calculations to reduce gas volumes to weights. Perhaps a display of precise measurement and calculation of the kind Lavoisier had mounted in 1785 was not necessary in the context in which van Marum performed his demonstrations. More important for winning over a more extensive audience was a version of the experiment that could readily be replicated in a large number of locations.

To this extent van Marum had succeeded by departing from Lavoisier's own rather uncompromising attitude to replication of his apparatus. In the *Traité* Lavoisier had written:

> In the present advanced state of chemistry, very expensive and complicated instruments are become indispensably necessary for ascertaining the analysis and synthesis of bodies with the requisite precision as to quantity and proportion; it is certainly proper to endeavour to simplify these, and to render them less costly; but this ought by no means to be attempted at the expense of their conveniency of application, and much less of their accuracy.[44]

In the event, simplifying and cheapening instruments did usually mean sacrificing their accuracy. But to extend acceptance of Lavoisier's chemical theories, the price was worth paying.

IV. CONCLUSION

Historians have noted how the "Chemical Revolution" brought with it new experimental apparatus and new standards of accurate measurement. What might not have been appreciated and certainly deserves further investigation is how integral these developments were to the achievement of Lavoisier and his allies. The lengthy and wide-ranging controversy he provoked inevitably raised numerous issues about the nature of scientific practice. The process of reaching a decision as to the facts of chemical phenomena was at the same time the formation of a consensus concerning how chemistry should be done. The program comprised a nomenclature and a textbook rhetoric; it was promulgated through social structures of communication and discipline; and it was embodied in new instrumentation and skills. Having triumphed, the revolution determined how history should be written: van Marum's

[42] Van Marum, "Letter to M. Berthollet" (cit, n. 41), p. 242.

[43] *Ibid.*, pp. 250, 251, 255.

[44] Lavoisier, *Traité élémentaire* II, pp. 359–360, trans. Robert Kerr, in *Elements of Chemistry in a New Systematic Order, Containing all the Modern Discoveries* (Edinburgh: William Creech, 1790), p. 319.

experiments were warranted as valid replications of Lavoisier's, while Priestley's were condemned as crude and contaminated.

Clearly, one task that historians should set themselves is to overcome this retrospective shaping of events. Another should be to situate material technology in such a reconstructed history. Study of instrumentation can lead to more than sterile antiquarianism; it can open the door to a broader appreciation of all the dimensions of scientific practice. A focus on controversies enables us to grasp the importance of the material culture of science, because when apparatus is disputed the connections with specific forms of practice and discourse are exposed. The links between instruments and their usage and interpretation are explicated in the course of attack or defense. We learn both that science is embodied in firmly material things and that it is nonetheless socially negotiated and historically variable.

The Ohm Is Where the Art Is: British Telegraph Engineers and the Development of Electrical Standards

*By Bruce J. Hunt**

THE HUMBLE RESISTANCE BOX has been central to the daily work, and crucial to many of the most outstanding achievements, of both physicists and electrical engineers since the middle of the nineteenth century. Without the ready availability of accurate and reliable resistance standards, the precision electrical measurement that underlies so many areas of science and technology would be virtually impossible. As Fleeming Jenkin, himself a pioneer in the development of the standard ohm, observed in 1865, "Resistance-coils . . . are now as necessary to the electrician as the balance to the chemist."[1] Standard resistances—ohms—have become so ubiquitous as to be almost invisible; they are now simply taken for granted. But how did they achieve this virtually unquestioned authority? Who undertook the immense labor necessary to secure the first reliable standards, and why? In particular, what role did the practical concerns of British telegraph engineers play in the origin of the ohm, and to what extent did the early standard coils in fact embody their aims and expertise?

Two quotations from William Thomson (later Lord Kelvin) and James Clerk Maxwell will shed some light on these questions. In a lecture on "electrical units of measurement" at the Institution of Civil Engineers in 1883, Thomson declared that British cable engineers had been ahead of physicists in the practice of accurate electrical measurement from about 1860 to the early 1870s. "Resistance coils and ohms," he said, "and standard condensers and microfarads, had been for ten years familiar to the electricians of the submarine-cable factories and testing-stations, before anything that could be called electric measurement had come to be regularly practised in almost any of the scientific laboratories of the world."[2] And in a review of Jenkin's 1873 textbook *Electricity and Magnetism,* Maxwell said that "at the present time there are two sciences of electricity—one that of the lecture-room and the popular treatise; the other that of the testing-office and the engineer's specification. The first

* Department of History, University of Texas, Austin, Texas 78712.

[1] Fleeming Jenkin, "Report on the New Unit of Electrical Resistance Proposed and Issued by the Committee on Electrical Standards Appointed in 1861 by the British Association," *Proceedings of the Royal Society,* 1865, *14:*154–164, repr. in *Philosophical Magazine,* June 1865, *29:*477–486, on p. 478. Resistance coils are perhaps better compared to a chemist's graduated weights, with the balance itself corresponding to a Wheatstone bridge.

[2] William Thomson, "Electrical Units of Measurement" (1883), *Popular Lectures and Addresses,* 3 vols. (London, 1891), Vol. I, pp. 82–83.

 0369-7827/94/0008-0004$01.00

deals with sparks and shocks which are seen and felt, the other with currents and resistances to be measured and calculated. The popularity of the one science depends on human curiosity; the diffusion of the other is a result of the demand for electricians as telegraph engineers."[3] An examination of the work of British telegraph engineers in the 1850s and 1860s will help bring out the context in which an effective demand for resistance standards first appeared, and will show how closely science and technology were intertwined in the making of the first ohms.

Most previous discussions of the development of electrical measurement in Britain have understandably focused on Thomson, who was a central figure among both scientists and engineers. The present account, while including Thomson and Maxwell, will broaden its focus to include those who were more firmly on the engineering side of the spectrum, particularly Latimer Clark and Fleeming Jenkin. As working engineers and as writers on electrical measurement (and as members, along with Thomson and Maxwell, of the British Association Committee on Electrical Standards), Clark and Jenkin laid the groundwork for much that was to follow, and their writings provide a valuable window into the aims and practices of British telegraph engineers in the mid-nineteenth century. They enable us to see how the demands and opportunities presented by the submarine telegraph industry helped shape the techniques, and literally set the standards, that were to become the ordinary working tools of both physicists and electrical engineers in the last quarter of the nineteenth century. It was through the work of these telegraph engineers that the art of electrical measurement was reduced to a set of standard practices, and that their own expertise and authority came to be embodied in the resistance boxes they calibrated and distributed.

I. LATIMER CLARK AND EARLY TELEGRAPHIC MEASUREMENT

Josiah Latimer Clark (1822–1898) worked as a chemist and a civil engineer in the 1840s before following his brother Edwin in 1850 into the Electric Telegraph Company, the first and long the largest telegraph company in Britain. At first Clark worked mostly on the ETC's system of land lines, but after helping lay the first Anglo-Dutch cables in 1853, he turned more and more to submarine telegraphy and eventually made it his main work as a consulting engineer.[4]

The network of telegraph lines that began to spread across Britain in the mid 1840s was a mix of overhead wires strung on poles and underground lines insulated with tarred cotton or (after 1850) gutta-percha. The overhead lines were electrically quite simple; as long as there were no actual breaks in the wire and the leakage of current at the supporting poles was kept within limits, operators could signal along them quite satisfactorily without worrying about measuring much of anything. As Clark later noted, "no very exact measurements are required to be made of overhead lines," and their spread gave little stimulus to the development of precision

[3] [James Clerk Maxwell], "Longman's Text-Books of Science" (rev. of Fleeming Jenkin, *Electricity and Magnetism*), *Nature,* 15 May 1873, *8:*42–43. Although this review is unsigned, its style and content (including some characteristic puns and several comments that closely parallel ones in the preface Maxwell had just written for his own *Treatise on Electricity and Magnetism*) make it clear that it is by Maxwell.

[4] On Clark see A. F. Pollard, s.v., *Dictionary of National Biography;* Bernard Finn, "Josiah Latimer Clark," *Dictionary of Scientific Biography* (New York: Scribners, 1970–1980), Vol. III, pp. 288–289; and the obituaries cited in the latter.

measurement techniques.[5] Simple vertical galvanometers (often just ordinary needle telegraph receivers) sufficed to show that enough current was getting through to produce readable signals, and that was the only real concern of the operators of the early overhead lines.

Underground lines were more complicated electrically and gave a first hint of some of the problems that were later to bedevil submarine cables. Long underground lines were troubled by electrostatic induction and retardation—obstacles to rapid signaling that were to play a major role in British thinking about electrical propagation.[6] They also suffered from flaws in their insulation (which led to the replacement of most long underground lines by overhead wires by the late 1850s), and such "faults" were much harder to find and repair on buried lines than were simple breaks on overhead wires. The main concern on the early underground and submarine lines was the "continuity" of the conductor, and very simple tests were used to check that current was getting through: Charlton Wollaston, the engineer on the first Channel cable of 1851, reportedly sometimes used his *tongue* to detect the current, a procedure that was apparently fairly common on land lines as well.[7] Clark did not use anything quite so primitive, but he and his colleagues generally tested the continuity of a cable by simply connecting it to a single cell and seeing whether it could pass enough current to deflect the needle of an uncalibrated vertical galvanometer. They checked the leakage through the insulation by connecting the same galvanometer across the gutta-percha in series with one or two hundred cells; if the deflection of the needle remained fairly small and steady, the insulation was judged to be "good enough."[8]

More precise electrical measurement first acquired a commercial value in connection with methods for locating faults in insulated wires. When the first underground lines were laid in the late 1840s, such faults—breaks in the wire or gaps in the insulation—were found by pulling up the wire, cutting it, and testing which side the fault was on. This was repeated, halving the distance each time, until the fault was isolated. The procedure was slow and expensive, and the many cuts and splices inevitably damaged the wire.[9] If a way could be found to locate faults, even approximately, by electrical tests on the ends of the line, the saving in time and money would be enormous.

Several people worked out electrical methods for locating faults in the late 1840s and early 1850s, first on underground lines and then on submarine cables. C. F. Varley of the ETC was one of the first. In 1846–1847 he was put in charge of a new underground line in London that was plagued by bad joints. He soon found that by connecting a battery and galvanometer to existing "good" wires, he could determine

[5] Latimer Clark and Robert Sabine, *Electrical Tables and Formulae for the use of Telegraph Inspectors and Operators* (London, 1871), p. 7; cf. Robert Rosenberg, "Academic Physics and the Origins of Electrical Engineering in America" (Ph.D. diss., Johns Hopkins Univ., 1990), pp. 6–8, 39–43.

[6] See Bruce J. Hunt, "Michael Faraday, Cable Telegraphy, and the Rise of Field Theory," *History of Technology,* 1991, *13:*1–19.

[7] Charles Bright, *Submarine Telegraphs* (London, 1898), p. 6n; cf. G. B. Prescott, *History, Theory, and Practice of the Electric Telegraph* (Boston, 1863), pp. 268, 282. Werner von Siemens, *Inventor and Entrepreneur: Recollections of Werner von Siemens* (London: Lund Humphries, 1966), p. 72, describes a method in which workmen located faults in cables during manufacture by putting their fingers into a tub of water and taking shocks as faulty sections were passed through the tub.

[8] Willoughby Smith, *The Rise and Extension of Submarine Telegraphy* (London, 1891), p. 23, describing the "recognised test for insulation" at the Gutta-Percha Co. in the early 1850s.

[9] Siemens, *Inventor and Entrepreneur* (cit. n. 7), pp. 78, 89.

what length of wire gave the same deflection (that is, the same current with the same battery power) as the faulty line, and by simply stepping out this distance along the course of the faulty line, he was able, he later said, to identify the bad joint "9 times out of 10."[10]

Varley's method did not require a deep understanding of currents and resistances; he simply set up a full-scale copy of the faulty circuit and walked off the corresponding length of wire. But while it was conceptually simple, Varley's method was cumbersome in practice (requiring as it did the use of enormous lengths of "good" wire) and not very precise. Werner Siemens worked out a somewhat better method for use on the Prussian underground lines in 1850, and in 1852 Charles Bright developed another, using a set of graduated resistance coils, for use on underground lines in England.[11] Resistance coils calibrated in feet of copper wire or similar small units had been introduced for laboratory use in the 1840s by Charles Wheatstone, M. H. Jacobi, and other scientists, but Bright and other engineers required much larger units; as Jenkin later noted, "the first effect of the commercial use of resistance was to turn the 'feet' of the laboratory into 'miles' of telegraph wire," and Bright's coils were indeed calibrated in equivalents of a mile of wire.[12] The replication and refinement of such resistance coils in the 1850s and 1860s was crucial to the spread of precision electrical measurement among both engineers and physicists. It is hard to gauge the precision these early fault-location methods achieved in practice, but it apparently was not very great (perhaps within 5 or 10 percent of the true position), partly because of ignorance of the true resistance per mile of the cable and partly because the resistance of the faults themselves varied unpredictably.[13]

The earliest submarine cables were built and operated without much reference to precision measurement, as the story of Wollaston's tongue indicates. But a string of failures on the second wave of cables laid in 1853–1854 led those in charge to begin taking somewhat more care, both in locating faults on laid cables and in testing the quality of the wire and insulation during manufacture. Willoughby Smith of the Gutta-Percha Company, makers of virtually all of the insulated wire used in cables, and F. C. Webb of a cable-laying subsidiary of the ETC were among the first to institute more careful electrical tests during manufacture—though by later standards these were still quite primitive, consisting mainly in using somewhat more sensitive horizontal galvanometers and recording the readings. At first some of the more rough-and-ready "practical men" derided these efforts as electrical "high farming" and a waste of time, but the tests proved of considerable use in detecting and eliminating faults and were taken up by others in the later 1850s.[14]

Most of these tests involved gauging the resistance of the wire (and sometimes the insulation), at first simply by noting the deflection the current from a given battery produced on a given galvanometer. Later, and more reliably, the resistance of the

[10] Testimony of C. F. Varley, in *Report of the Joint Committee on the Construction of Submarine Telegraphs,* British Parliamentary Papers, 2744 (1860), Vol. LXII (London, 1861), par. 2900.

[11] Siemens, *Inventor and Entrepreneur* (cit. n. 7), p. 80; and Prescott, *Electric Telegraph* (cit. n. 7), pp. 286–288.

[12] Jenkin, "Report on the New Unit" (cit. n. 1), p. 479.

[13] Siemens, *Inventor and Entrepreneur* (cit. n. 7), pp. 132–133; and William Thomson, "On the Forces Concerned in the Laying and Lifting of Deep-Sea Cables" (*Proceedings of the Royal Society of Edinburgh,* 1865, *5*:495–509), *Mathematical and Physical Papers* (Cambridge, 1884), Vol. II, pp. 153–167, on p. 158.

[14] Smith, *Rise and Extension* (cit. n. 8), pp. 25, 331.

wire was compared to that of an arbitrarily chosen standard by using a differential galvanometer or Wheatstone bridge.

As better gutta-percha came into use later in the 1850s, the leakage current on well-made cables became too small to deflect the needle of even a fairly sensitive galvanometer, and other ways had to be found to test the resistance of the insulation. In one widely used method, an insulated length of cable was charged to a high potential (as indicated on an electrometer) and then allowed to discharge by leakage; by noting the time it took to fall to half its original potential, the leakage rate and so the insulation resistance could be readily calculated. A related technique involved partially discharging the cable at intervals into short lengths of insulated cable or calibrated condensers. The electrostatic capacity of cables (an important determinant of the maximum signaling rate they could handle) was sometimes also measured, either by discharging the cable through a galvanometer and noting the "throw" of the needle, or by partially discharging it into calibrated condensers or lengths of insulated cable.[15]

By the late 1850s, Clark, Varley, Bright, Smith, and other leading British cable engineers were using calibrated resistance coils on a regular basis and were beginning to use calibrated condensers as well. To that extent Thomson's 1883 remark was correct. The engineers had developed some of their techniques on their own and had borrowed others from the small number of laboratory physicists who had worked on electrical measurement in the 1820s, 1830s, and 1840s—particularly Georg Ohm and Wheatstone—and then adapted them to fit their own needs, especially for the comparison of relatively large resistances. By the late 1850s cable engineers were doing fairly precise electrical measurements far more often and on a far larger scale than laboratory physicists ever had. But neither engineers nor physicists had yet begun to state their results in "ohms" or "microfarads." The move to these standardized units did not come until a few years later, after the failure of the first Atlantic cable in 1858.

II. SPECIFICATIONS AND STANDARDS

The attempt to lay a cable across the Atlantic in 1857–1858 carried telegraph engineering into many new areas; its failure raised questions that went to the heart of the future development of submarine telegraphy. One of the most important questions—one that did much to shape the interaction between electrical physics and telegraph engineering in the 1860s—was that of engineering specifications and standards.

William Thomson's involvement with the Atlantic cable is well known: he was a director of the Atlantic Telegraph Company, he sailed on every expedition, and his theory of signal propagation and his new instruments—particularly his mirror galvanometer—contributed greatly to the eventual success of the project in 1866. Less well known, though nearly as important, was an investigation he undertook in 1857 of the conductivity of "commercial copper." Using a Wheatstone bridge and a tangent galvanometer, Thomson and several of his Glasgow students compared various

[15] See Latimer Clark, "Appendix 2," in *Joint Committee Report* (cit. n. 10), p. 310; Clark, *Elementary Treatise on Electrical Measurement* (London, 1868), p. 113; and Siemens, *Inventor and Entrepreneur* (cit. n. 7), p. 132.

samples of wire made for use in the cable with one chosen as a standard. They soon found that the resistance of supposedly identical "pure" copper wires sometimes differed by nearly a factor of two.[16] Subsequent analysis showed that small impurities, particularly arsenic, greatly impaired the conductivity of copper, a point the London chemist Augustus Matthiessen later followed up in detail.

Thomson's discovery did not require unusually precise measurements. The differences in conductivity were large enough to have been found by anyone who undertook a systematic comparison of different samples—that is, if they trusted their standards and instruments. But before reliable standards were established, such trust was rare; it was easy, when anomalous results were found, to ascribe them to some unspecified fault in one's procedures or instruments. The real significance of Thomson's discovery lay in the stimulus it gave to electrical measurement as a means of quality control. Before 1857 cable contracts had specified only the weight or gauge of the wire, with perhaps a vague reference to its chemical purity; they had said nothing about its electrical characteristics. By the time Thomson discovered the variation in the conductivity of commercial copper, the main length of the original Atlantic cable had already been manufactured, but when the Atlantic Telegraph Company made plans in the autumn of 1857 to order an additional six hundred miles of cable to replace the length lost in that summer's abortive expedition, Thomson asked that high-conductivity copper be specified in the contract. According to Thomson's propagation theory, higher conductivity would yield a higher signaling rate, making the conductivity of the wire used in the cable a matter of direct economic importance.[17] The other directors of the Atlantic Telegraph Company hesitated at first: their electrician, Wildman Whitehouse, assured them that the conductivity of the wire was not nearly as important to rapid signaling as use of his patented induction-coil sending and receiving apparatus. But after a determined campaign, Thomson secured insertion of a clause calling for the use of only high-conductivity wire. Tests against a standard wire were instituted at the Gutta-Percha Company and manufacturers began, for a small premium, to supply copper of the required conductivity.[18]

The failure of the first Atlantic cable in September 1858 after just a month of fitful service was blamed on many things, most dramatically the high-voltage jolts Whitehouse had given it with his huge induction coils.[19] But poor materials, hasty manufacture, and the rough handling the cable had received since 1857 had evidently contributed to its demise. In 1859 the British government and the Atlantic Telegraph Company set up a joint committee to investigate the causes of the failure (and the costly collapse of the Red Sea cable a short time later) and to advise on how such disasters might be avoided in the future. The committee (made up mainly

[16] William Thomson, "On the Electric Conductivity of Commercial Copper of Various Kinds" (*Proc. Roy. Soc.*, 1857, *8*:550–555), *Mathematical and Physical Papers,* Vol. II, pp. 112–117. Willoughby Smith, *Rise and Extension* (cit. n. 8), pp. 116–117, later found that the conductivity of wire used in various early cables ranged from 17% to 75% of that of pure copper.

[17] Thomson, "Commercial Copper," p. 113.

[18] William Thomson, "Analytical and Synthetical Attempts to Ascertain the Cause of the Differences of Electric Conductivity Discovered in Wires of Nearly Pure Copper" (*Proc. Roy. Soc.,* 1860, *10*:300–309), *Mathematical and Physical Papers,* Vol. II, pp. 118–128, on p. 125n; see also Smith, *Rise and Extension* (cit. n. 8), pp. 47, 66.

[19] Vary Coates and Bernard Finn, *A Retrospective Technology Assessment: Submarine Telegraphy—The Transatlantic Cable of 1866* (San Francisco: San Francisco Press, 1979), p. 30; and Donard de Cogan, "Dr. Whitehouse and the 1858 Trans-Atlantic Cable," *Hist. Tech.,* 1985, *10*:1–15.

of eminent engineers, including Varley and Edwin and Latimer Clark) set out to identify and promote "best practice" in submarine telegraphy and to lay out clear procedures, particularly on quality control and the specification of materials, that would help secure the reliability of future cable projects and restore confidence in what had come to be seen as a great *failed* technology.[20]

The committee held hearings from December 1859 to September 1860, calling as witnesses most of the principal figures in cable telegraphy and taking supplementary reports from many of them. The committee also commissioned new experiments, many of a scale and expense far beyond anything laboratory scientists could hope to do on their own. Latimer Clark's long report, for example, was based on months of careful experimentation on enormous stocks of copper and gutta-percha and constituted a virtual treatise on electrical measurement and the properties of conducting and insulating materials. Many of his electrostatic measurements were made in a special "dry room," which Clark himself described as a "luxury," and which would have been beyond the means of any purely scientific laboratory at the time.[21] Besides appearing as an appendix to the joint committee's massive "Blue Book," Clark's report was published separately in 1861, and it provided much of the basis for his 1868 *Elementary Treatise on Electrical Measurement, for the Use of Telegraph Inspectors and Operators,* which soon became a standard reference work.[22]

Another important contributor to the committee hearings was Fleeming Jenkin, a young English civil engineer who had begun work at R. S. Newall's Birkenhead cable factory in 1857. There Jenkin learned the usual techniques of cable testing. He also assisted Werner Siemens with electrical tests on the Malta-Corfu cable, built by Newall's, and early in 1859 he began an intensive correspondence with Thomson on how best to measure the resistance of copper and gutta-percha.[23] That summer Jenkin carried out extensive tests on coils of the Red Sea cable being readied at Newall's, which he wrote up in an elaborate report for the joint committee, "On the Insulating Properties of Gutta-Percha."[24]

Rather than simply comparing the insulating power of one sample of gutta-percha with that of another, as had been the usual practice before, Jenkin set out to measure the specific resistivity of gutta-percha on the same scale as that of copper. The enormous disparity between such a good insulator and such a good conductor made this a difficult task, but by using an ordinary galvanometer to measure the current flowing through the copper core of a long cable and a far more delicate one to measure the leakage through the gutta-percha covering, Jenkin was able to find the ratio of their resistivities, and so, for any given length of cable, to calculate how much of the total

[20] *Joint Committee Report* (cit. n. 10); and C. A. Hempstead, "The Early Years of Oceanic Telegraphy: Technology, Science, and Politics," *Proceedings of the Institution of Electrical Engineers,* 1989, *136A:*297–305.

[21] Clark, "Appendix 2," *Joint Committee Report* (cit. n. 10), p. 333. The Cavendish Laboratory acquired a set of "Latimer Clark's drying apparatus for electric room" in 1875/76; see Maxwell's report in *Cambridge University Reporter,* 20 May 1876, p. 496.

[22] Clark, *Elementary Treatise* (cit. 15). On its popularity among telegraphists see Clark and Sabine, *Electrical Tables* (cit. n. 5), p. v; and Bright, *Submarine Telegraphs* (cit. n. 7), p. 60n.

[23] See Siemens, *Inventor and Entrepreneur* (cit. n. 7), p. 130; and Jenkin's letters to Thomson, 4 Feb.–4 Sept. 1859, J3–J16, Kelvin Collection, University Library, Glasgow.

[24] Fleeming Jenkin, "On the Insulating Properties of Gutta-Percha," in *Joint Committee Report* (cit. n. 10), pp. 464–481.

current would leak out through the gutta-percha and how much would pass to the far end through the copper.

In comparing the resistivities of two such different materials, Jenkin had to decide both what units to state his results in and what standards to calibrate his measurements against. These considerations were especially important because parties to a contract required a common reference point, and engineers planning a project required a set standard against which to gauge the quality of the materials they might select.

A welter of resistance "standards" were available in the late 1850s, but none were entirely satisfactory. Wheatstone had proposed a foot of copper wire weighing 100 grains in 1843, and M. H. Jacobi had sent copies of a longer "étalon" to various physicists in 1848, but neither of these standards had come into wide use—in part because there was little effective demand for them at the time they were proposed. As noted above, when telegraph engineers began to make resistance measurements, they preferred coils calibrated in larger units: miles of copper wire in Britain (standardized at the ETC as "Varley units"), kilometers of iron wire in France, and German miles of iron wire in Germany. These standards came into fairly wide use in the 1850s, but they were called into question by Thomson's discovery of the variation in the resistance of "commercial copper," and comparisons soon showed that supposedly identical standard coils often differed by several percent.[25] In his report to the joint committee Jenkin complained that the only resistance coils available to him at Newall's were three "very imperfect" ones (made by Siemens) that were nominally equivalent to 30, 60, and 90 German miles of telegraph wire, but which, when compared on a Wheatstone bridge, turned out actually to be in the ratios 30 : 59.15 : 88.27.[26]

An alternative to arbitrary material standards of resistance already existed, at least on paper, in the "absolute" system based on units of force and motion that Wilhelm Weber, building on C. F. Gauss's earlier magnetic work, had published in 1851. The absolute system was comprehensive, and as Thomson emphasized, it tied electrical quantities directly to those of work and energy, but it was difficult to explain in an elementary way how its units were defined, and even more difficult to embody them accurately in material standards. Rather than simply defining a certain piece of wire as the unit of resistance, the absolute system required a delicate measurement with special apparatus to determine the resistance of a given wire in terms of a velocity. But Weber's measurements were uncertain: some of his own determinations differed among themselves by several percent, and his unit of resistance, the meter per second, was ludicrously small (less than 1/100 millionth of the resistance of a mile of ordinary wire). It is therefore not surprising that the few telegraph engineers who knew about the absolute system in the 1850s regarded it as having little practical value.[27]

In 1860 Werner Siemens attempted to resolve the resistance standard problem by

[25] See the comparative table of standards in Jenkin, "Report on the New Unit" (cit. n. 1), facing p. 480.

[26] Jenkin, "Insulating Properties" (cit. n. 24), p. 469.

[27] [William Thomson], "Second Report—Newcastle-on-Tyne, 1863" (*British Association Report,* 1863), *Reports of the Committee on Electrical Standards Appointed by the British Association for the Advancement of Science,* ed. F. E. Smith (Cambridge: Cambridge Univ. Press, 1913), pp. 58–59.

introducing a new unit based on the resistance, at a temperature of 0°C, of a tube of pure mercury one meter long and one square millimeter in section. This unit was arbitrary, but it was of a convenient magnitude (about 1/20 the resistance of a mile of ordinary wire), and standards representing it could be measured and reproduced quite precisely—within a small fraction of one percent, according to Siemens. Matthiessen later criticized the standard, claiming that impurities dissolved from the connecting wires would alter the conductivity of the mercury, but when Siemens began to issue wire coils calibrated in terms of his mercury unit in 1860, he had every reason to think they would soon came into general use.[28]

By the time Siemens proposed his mercury unit in 1860, the idea of a single universal standard of resistance was very much in the air. Thomson had advocated the use of absolute standards in his testimony to the joint committee in December 1859, and Jenkin, appearing a few days later, went even further, calling for a "standard coil with which any resistance might be compared" to be deposited in "some public institution," much as the standard meter was kept in Paris and the standard yard in London.[29] A standard of resistance should, he said, have the same official status and authority.

Sets of standard resistance coils were of considerable use by themselves for locating faults and testing lines, but the real advantages of standardization came into play only when the process was extended beyond the coils and made part of a general policy of quality control. Resistance standards could then be used to measure, control, and record the properties of the copper and gutta-percha in a cable while it was being made and laid. By securing the ability to impose strict control over the quality of materials, engineers equipped with standard coils could ensure that only materials that met a specified standard were used—an important consideration, since a single tiny flaw could put an entire cable out of operation. Moreover, standardization of materials would allow faults to be located more exactly, since the engineer could then be sure that the resistance of each mile of his cable was strictly comparable to that of his coils, rather than subject to the large variations typical in untested materials.

Such standardization—first of resistance coils, then of production materials—is a good example of the process Bruno Latour discusses in the section "Metrologies" in *Science in Action.* Standardization of instruments and materials enables scientists and engineers to extend their networks of calculation and control by simply making and sending out what are, in effect, little pieces of their laboratories and testing rooms.[30] They can then travel around the world without, in a sense, ever having to leave their laboratories—as long as they are able to put certified copies or extensions of their instruments wherever they have to go. The engineer sought to make his entire cable as nearly as possible a chain of little standard resistances strung end to

---

On Thomson's authorship see S. P. Thompson, *Life of Sir William Thomson, Baron Kelvin of Largs* (London: Macmillan, 1910), Vol. I, p. 419.

[28] Werner Siemens, "Proposal for a New Reproducible Standard Measure of Resistance to Galvanic Currents," *Phil. Mag.,* Jan. 1861, *21:*25–38 (trans. from *Annalen der Physik,* Jan. 1860); Augustus Matthiessen and C. Vogt, "On the Influence of Traces of Foreign Metals on the Electrical Conducting Power of Mercury," *Phil. Mag.,* March 1862, *23:*171–179; see also Siemens, *Inventor and Entrepreneur* (cit. n. 7), pp. 133, 160.

[29] Thomson, par. 2524, and Jenkin, par. 2837, in *Joint Committee Report* (cit. n. 10).

[30] Bruno Latour, *Science in Action: How to Follow Scientists and Engineers through Society* (Cambridge: Harvard Univ. Press, 1987), pp. 247–257, esp. p. 251.

end, with the resistance of each mile and yard of it known, controlled, and recorded. Guesswork would be eliminated, and he would be able to dazzle his lesser brethren by specifying the location of a distant and unseen fault more precisely than the repair-ship captain could navigate to it—something cable engineers in fact began to do in the 1860s and 1870s.[31]

As useful as the precision and control afforded by standardization was within a single company's system, it became even more important when an exchange of materials was involved—when standardization became a part of contract specifications. By providing fixed and agreed reference points in which both parties could have confidence, standard resistances were crucial in settling or heading off possible disputes. By enabling engineers to secure the comparability and even uniformity of their copper and gutta-percha, to identify and police deviations, and to reproduce the properties of successful cables in a predictable way, reliable standards were crucial to the growth of the cable manufacturing industry and to the efficient operation and extension of the world cable system.

The prospective advantages of this kind of standardization became increasingly clear after the failure of the Atlantic and Red Sea cables, and in its final summary report the joint committee called for standard resistances to be used in all future contract specifications.[32] By the time the committee report was issued in April 1861, the demands of cable telegraphy were rapidly bringing the movement toward a common standard of electrical resistance to a head.

III. MANCHESTER, 1861

As is well known, this movement culminated in the formation of a Committee on Electrical Standards at the Manchester meeting of the British Association for the Advancement of Science in September 1861. Over the next few years, this committee (which initially included Thomson, Jenkin, Wheatstone, and several eminent chemists, and later added Maxwell and a number of other physicists and cable engineers) produced essentially the system of ohms, amps, and volts that we still use. It was the most important point of intersection between physicists and telegraph engineers in the 1860s, and its work, especially on the ohm, had a far-reaching effect on virtually all later work in precision electrical measurement.

The story of the formation of the British Association committee has been told many times, though perhaps never as fully as it deserves.[33] Previous accounts have contained several errors and have left a minor mystery, but guided by some previously overlooked documents, we can now clear these up.

Most accounts of the formation of the committee have focused on a paper, "On the Formation of Standards of Electrical Quantity and Resistance," presented at the meeting by Latimer Clark and Sir Charles Bright, who had recently become partners

[31] Bright, *Submarine Telegraphs* (cit. n. 7), p. 182n; and Siemens, *Inventor and Entrepreneur* (cit. n. 7), pp. 132–133.

[32] *Joint Committee Report* (cit. n. 10), pp. xvi–xvii.

[33] A. C. Lynch, "History of the Electrical Units and Early Standards," *Proc. IEE,* 1985, *132A:*564–573; Bright, *Submarine Telegraphs* (cit. n. 7), p. 61; Thompson, *Life of Kelvin* (cit. n. 27), Vol. I, pp. 417–418; Graeme Gooday, "Precision Measurement and the Genesis of Physics Teaching Laboratories in Victorian Britain," *British Journal for the History of Science,* 1990, *23:*25–51, on p. 34; and Crosbie Smith and Norton Wise, *Energy and Empire: A Biographical Study of Lord Kelvin* (Cambridge: Cambridge Univ. Press, 1990), p. 687.

in a cable consulting firm. Although the paper bore both men's names, Clark later said the "original ideas" in it were his.[34] Those ideas were certainly remarkable: after noting that "the science of Electricity and the art of Telegraphy have both now arrived at a stage of progress at which it is necessary that universally received standards of electrical quantities and resistances should be adopted, in order that precise language and measurement may take the place of the empirical rules and ideas now generally prevalent," the paper outlined a connected system of units of electrical tension, "quantity" (i.e., charge), current, and resistance; proposed that these units be given magnitudes suited to the needs of cable engineers, with prefixes to indicate multiples and submultiples; and suggested that they be named after "some of our most eminent philosophers": "ohma" for tension, "farad" for quantity, "galvat" for current, and "volt" for resistance. The practical introduction of such a system would, the paper concluded, be a "great a boon to science and to the art of telegraphy," and the authors called on the British Association to appoint a committee to promulgate such a system and to prepare electrical standards "for public use and reference."[35]

Given all this, it is not surprising that Clark and Bright's paper has generally been cited as the direct stimulus for the formation of the standards committee at the same meeting. But several facts call this straightforward story into question. First, neither Bright nor Clark was named to the committee in 1861, which would be odd if they were indeed its instigators. (Bright was added to it in 1862, Clark in 1866.) Second, the committee as originally formed was asked to report only on "standards of electrical resistance"—though Clark and Bright's main point was their call for a connected system of standards of tension, current, and quantity as well. (The committee's terms of reference were broadened to include these other electrical standards in 1862.)[36] Finally, several of the best-informed accounts from the time explicitly state that the committee was set up at *Thomson's* suggestion—and Thomson was not at the Manchester meeting (he was in Scotland nursing a badly broken leg), and so could not have been responding to Clark and Bright's paper.[37] What, then, was the relationship between Clark and Bright's paper and the formation of the standards committee?

The answer, or most of it, can be found in two letters from Jenkin to Thomson, now held in Thomson's papers at the Cambridge University Library.[38] Although they are undated, their contents show that Jenkin wrote them during the Manchester meeting, at which he acted on Thomson's behalf. Jenkin's remarks show that Thomson (no doubt motivated by concerns similar to Clark and Bright's) had launched his own effort to get the British Association to set up a committee on electrical units, or at least resistance standards, well before the Manchester meeting began, and that he was already campaigning for the adoption of an "absolute" system based on

[34] Latimer Clark to William Thomson, 3 May 1883, quoted in Smith and Wise, *Energy and Empire,* p. 687.

[35] Latimer Clark and Sir Charles Bright, "Measurement of Electrical Quantities and Resistance," *Electrician,* 9 Nov. 1861, *1:*3–4. A brief abstract, "On the Formation of Standards of Electrical Quantity and Resistance," appeared in *Brit. Assoc. Report,* 1861, pp. 37–38.

[36] The texts of the relevant resolutions are given in *Brit. Assoc. Report,* 1861, p. xl, and 1862, p. xxxix.

[37] Jenkin, "Report on the New Unit" (cit. n. 1), p. 480; Thompson, *Life of Kelvin* (cit. n. 27), Vol. I, p. 418; and Smith and Wise, *Energy and Empire* (cit. n. 33), p. 687.

[38] Fleeming Jenkin to William Thomson, J36 and J37 (undated, but between 4 and 11 Sept. 1861; J37 was written a little before J36), Kelvin Collection, University Library Cambridge (**ULC**).

Weber's. Thomson had written to Matthiessen and apparently also to Wheatstone, who drafted the resolution ("not altogether a bad one," according to Jenkin) to set up the committee.[39] The committee had in fact already been appointed before Clark and Bright arrived at the meeting with their own proposal, which they had apparently prepared without knowing of Thomson's efforts. (Bright's brother and son later wrote that Clark and Bright's paper "formed the sequel to a letter addressed by Bright to Prof. J. Clerk Maxwell, F. R. S., some months previously, on the whole subject of electrical standards and units," but I have found no further evidence of this. It was at just this time—summer 1861—that Maxwell first used measurements of electric and magnetic constants to work out the speed with which electromagnetic disturbances would propagate in his vortex ether; it is conceivable that he wrote to Bright for information on such measurements, and the biographer Brights' reference was to Bright's reply.)[40]

Jenkin immediately tried to enlist Clark and Bright behind Thomson's effort, apparently with some success—he wrote Thomson that "Latimer Clark looked delighted" when told of the plan and was "eager to have it all explained." But it was not all smooth sailing: Jenkin said later in this first letter to Thomson that he was "writing in a very great hurry after hot argument with Sir C. and Latimer."[41] In his next letter he said, "They will no doubt be easily converted," but did not say what Clark and Bright had to be converted *from;* he presumably meant they were skeptical about simply adopting Weber's absolute units. Jenkin in fact told Thomson that it was only "by force of telling others about them" that he was "beginning really to understand" absolute units himself, and they certainly were not well understood or accepted by other telegraph engineers.[42] Practical men were mainly interested in securing a concrete material resistance standard that was of convenient magnitude and perhaps bore a simple relation to units of tension and current. (One of the arguments against Siemens's mercury unit was that it was not related systematically to units of tension or current.) Telegraph engineers were leery of Weber's abstract and seemingly unrealistic system—the idea of measuring resistance by a velocity seemed odd to them, and the "meter per second" was in any case far too small a unit to be of much practical use. In January 1862 Clark wrote a letter to the *Electrician* (a new "journal of telegraphy") in which he complained that those initially appointed to the British Association Committee were "but little connected with practical telegraphy, and there is a fear that while bringing the highest electrical knowledge to the subject, and acting with the best motives, they may be induced simply to recommend the adoption of Weber's absolute units, or some other units of a magnitude ill adapted to the peculiar and various requirements of the electric telegraph."[43]

[39] Jenkin to Thomson, J37, Kelvin Collection, ULC.

[40] E. B. Bright and Charles Bright, *The Life of Sir Charles Tilston Bright* (London: Lockwood, 1910), p. 213; Maxwell to Michael Faraday, 19 Oct. 1861, in *The Scientific Letters and Papers of James Clerk Maxwell*, ed. Peter Harman (Cambridge: Cambridge Univ. Press, 1990), Vol. I, pp. 683–688.

[41] Jenkin to Thomson, J37, Kelvin Collection, ULC.

[42] Jenkin to Thomson, J36, Kelvin Collection, ULC. Cf. Clark to Thomson, 3 May 1883, quoted in Smith and Wise, *Energy and Empire* (cit. n. 33), p. 687, where Clark says that in 1861 he was "not mathematician enough to see the enormous value of an absolute system" of units.

[43] Latimer Clark, letter to the editor, *Electrician,* 17 Jan. 1862, *1:*129, as quoted in Lynch, "Electrical Units" (cit. n. 33), p. 565.

Clark's fears were soon allayed, for the committee proved intent on choosing units whose magnitudes would suit the needs of telegraph engineers. This was in fact the first principle stated in the committee's initial report, drafted by Jenkin in 1862, and Jenkin drew some of the examples he used to illustrate the usefulness of the proposed system directly from cable telegraphy.[44] The committee members—especially Thomson and Jenkin, who took the lead in most of its activities—sincerely wished to serve the needs of the cable industry; they also knew they could not hope to win universal acceptance for their system of units unless they first satisfied the cable men. Telegraphy, especially cable telegraphy, was the main arena for electrical measurement in the mid-nineteenth century and so provided the main market for the work of the British Association committee; by devising a system that would meet the needs of both telegraphers and laboratory physicists, the committee sought to unify the practices of both groups on a single basis and so give their joint units and standards a broader and more comprehensive authority than either could have commanded on their own.

The committee met several times in 1861–1862, and led by Thomson, who was intent on making the electrical units fit with those for work and energy, it eventually settled on a version of the metric absolute system. But in place of Weber's inconveniently sized "meter per second" and other units, the committee recommended the adoption of what it called "practical units"—decimal multiples of Weber's units chosen to fit cable needs. The most important of these was, of course, the resistance unit; it acquired the name "ohm," a modification of Clark's suggestion, a few years later. It was to be equivalent to 10 million meters per second, making it just a few percent larger than Siemens's mercury unit, and so of a convenient size for cable use. Once a material standard as near as possible to 10 million meters per second had been constructed, this was to serve as the permanent "British Association unit" and was not to be changed as absolute measuring techniques improved, any more than the standard meter in Paris was altered to accommodate new geodetic determinations.[45]

The committee had originally hoped to be able to issue its standards fairly quickly, but the difficulty of performing absolute measurements of the required precision meant it was late 1863 before it could offer even tentative resistance standards (through Elliott Brothers, the London instrument makers). These early coils were issued in answer to the urgent demand for practical standards—and because "defective systems" (e.g., Siemens's unit) were, the committee said, "daily taking firmer root" in the absence of concrete exemplars of the British Association unit.[46] After nearly two years of painstaking measurements by Jenkin, Maxwell, Balfour Stewart, and Charles Hockin (a recent Cambridge graduate who later became a leading cable engineer), the committee finally issued its official resistance standards in February 1865. In a brief notice in the *Philosophical Magazine,* Jenkin, as secretary of the committee, announced that copies of the standard resistance were now available, and that "A unit coil and box will be sent on the remittance [to Jenkin] of £2 10s."[47]

[44] Fleeming Jenkin, "First Report—Cambridge 1862," *Electrical Standards,* ed. Smith (cit. n. 27), pp. 1, 5.

[45] *Ibid.,* pp. 7–8. This idea was soon dropped, and the value of the ohm was in fact adjusted several times over the next few decades.

[46] "Third Report—Bath 1864," *Electrical Standards,* ed. Smith (cit. n. 27), pp. 159, 162.

[47] Fleeming Jenkin, "Electrical Standard" (letter to the editor, 7 Feb. 1865), *Phil. Mag.,* March 1865, *29:*248.

What purchasers were really buying, of course, was not just bits of wire in a box, but the concentrated expertise the coils embodied and the certified authority of the British Association committee that stood behind them. A number of coils went to scientists in Britain and Germany, but most were sold or given to telegraph companies and government telegraph departments around the world. In its 1865 report the committee proudly noted that the new standard was now to provide the basis for resistance measurements at the ETC, the Magnetic Company, the Atlantic Telegraph Company, and various colonial telegraph departments, and that several instrument makers had ordered standard coils in order to make and sell copies for general use.[48] The distribution and replication of these standards provides an excellent example of Latour's process of extending a network of calculation by sending out certified bits of a laboratory or testing room.

The British Association "ohm," as it soon came to be known, had to compete with Siemens's mercury unit for some time, but by 1868 Latimer Clark could assert (perhaps with some exaggeration) that "the measures now universally adopted are those of the British Association."[49] The adoption of the ohm by the big British cable companies played a key role in this success, especially since it came at the beginning of a huge boom in cable laying. The successful completion of the new Atlantic cables in 1866 was soon followed by the laying of cables to India, Australia, China, Japan, and along the African and South American coasts—virtually all by British companies.[50] These cables provided a much-expanded arena for the practice of precision electrical measurement, and they were standardized and tested from the first mainly in terms of British Association ohms.

The provision of reliable standards put much of electrical measurement on a substantially new basis; electricians sensed that they had entered a new era, and Jenkin noted in 1865 that "we have now reached a point where we look back with surprise at the rough and ready means by which the great discoveries were made on which all our work is founded."[51] An enormous amount of expertise was now built into the ordinary electrician's instruments, where it remained largely invisible—nowhere more so than in that humble but crucial "instrument," the resistance box.

When Jenkin declared that the resistance coil had become "as necessary to the electrician as the balance to the chemist," he was expressing a fundamental shift in the working world of both physicists and telegraph engineers.[52] A box of ohms became a central part of the practice of precision electrical measurement as it began to spread more widely from the mid 1860s. It was no coincidence that the first physics teaching laboratories appeared in Britain in this period, or that almost all of them strongly emphasized electrical measurement. Telegraphy had provided the initial market and demand for such measurement, and it continued to stimulate the development of improved measurement techniques and tools. By 1886 the London physics professor Frederick Guthrie could declare that "electricity especially voltaic lends itself perhaps *more abundantly* to exact measurement than the other branches

[48] "Fourth Report—Birmingham, 1865," *Electrical Standards,* ed. Smith (cit. n. 27), pp. 193–194.

[49] Clark, *Elementary Treatise* (cit. n. 15), p. 43. On the later history of the ohm and other units, see Larry Randles Lagerstrom, "Constructing Uniformity: The Standardization of International Electromagnetic Measures, 1860–1912" (Ph.D. diss., Univ. California at Berkeley, 1992).

[50] Bright, *Submarine Telegraphs* (cit. n. 7), pp. 106–145.

[51] Jenkin, "Report on the New Unit" (cit. n. 1), p. 479.

[52] *Ibid.,* p. 478.

of physics."[53] It was a statement that would have seemed absurd thirty years before, and it reflected the way in which electrical measurement, and with it the position of electricity relative to the rest of physics, had been transformed in the intervening years in response to the demands and opportunities presented by telegraphy.

IV. CONCLUSION

The quotations from Thomson and Maxwell with which we began reflect the sentiments of three participants (three, since Maxwell was mainly paraphrasing and endorsing remarks Jenkin had made in his book[54]) in a major transition in electrical practice in the early 1860s—a transition involving first the cable industry and then the British Association Standards Committee. All three sought (though in somewhat different ways) to encourage the alliance that had grown up between physics and telegraph engineering and to reinforce the increasing emphasis on precision electrical measurement. This is especially clear in Thomson's case—he was, after all, addressing an audience of engineers about "electrical units of measurement"—and the enormous role practical measurement played in his thinking has been ably explored by Crosbie Smith and Norton Wise. Jenkin, too, was keenly aware of the importance of electrical standards to both scientists and engineers, as Colin Hempstead has recently emphasized.[55]

The immediate context of Maxwell's remarks is especially revealing. In 1863–1864, Maxwell served with Jenkin on the British Association subcommittee that had determined the value of the ohm and devoted great effort to working out, both experimentally and conceptually, the relationship between electrical measurements. This experience had a greater effect than has often been appreciated on his thinking at a crucial stage in the development of his electromagnetic theory.[56] By the time Maxwell wrote his review of Jenkin's book in the spring of 1873, he had just completed his own *Treatise on Electricity and Magnetism* and was busy supervising the preparation of the new Cavendish Laboratory. He was intent on installing *measurement,* particularly electrical measurement, as the chief activity of the laboratory, and even before the Cavendish opened early in 1874, he wrote to Jenkin about securing the transfer to it of the Standards Committee equipment for use in research and training.[57] Maxwell clearly intended the electricity studied at the Cavendish to be

[53] Quoted in Gooday, "Precision Measurement" (cit. n. 33), pp. 42–43, which emphasizes the connection with physics teaching laboratories.

[54] Fleeming Jenkin, *Electricity and Magnetism* (London, 1873); see esp. pp. v–vi.

[55] See Smith and Wise, *Energy and Empire* (cit. n. 33); M. Norton Wise and Crosbie Smith, "Measurement, Work and Energy in Lord Kelvin's Britain," *Historical Studies in the Physical Sciences,* 1986, *17:* 146–173; and C. A. Hempstead, "An Appraisal of Fleeming Jenkin (1833–1885), Electrical Engineer," *Hist. Tech.,* 1991, *13:*119–144.

[56] A possible connection is explored in Salvo d'Agostino, "Esperimento e teoria nell'opera di Maxwell: Le misure per le unita assolute elettromagnetiche e la velocita della luce" (Experiment and theory in Maxwell's work: The measurements for absolute electromagnetic units and the velocity of light), *Scientia,* 1978, *113:*453–480, but d'Agostino's discussion is vitiated by his misattribution to Maxwell of Thomson's 1863 BA Committee Report (see above, n. 27).

[57] J. C. Maxwell to Fleeming Jenkin, [17 March], 22 July, and 18 Nov. 1874, in Jenkin papers (uncatalogued), ULC; see also Maxwell's list of the equipment in his annual report on the Cavendish, *Camb. Univ. Reporter,* 27 April 1875, pp. 352–354. On the later history of electrical standards work at the Cavendish see Simon Schaffer, "Late Victorian Metrology and Its Instrumentation: 'A Manufactory of Ohms,'" in *Invisible Connections: Instruments, Institutions, and Science,* ed. Robert Bud and Susan E. Cozzens (Bellingham, Wash.: SPIE Optical Engineering Press, 1992).

not that of "sparks and shocks which are seen and felt," but that of "currents and resistances to be measured and calculated"—not that of "the lecture-room and the popular treatise," but ultimately that of "the testing-room and the engineer's specification." Thomson, Jenkin, Maxwell, and their collaborators had helped extend the physics laboratory into the cable industry, but they had also brought part of the cable industry—encapsulated in the resistance box—into the physics laboratory.

INSTRUMENTS & AUDIENCE

The Large Space Telescope, as it was conceived in the 1970s. See page 101. Courtesy of NASA.

Terrestrial Magnetism: For the Glory of God and the Benefit of Mankind

*By Deborah Warner**

I. NATURAL PHILOSOPHY AND MIXED MATHEMATICS

IN THE CATALOGUE OF OBJECTS belonging to the Royal Society that he published in 1681, Nehemiah Grew distinguished "Instruments relating to Natural Philosophy" from "Things relating to the Mathematicks." Philosophical instruments, as most of Grew's first group soon came to be known, were the new tools of the new experimental natural philosophy. The phrase was apparently already current in 1649 when Samuel Hartlib wrote to Robert Boyle about "models and philosophicall apparatus." Mathematical instruments, on the other hand, were the tools of mixed mathematics, used to weigh, measure, or otherwise attach numbers to nature and to fabricated items.[1] The contemporary distinction between science and engineering suggests that Grew's basic distinction endures to this day.

Grew's definition of mathematical instruments reflects a broad definition of mathematics—a definition that originated in antiquity and remained popular throughout the early modern period. To quote Descartes, "All sciences which have for their end investigations concerning order and measuring are related to mathematics."[2] In this conception, mathematics was understood to have two components. Pure mathematics encompassed geometry and arithmetic, while mixed mathematics encompassed such abstruse subjects as astronomy, optics, and statics, as well as such mundane subjects as surveying, navigation, and fortification.[3]

* Division of Physical Sciences, National Museum of American History, Smithsonian Institution, Washington, D.C. 20560.

[1] Nehemiah Grew, *Musaeum Regalis Societatis; or, a Catalogue & Description of the Natural and Artificial Rarities Belonging to the Royal Society and Preserved at Gresham Colledge* (London, 1681), pp. 357–368. Hartlib's letter is quoted in Charles R. Weld, *A History of the Royal Society* (London, 1848; repr. New York: Arno, 1975), Vol. I, p. 53. For an overview of mathematical instruments see J. A. Bennett, *The Divided Circle: A History of Instruments for Astronomy, Navigation, and Surveying* (Oxford: Phaidon, for Christie's, 1987).

[2] René Descartes, *Rules for Direction of the Mind* (1628), as quoted in Charles Singer, *A Short History of Scientific Ideas to 1900* (Oxford: Clarendon Press, 1960), p. 227.

[3] This diversity appears in the range of subjects that the Lucasian Professor of mathematics was required to cover: see Richard Westfall, *Never at Rest: A Biography of Isaac Newton* (New York: Cambridge Univ. Press, 1980), p. 208. See also John Dee, preface to Euclid, *The Elements of Geometrie,* trans. H. Billingsley (London, 1570); Jonas Moore, *A New System of the Mathematics* (London, 1681); and the description of a text by Peter Galtruchius widely used in Cambridge during Newton's tenure, in John Gascoigne, "The Universities and the Scientific Revolution: The Case of Newton and Restoration Cambridge," *History of Science,* 1985, *23:*391–434, on p. 413.

 0369-7827/94/0008-0005$01.00

Historians have long been aware of this broad definition of mathematics, but have seldom given it the attention it deserves. In his provocative essay "Mathematical versus Experimental Traditions in the Development of Physical Science," Thomas S. Kuhn recognized that "among the large number of topics now included in the physical sciences," astronomy, optics, statics, harmonics, and mathematics "were already in antiquity foci for the continuing activity of specialists." These topics were empirical rather than a priori; they were based on observations, but not on experiments; and the "men who contributed to one of these subjects, almost always made significant contributions to others as well." Kuhn noted further that "from some significant points of view" these topics "might better be described as a single field, mathematics." He failed, however, to recognize the historical connection between these "classical sciences" and the larger universe of mathematics in general, and mixed mathematics in particular.[4]

While both mixed mathematics and natural philosophy were concerned with the natural world, the two approaches differed in several important ways. The most obvious difference pertained to quantification. Galileo's assertion that the book of nature was written in the language of mathematics may have been news to natural philosophers, but his fellow mathematicians were united in believing that "all natural bodies and qualities be, as far as is possible, reduced to number, weight, measure, and precise determination," as Francis Bacon phrased it.[5] Galileo, of course, earned his living as a mathematician and, despite his forays into natural philosophy, he is said to have remained "a mathematician in his method and outlook." Most natural philosophers, however, even after the advent of the experimental program, were reluctant to quantify their investigations. Boyle went so far as to state that "there are few physical experiments, wherein mathematical preciseness is necessary, and fewer wherein it is to be expected."[6]

Another important element of the distinction between mixed mathematics and natural philosophy pertained to the question of truth. Natural philosophy concerned the properties of natural bodies, the reasons and causes of the effects that nature produces; mathematics did not. This attitude informs the anonymous letter to the reader with which Andreas Osiander prefaced Copernicus's *De revolutionibus.* Bacon echoed it when he described Copernicus and Ptolemy as mathematicians who produced mere "calculations and predictions," and distinguished them from philosophers who sought to understand "what is found in nature herself, and is actually and really true." Isaac Newton was clearly aware of the distinction. According to R. S. Westfall, the *Principia* "repeatedly insisted that the mathematical treatment of attraction asserted nothing about its physical cause."[7]

[4] Thomas S. Kuhn, *The Essential Tension: Selected Studies in Scientific Tradition and Change* (Chicago: Univ. Chicago Press, 1977), pp. 31–65, on p. 65. See also J. A. Bennett, "The Mechanics' Philosophy and the Mechanical Philosophy," *Hist. Sci.,* 1986, *24:*1–28.

[5] William R. Shea, *Galileo's Intellectual Revolution: Middle Period, 1610–1632* (London: Macmillan, 1972), p. 10; and Francis Bacon, *Parasceve,* in *Works,* Vol. I, p. 400, as quoted in Charles Webster, *The Great Instauration: Science, Medicine and Reform 1626–1660* (London: Duckworth, 1975), p. 351.

[6] Robert Boyle, *The General History of the Air* (London, 1692), p. 99. See also Steven Shapin, "Robert Boyle and Mathematics: Reality, Representation, and Experimental Practice," *Science in Context,* 1988, *2:*23–58.

[7] Francis Bacon, *Descriptio globi intellectus* (1612), in *Works,* Vol. III, p. 734, as quoted in Mary Hesse, "Francis Bacon," *Dictionary of Scientific Biography* (New York: Scribners, 1970–1980), Vol. I, p. 373; and Westfall, *Never at Rest* (cit. n. 3), p. 422.

The status of practitioners represents a third element of the distinction. As we learn from John Wallis, mathematics in seventeenth-century England was often seen "as the business of Traders, Merchants, Seamen, Carpenters, Surveyors of Land, or the like, and perhaps some Almanack Makers in London."[8] Wallis's complaint that mathematics "were scarce looked upon as Academical Studies" can hardly be taken at face value—from the medieval quadrivium (arithmetic, geometry, astronomy, and music) to an increasing response to the growing practical concerns in the early modern period, mathematics was well ensconced in university curricula—but it may reflect the fact that academic mathematicians did not often enjoy the salary or status of their philosophical colleagues. Newton complained that "mathematicians, that find out, settle, and do all the business, must content themselves with being nothing but dry calculators and drudges."[9]

Although we lack a similar account of natural philosophers, Thomas Sprat tells us that the early members of the Royal Society tended to be "Gentlemen, free, and unconfin'd," and Steven Shapin and Simon Schaffer detail the ways in which Boyle and his colleagues used social and cultural norms to raise the status of the laboratory and the labor performed within it. Boyle's own nobility played a role here, as did his argument that while tradesmen might have a vested interest in the outcome of an experiment, men of gentle birth could be counted on to be reliable witnesses of instrumentally produced matters of fact.[10]

A fourth element of the distinction pertained to practicality. Philosophical instruments might be used for practical purposes, but they were never intended to be so. Mathematical instruments, however, were the means by which the mathematical sciences were "rendered useful in the Affairs of Life." This practical emphasis pertained to the most sophisticated astronomical instruments as well as to the relatively simple instruments used for surveying and navigation. Some observatories might have been sites for the production of astronomical knowledge, but most had intensely practical mandates. The Royal Observatory at Greenwich was established "to the finding out of the longitude of places for perfecting navigation and astronomy."[11]

Although mathematical instruments originated in antiquity, if not before, the early modern period saw a proliferation of measurement corresponding with, and undoubtedly attributable to, the expansion of commerce and industry and, increasingly, the imperial demands of the state. Indeed, the first capital intensive efforts at quantification pertained to matters of major importance to states whose wealth and power depended on their access to and control of foreign markets. These matters included

[8] John Wallis as quoted in E. G. R. Taylor, *The Mathematical Practitioners in Tudor and Stuart England* (Cambridge: Cambridge Univ. Press, 1954), p. 4.

[9] Newton as quoted in Charles Gillispie, *The Edge of Objectivity: An Essay in the History of Scientific Ideas* (Princeton: Princeton Univ. Press, 1960), p. 139. See also Mordechai Feingold, *The Mathematician's Apprenticeship: Science, Universities, and Societies in England, 1560–1640* (New York: Cambridge Univ. Press, 1984); and John Gascoigne, "A Reappraisal of the Role of Universities in the Scientific Revolution," in *Reappraisals of the Scientific Revolution,* ed. David Lindberg and Robert Westman (New York: Cambridge Univ. Press, 1990), pp. 207–260.

[10] Thomas Sprat, *The History of the Royal Society of London* (London, 1667), p. 67; and Steven Shapin and Simon Schaffer, *Leviathan and the Air-Pump: Hobbes, Boyle, and the Experimental Life* (Princeton: Princeton Univ. Press, 1985).

[11] Quoting from Edmund Stone, *The Construction and Principal Uses of Mathematical Instruments* (London, 1723), preface; and the warrant of Charles II, 22 June 1675, quoted in Eric G. Forbes, *Greenwich Observatory,* Vol. I: *Origins and Early History (1675–1835)* (London: Taylor & Francis, 1975), p. 22.

the size and shape of the earth as well as the latitude and longitude of places of military or economic significance—and of prominent landmarks en route. A British sea captain spoke truly when he told Charles II that "the Amplitude and Prosperousness of Your Majesties Dominions are chiefly maintained by Your Majesties Countenance upon the Mathematical Arts."[12]

Mixed mathematics and natural philosophy may have been relatively stable categories, but they were hardly immutable. Dissolution of the boundaries between the two realms, already evident in the sixteenth century, advanced rapidly during the course of the seventeenth century. The Royal Society may have been dedicated to "the Advancement of Experimental Philosophy," but experimental philosophy was never its exclusive concern. The society's *Philosophical Transactions* published numerous reports of and encouragements for useful arts and sciences, and the frontispiece to Sprat's 1667 *History of the Royal Society* shows such emblematic philosophical instruments as an air pump and an aerial telescope in the midst of a wide variety of common mathematical instruments. This broad approach was already evident in the 1650s, when the members of the Philosophical Club at Oxford went "over all the heads of naturall philosophy and mixt mathematics" and agreed "to pass Conjectures, and propose Problems, on any Mathematical, or Philosophical Matter, which comes in their way."[13]

The broad approach was evident in the world of commerce as early as the 1720s, when Richard Glynne began advertising "Mathematical Instruments, either for Land or Sea, & Apparatus for Experimental Philosophy." By mid century several London dealers were offering a wide range of mathematical and philosophical instruments, and often optical ones as well, and some wealthy individuals were amassing collections that included examples from each category. The broad approach also pertained in Paris, where the Académie Royale des Sciences embraced mixed mathematics as well as natural philosophy.[14]

One factor blurring the boundaries between mixed mathematics and natural philosophy was "the Genius of Experimenting" that was said to be "so much dispers'd" in Restoration London that, according to Sprat, "All places and corners are now busie, and warm about this work." The early experimental philosophers provided some of this warmth, but much of it came from apothecaries, mechanics, and other "Artists, such as are indeed skilfull in the Mathematicks, and lovers and inquirers of the truth."[15] Many of the experiments and observations made by these practical men

[12] Samuel Sturmy, *The Mariners Magazine* (London, 1679), dedication.

[13] Seth Ward to Justinian Isham, Feb. 1652, as quoted in Webster, *The Great Instauration* (cit. n. 5), p. 157. See also K. T. Hoppen, "The Nature of the Early Royal Society," *British Journal for the History of Science,* 1976, *9:*1–24, 243–273; and Sprat, *History of the Royal Society of London* (cit. n. 10), p. 56.

[14] For Glynne's advertisement (which appeared in G. Gordan, *An Introduction . . . to Geography,* London, 1726) see H. R. Calvert, *Scientific Trade Cards in the Science Museum Collection* (London: Her Majesty's Stationery Office, 1971), p. 23. See also G. L'E. Turner, "The Auction Sales of the Earl of Bute's Instruments, 1793," *Annals of Science,* 1967, *23:*213–242. For the Académie see David S. Lux, "The Reorganization of Science, 1450–1700," in *Patronage and Institutions: Science, Technology and Medicine at the European Court, 1500–1750,* ed. Bruce Moran (Rochester, N.Y.: Boydell Press, 1991), pp. 185–194, on p. 192.

[15] Sprat, *History of the Royal Society* (cit. n. 10), p. 71; and William Oughtred, *The Circles of Proportion* (London, 1633), Pt. 2, pp. 55–56, as quoted in Margaret Deacon, "Founders of Marine Science in Britain: The Work of the Early Fellows of the Royal Society," *Notes and Records of the Royal Society of London,* 1965, *20:*29. See also W. H. G. Armytage, "The Royal Society and the Apothecaries, 1660–1722," *ibid.,* 1954, *11:*22–37.

were extensions of traditional practical concerns, but some encroached on philosophical domains. One might not consider compass makers and navigators to be natural philosophers, but what about a compass maker who discovered magnetic dip, or a navigator who charted magnetic variation as it changed with time and place? It has even been suggested that these mundane investigations actually stimulated the emerging experimental program—in J. A. Bennett's felicitous phrase, the mechanical philosophy derived from the mechanics' philosophy.[16]

It may no longer be fashionable to view the mathematicization of physics as "the crucial act of the scientific revolution," or even to admit that there was such a thing as a scientific revolution.[17] It is, however, useful to remember that the quantification of natural philosophy began in the seventeenth century and, as John L. Heilbron has argued, this quantification proved to be an "uncomfortable process" that necessitated "a radical readjustment of the divisions of knowledge." According to this reading, the "quantifying spirit" that swept over Europe in the latter decades of the eighteenth century brought about "the completion of the Scientific Revolution."[18] With the advent of quantification, some mathematical practitioners felt that natural philosophers were encroaching onto their terrain. Jesse Ramsden, the leading instrument maker in London, if not the world, at the end of the eighteenth century, was highly critical of the Royal Society's extensive and precise excise investigation, arguing that "work of this nature is more the work of an instrument maker than a philosopher."[19]

The second important factor blurring the boundaries concerns the question of truth. Dissatisfaction with the distinction between prediction and explanation appeared first in the field of astronomy. Copernicus believed that his calculations described what was "actually and really true," and Kepler argued that astronomy must predict celestial positions and explain the physics behind celestial motions. This dissatisfaction soon spread to other areas of investigation. Newton argued that many "mathematical sciences" actually "treat of physical things," and that optics depended "as well on Physicall Principles as on Mathematicall Demonstrations."[20]

The status differential between mathematicians and natural philosophers never dissolved completely—engineers are still less classy than scientists—but it lessened

[16] Bennett, "The Mechanics' Philosophy and the Mechanical Philosophy" (cit. n. 4). See also J. A. Bennett, "The Challenge of Practical Mathematics," in *Science, Culture and Popular Belief in Renaissance Europe,* ed. Stephen Pumfrey, Paolo Rossi, and Maurice Slawinski (Manchester: Manchester Univ. Press, 1991), pp. 176–190.

[17] Gillispie, *The Edge of Objectivity* (cit. n. 9), p. 54. See also David Lindberg and Robert Westman, introduction to *Reappraisals of the Scientific Revolution,* ed. Lindberg and Westman (cit. n. 9).

[18] John L. Heilbron, *Electricity in the Seventeenth and Eighteenth Centuries: A Study of Early Modern Physics* (Berkeley: Univ. California Press, 1979), p. 11. See also Roger Hahn, "New Considerations on the Physical Sciences of the Enlightenment," *Studies on Voltaire and the Eighteenth Century,* 1989, *264:*789; and Tore Frangsmyr, J. L. Heilbron, and Robin E. Rider, eds., *The Quantifying Spirit in the Eighteenth Century* (Berkeley: Univ. California Press, 1990).

[19] Jesse Ramsden, *Account of Experiments to Determine the Specific Gravity of Fluids, Thereby to Obtain the Strength of Spiritous Liquors* (London, 1792).

[20] For Copernicus see Robert S. Westman, "The Astronomer's Role in the Sixteenth Century: A Preliminary Study," *Hist. Sci.,* 1980, *18:*105–147, esp. pp. 107–116. For Kepler see Westman, "Kepler's Theory of Hypothesis and the 'Realist Dilemma,'" *Studies in the History and Philosophy of Science,* 1972, *3:*233–264. Newton is quoted in Henry Guerlac, "'Newton's Mathematical Way': Another Look," *British Journal for the History of Science,* 1984, *17:*61–63, on p. 61. Francis R. Johnson, *Astronomical Thought in Renaissance England* (Baltimore: Johns Hopkins Univ. Press, 1937), p. 235, points to "a prevailing eagerness to evolve some new physical explanation of the movements of the planets in their orbits." See also J. A. Bennett, "Cosmology and the Magnetical Philosophy," *Journal of the History of Astronomy,* 1981, *12:*165–177.

markedly in the early modern period. This change came about, in large part, because of the high cost of doing mathematics and natural philosophy. Robert S. Westman has shown that patronage by princely courts helped to raise the status of mathematicians in Germany and Denmark in the sixteenth century.[21] A similar trend can be seen in seventeenth-century England and Scotland, as men of gentle birth and deep pockets came to appreciate the extent to which their way of life depended on astronomy, navigation, cartography, and other parts of mixed mathematics. The great observatory at Greenwich would not have been possible without royal patronage. Nor would those at Paris or Hven. By 1723 Edward Stone could write, with some justification, that "Mathematicks are now become a popular Study, and make a part of the Education of almost every Gentleman."[22]

Bacon's call for useful knowledge pertained more to mathematics and other practical arts and sciences than to natural philosophy. Nevertheless, the merging of the traditional categories led, albeit slowly, to an awareness that philosophical investigations might yield practical benefits. It also occasionally led natural philosophers to appropriate the techniques of practitioners and then proclaim that improvements of these techniques would yield practical benefits.

II. TERRESTRIAL MAGNETISM: INSTRUMENTS AND OBSERVATIONS

Astronomy was the first important subject straddling the divide between mixed mathematics and natural philosophy, but terrestrial magnetism was not far behind. Entailing measurement and made for practical purposes, magnetic investigations originated within the realm of mixed mathematics. But by the seventeenth century many investigators were as interested in understanding the earth's magnetic field as in using that knowledge for practical purposes.[23]

As magnetic phenomena came to be seen as aspects of nature rather than simply artifacts of practice, terrestrial magnetism became a hot topic. Instruments were invented, refined, and made commercially available, and many observers were trained in their use. Often supported with public funds, investigators mapped variation and dip with ever greater precision in ever more remote locales, and they began to correlate changes in variation and dip with time, temperature, and other meteorological phenomena. These activities escalated throughout the eighteenth century, reaching a crescendo in the magnetic crusades of the middle decades of the nineteenth century.[24]

Interest in terrestrial magnetism began with the invention of the magnetic compass in the thirteenth century and kept pace with the growing importance of that instru-

[21] Westman, "Astronomer's Role in the Sixteenth Century" (cit. n. 20), esp. pp. 121–127.

[22] Stone as quoted in A. J. Turner, "Mathematical Instruments and the Education of Gentlemen," *Ann. Sci.*, 1973, *30:*51–65. For a discussion of Roger North's interest in mathematics see Robert T. Gunther, *Early Science in Cambridge* (Oxford: Clarendon Press, 1937), pp. 46–49. John Skene's 1599 description of John Napier as "a gentleman of singular judgement and learning, especially in mathematic sciences" is quoted in Margaret Baron, "John Napier," *Dictionary of Scientific Biography* (cit. n. 7), Vol. IX, p. 609.

[23] The standard source remains Heinz Balmer, *Beiträge zur Geschichte der Erkenntnis des Erdmagnetismus* (Aarau: Sauerlander, 1956). See also Stephen Pumfrey, " 'O Tempora, O Magnes!' A Sociological Analysis of the Discovery of Secular Magnetic Variation in 1634," *Brit. J. Hist. Sci.*, 1989, *22:*181–214.

[24] John Cawood, "Terrestrial Magnetism and the Development of International Collaboration in the Early Nineteenth Century," *Ann. Sci.*, 1977, *34:*551–587.

ment. By the Renaissance the compass was being hailed as one of the three technologies that indicated how far the moderns had advanced beyond the ancients. Seventeenth-century English propagandists noted that because the compass enabled England to procure trade and transmit the gospel beyond the seas, it "justly ranked amongst the greatest wonders that this world affords."[25]

But there were problems. By the fifteenth century it was well known that magnetic north seldom coincides with the meridian, and that this "variation" (declination) varies from one place to another. Sixteenth-century navigational manuals routinely published information about the spatial distribution of variation, and they urged seafarers to make frequent observations of magnetic variation. The immediate goal was to correct compass readings. The longer-term goal was to use variation to determine longitude at sea. To measure the "northeasting" and "northwesting" of their needles, navigators might use a simple compass with a shadow line, as shown in William Borough's *A Discours of the Variation* (London, 1581), or a gimball mounted compass equipped with vertical sights, as shown in William Barlow's *The Navigators Supply* (London, 1597).[26] Very few early navigational compasses are available for us to study, but elegant diptych dials from the sixteenth and seventeenth centuries survive in relatively great profusion. On many of these instruments the inset compass has a north-south line offset for variation at the place of manufacture.[27]

Interest in terrestrial magnetism was especially apparent in states with expanding overseas empires. Moreover, although this interest originated with navigators and instrument makers, it soon spread to universities. By the early seventeenth century, as Stephen Pumfrey has shown, magnetic investigators in London had formed a particularly strong community with shared rules of practice and theory—rules that enabled Henry Gellibrand, a professor of astronomy at Gresham College, to trust the observations made by his predecessors and thus "discover" that variation changes over time. Pierre Petit in Paris made observations similar to those of Gellibrand, but he attributed the discrepancies to erroneous observations made by such predecessors as Oronce Finé, a professor at the Collège Royale in Paris, who had begun observing magnetic variation in the middle decades of the sixteenth century.[28] By the 1650s there were plans to erect at Oxford a "Magneticall, Mechanicall and Optick Schoole, furnished with the best instruments, and Adapted for the most usefull experiments in all those faculties."[29]

The Royal Society undertook a variety of measures to help stabilize the practice of magnetic observations. It issued an order "that precise Meridians be made in several places of England; for observing the present Declination of the Needle from them." It encouraged those undertaking distant sea voyages to observe the magnetic

[25] John Seller, *Practical Navigation* (London, 1680), p. 129.

[26] Ernst Crone, "The Measurement of the Variation of the Compass," in *The Principal Works of Simon Stevin,* Vol. III, ed. A. Pannekoek and E. Crone (Amsterdam: Swets & Zeitlinger, 1961), pp. 380–413; and David W. Waters, *The Art of Navigation in England in Elizabethan and Early Stuart Times* (Greenwich: National Maritime Museum, 1978), pp. 70–71, 228.

[27] Penelope Gouk, *The Ivory Sundials of Nuremberg, 1500–1700* (Cambridge: Whipple Museum of the History of Science, 1988), pp. 14–15; and Steven A. Lloyd, *Ivory Diptych Sundials 1570–1750* (Cambridge, Mass.: Collection of Historical Scientific Instruments, Harvard University, 1992), p. 18.

[28] Pumfrey, "O Tempora, O Magnes!" (cit. n. 23). For the early French observations see "Aiguille aimentée," in *Encyclopédie; ou, Dictionnaire raisonné des sceinces, des artes et des métiers,* ed. D. Diderot, et al., (Paris, 1751), Vol. I, pp. 199–202.

[29] Webster, *Great Instauration* (cit. n. 5), p. 171.

variation in different locales. It procured a magnetic needle for John Flamsteed and urged him to measure the variation at Greenwich. Its Magnetics Committee measured variation at Whitehall.[30] The Académie Royale des Sciences offered similar encouragement to French investigators, both at home and abroad, and published the annual observations of magnetic variation begun by Jean Picard at the Paris Observatory in 1666 and continued by others throughout the eighteenth century.[31]

Philosophical speculation concerning terrestrial magnetism paralleled practical concerns.[32] Individual investigators may have leaned towards theoretical understanding or practical application, but most did not distinguish between the two in any meaningful way. The Royal Society recognized that magnetism was both an important aid to navigation and "one of the Noblest and most abstruce Phaenomena, that falls under the cognizance of humane Reason." Daniel Bernoulli told the Académie Royale des Sciences that magnetic observations were of great utility for "la Physique générale" as well as for "la Navigation en particulier," and thus merited the attention "des Physiciens." Petrus van Musschenbroek penned a similar argument.[33]

As observations accumulated, however, indicating that magnetic variation varies in an apparently irregular manner from one location to another in the same time, and from one time to another in the same location, the possibility of a simple theory receded from view. In the 1690s Philippe de La Hire called for "exact Observations of the irregularities of this Variation, made all over the Earth, and at a considerable interval of time" in order to "discover some Period of this motion, and establish a System which might be of great use in Navigation."[34]

The first to rise to this challenge was Edmund Halley, who comfortably straddled the divide between natural philosophy and mixed mathematics, and who believed that large quantities of additional data would advance both causes. Halley had published two papers speculating on the positions and numbers of the earth's magnetic poles and the direction and speed of their movements. He had also spent considerable time on the docks and understood the practical benefits of a reliable magnetic chart. It was he, for instance, who reported that faulty information about variation had sent many ships into the Bristol Channel when they were bound up the English Channel, often with disastrous consequences.[35]

[30] For the quotation from Sprat see Robert K. Merton, *Science, Technology and Society in Seventeenth-Century England* (New York: H. Fertig, 1970), pp. 163–164. For the "Directions for Observations and Experiments to be made by Masters of Ships, Pilots, and other fit Persons in their Sea-Voyages . . . ," see *Philosophical Transactions of the Royal Society,* 1667, *2:*433–438. For Flamsteed see Weld, *History of the Royal Society* (cit. n. 1), Vol. I, p. 269. For the Magnetics Committee notice see *Phil. Trans.*, 1670, *5:*1187.

[31] *Histoire de l'Académie Royale des Sciences (1666–1699),* 2 vols. (Paris, 1733), Vol. I, pp. 176–178; and *Mémoires de l'Académie Royale des Sciences, depuis 1666 jusqu'à 1699,* Vol. VII (Paris, 1729), pp. 232–233, 394, and 508.

[32] Stephen Pumfrey, "Mechanizing Magnetism in Restoration England: The Decline of Magnetic Philosophy," *Ann. Sci.,* 1987, *44:*1–21.

[33] "Directions for Observations and Experiments" (cit. n. 30), p. 431; Daniel Bernoulli, "Mémoire adressé à messieurs les auteurs du Journal des Sçavans," *Journal des Sçavans,* 1757, *25:*27; and Petrus van Musschenbroek, *Essai de physique* (Leiden, 1751), pp. 273, 291.

[34] "A Letter of Mr. De la Hire of the Royal Academy of Sciences at Paris, concerning a new sort of Magnetical Compass, with several curious Magnetical Experiments," *Phil. Trans.,* 1686–1692, *16:*344–351.

[35] Edmund Halley, "A Theory of the Variation of the Magnetical Compass," *Phil. Trans.,* 1683, *13:*208–221; Halley, "An Account of the Cause of the Change of the Variation of the Magnetickal Needle, with an Hypothesis of the Structure of the Earth," *Phil. Trans.,* 1692, *16:*563–578; and [Halley], "An Advertisement Necessary to be Observed in the Navigation Up and Down the Channel of

Halley suggested that since magnetic variation varies in a continuous manner from one location to the next, it should be possible to trace lines of equal variation on the surface of the earth. In 1698 the Royal Navy gave him command of a ship and instructed him to chart the "Longitude and Variations of the Compasse." This endeavor yielded two isogonic maps: *A New and Correct CHART Shewing the VARIATIONS of the COMPASS in the WESTERN AND SOUTHERN OCEANS as observed in ye YEAR 1700 by his Maties Command,* and *A New and Correct Sea Chart of the Whole World Shewing the Variations of the Compass as they were found in the year MDCC.*[36]

Halley was undoubtedly aware of the discrepancies between one magnetic observation and another, and concerned that his observations be accepted as matters of fact. Yet he never published any accounts of his methods for determining position or variation, and he apparently left no manuscript reports. The only clue to his instrumentation appears on his Atlantic chart, where figures above the dedicatory cartouche in central west Africa hold a backstaff (used for determining latitude), an azimuth compass (for variation), and an armillary sphere (for education?). Longitude would have been found by dead reckoning. Halley's behavior, although typical of mathematical practitioners, stands in sharp contrast to that of Boyle, who recognized that his philosophical instruments were often as contested as the observations and experiments they made possible. Boyle in consequence developed sophisticated literary technologies designed to make readers think themselves "virtual witnesses" of his investigations.[37]

Halley's maps were immediately and repeatedly hailed as a monumental achievement for both philosophical and practical purposes. For John Wallis they showed that "a true Natural History of matter of Fact, is certainly the surest Foundation on which to ground a Physical Hypothesis, to explain the Causes."[38] In France, where the state had invested heavily in the advancement of geography and geodesy, the Académie Royale des Sciences described Halley's "sistême" as "very beautiful and worthy to be followed with great attention." Jacques Cassini found much conformity between Halley's observations and those made on French expeditions to China, the Americas, and elsewhere; the minor discrepancies he attributed to "the extreme difficulty in making these sorts of observations at sea with exactitude." Halley himself recognized that his maps were simply the first firm step in the right direction, and he asked "all knowing Mariners . . . to lend their Assistance and Informations, towards the perfecting of this useful Work."[39] The necessity for more and better data

---

England," reproduced in Colin Ronan, *Edmund Halley: Genius in Eclipse* (London: Macdonald & Co., 1970). See also [Halley], "An Advertisement Necessary for all Navigators Bound up the Channel of England," *Phil. Trans.,* 1700–1701, *22:*725–726.

[36] Norman Thrower, "Edmond Halley as a Thematic Geo-Cartographer," *Annals of the Association of American Geographers,* 1969, *59:*652–676; Thrower, "Edmond Halley and Thematic Geocartography," in *The Compleat Plattmaker,* ed. Norman Thrower (Berkeley: Univ. California Press, 1978), pp. 195–228; and Halley, *The Three Voyages of Edmond Halley in the Paramore, 1698–1701,* ed. Thrower (London: Hakluyt Society, 1981).

[37] Shapin and Schaffer, *Leviathan and the Air-Pump* (cit. n. 10).

[38] "A Letter of Dr Wallis to Captain Edmund Halley; Concerning the Captains Map of Magnetick Variation; and Some Other Things Relating to the Magnet," *Phil. Trans.,* 1702–1703, *23:*1106.

[39] "Sur l'aiman et sur l'aiguille aimantée," *Histoire de l'Académie Royale des Sciences, 1705* (1707), pp. 5–10; and "Sur la déclinaison de l'aiman," *ibid., 1701* (1703), pp. 9–11. See also Edmund Halley, "The Description and Uses of a New and Correct Sea-Chart of the Western and Southern

became painfully apparent in 1707 when faulty compasses and misinformation about variation off the southwest coast of England caused the destruction of a squadron of ships commanded by Admiral Clowdisley Shovell.[40]

As the eighteenth century advanced, magnetic data mushroomed—refinements in the Atlantic Ocean, and also basic data for those areas Halley had not reached. Although the Royal Navy did not sponsor any more dedicated voyages, it encouraged those who went abroad to chart variation around the world and to share this information with those who stayed at home. Thus the London mathematicians William Mountaine and James Dodson were able to develop an isogonic chart, correct for 1744, from some 1,100 observations of variation made by Robert Douglas, a teacher of navigation in the Royal Navy. A later edition, correct for 1756, incorporated observations made by or for the Royal Society, the Royal Navy, the East India Company, the Hudson's Bay Company, the Royal Observatory at Greenwich, and several private individuals. This latter chart was accompanied by a table of 50,000 figures adapted to every 5 degrees of latitude and longitude over the most frequented oceans, and dating from 1700 to 1756.[41] Samuel Dunn, another teacher of mathematics in London, brought out yet another variation chart in 1775.[42]

French mathematicians recognized that Halley's charts needed improvement, but were slow to get a French product on the market. Charles-Marie de La Condamine noted the lack of data for the Pacific and Indian Oceans, for most land areas, and for the Mediterranean.[43] Guillaume Delisle, the "Premier Géographe du Roi," noted that the English charts sported numerous geographical errors not found on the best French ones. Delisle also noted discrepancies between Halley's values of variation and those measured by French navigators in the years 1706–1709, and he correctly attributed these discrepancies to the secular change in variation.[44] Delisle made preparations for an improved isogonic chart, gathering more than 15,000 readings

---

Ocean, Shewing the Variations of the Compass," appended to the sides of some copies of the Atlantic chart; repr. in Halley, *Three Voyages* (cit. n. 36), pp. 365–366.

[40] Charles-Marie de La Condamine, "Nouvelle manière d'observer en mer la déclinaison de l'aiguille aimantée," *Mémoires de l'Académie Royale des Sciences, 1733* (1735), pp. 446–456; and M. Meynier, *Mémoire sur le suject du prix proposé par l'Académie Royale des Sciences en l'année 1729* (Paris, 1732).

[41] The first state of their map was entitled "Accuratissima totius Terrarum Orbis tabula nautica, celeberrimo viro E. Halley anno 1700 constructa . . . Ad observationes circiter Annum 1744 habitas, renovata Gulielmo Mountaine et Jacobo Dodson." The later edition was "Ad observationes circiter annum 1756 habitas, renovata." See also William Mountaine and James Dodson, "An Attempt to Point Out . . . the Advantages Which Will Accrue from a Periodic Review of the Variation of the Magnetic Needle, Throughout the Known World," *Phil. Trans.*, 1754, *48:*875–880; Mountaine and Dodson, "A Letter . . . Concerning the Variation of the Magnetic Needle; with a Sett of Tables Annexed, Which Exhibit the Result of Upwards of Fifty Thousand Observations, in Six Periodic Reviews, from the Year 1700 to the Year 1756, Both Inclusive; and are Adapted to Every Five Degrees of Latitude and Longitude in the More Frequented Oceans," *Phil. Trans.*, 1757, *50:* 329–349; and Mountaine and Dodson, *An Account of the Methods Used to Describe Lines on Dr. Halley's Chart of the Terraqueous Globe Showing the Variation of the Magnetic Needle About the Year 1756 in All the Known Seas, &c.* (London, 1758).

[42] Samuel Dunn, *A Variation Chart of the Atlantick, Ethiopick & Southern Oceans, for the year 1776* (London, 1775).

[43] Charles-Marie de La Condamine, "Observations mathématiques et physiques faites dans un voyage de Levant en 1731 & 1732," *Mém. Acad. Roy. Sci., 1732* (1735), pp. 295–322, on pp. 297–300.

[44] Guillaume Delisle, "Observations sur la variation de l'aiguille par rapport à la carte de M. Halley," *Mém. Acad. Roy. Sci., 1710* (1712), pp. 353–365; see also "Some Remarks on the Variations of the Magnetical Compass published in the Memoirs of the Royal Academy of Sciences, with regard to the General Chart of those Variations made by E. Halley . . . ," *Phil. Trans.*, 1714–1716, *29:*165–168.

of variation made by some 150 individuals. Philippe Buache, Delisle's son-in-law and successor, planned a series of isogonic charts correct for different years, and he actually drew isogonic charts of the Atlantic and Pacific. As it happened, however, the first isogonic charts published in France were based on the data collected by Mountaine and Dodson.[45]

Unlike magnetic variation, magnetic dip changed little over time, had little effect on navigation, and attracted relatively little attention. After the first publication of dip by Robert Norman, a compass maker in London, there was some hope that the "inclinatory needle" would "Shew you the Latitude of any Place without Observation either of Sun or Stars." But this hope was generally abandoned after Jean Richer at Cayenne found no easy correlation between dip and latitude. And although the Royal Society had two dip needles "designed for the taking of longitudes," that hope was also soon seen as illusory.[46] Nevertheless, observations of dip did begin to accumulate. At the turn of the century, while Halley was charting variation in the Atlantic, James Pound and James Cunningham were making a few observations of dip in the Indian Ocean.[47]

During the eighteenth century, countless measurements of variation and dip were made by global explorers of the stature of Lord George Anson, Louis de Bougainville, James Cook, and Antonio de Ulloa; by astronomers and natural philosophers who lived in places not yet charted—Anders Celsius in Uppsala, Musschenbroek in Utrecht, and Johan Wilcke in Stockholm; by geodesists traveling to distant lands to measure the size and shape of the earth; by astronomers undertaking distant voyages to chart the southern stars or to observe the transit of Venus; and by a host of navigators and travelers, whose names are now forgotten but who once had a lively interest in philosophical and practical knowledge.

As magnetic observations proliferated, so did concern with the design and construction of magnetic instruments. Investigators had long known that different compasses gave different measures of variation, and that even the best compasses gave different readings from one moment to the next. And makers had long touted the advantages of their particular instruments. In the eighteenth century the design and construction of compasses became a matter of public concern. From this remove it is difficult to evaluate the various innovations mentioned in the historical record, but one is struck by the repetition of complaints continuing from year to year.

[45] "CARTE qui indique avec la direction des Vents géneraux les Variations qu'avoit la Boussole, en 1700 et 1744 dans presque tous les lieux de la Mer," in Pierre Bouguer, *Nouveau traité de navigation, contenant la théorie et la pratique du pilotage* (Paris, 1753); and Jacques Nicolas Bellin, *Carte des variations de la boussole et des vents généraux que l'on trouve dans les mers les plus fréquentés* (1765). On Delisle's efforts see Josef Konvitz, *Cartography in France, 1660–1848* (Chicago: Univ. Chicago Press, 1987), p. 70; see also [Philippe] Buache, "Construction d'une nouvelle boussole," *Mém. Acad. Roy. Sci., 1732* (1735), pp. 377–384.

[46] Robert Norman, *The New Attractive* (London, 1581). For the putative connection between longitude and dip see Seller, *Practical Navigation* (cit. n. 25), pp. 246–250; "The Undertakings of Mr. Henry Bond Senior, a Famous Teacher of the Art of Navigation in London, Concerning the Variation of the Variation of the Magnetical Compass and the Inclination of the Inclinatory Needle; as the Results and Conclusion of 38 Years Magnetical Study," *Phil. Trans.*, 1673, *8:*6065–66; and William Whiston, *The Longitude and Latitude Found by the Inclinatory or Dipping Needle* (London, 1721). For Richer see *Histoire de l'Académie (1666–1699),* Vol. I, pp. 176–178; and *Mémoires de l'Académie (1666–1699),* Vol. VII, pp. 322–323 (both cit. n. 31).

[47] "Part of a Letter from Mr. James Cunningham . . . from the Cape of Good Hope, Ap. 6, 1700, Giving an Account of his Observations on the Thermometer and Magnetick Needle in his Voyage Thither," *Phil. Trans.*, 1700, *22:*577–578.

Even as they became available, however, reliable compasses did not ensure reliable observations. The numerous manuals on the market suggest that mariners had to be taught how to use an azimuth compass.[48] Moreover, the stories about Captain William Bligh keeping pistols in his binnacle, or Captain Cook keeping an iron key in his, although perhaps apocryphal, suggest that many mariners did not understand the way that iron affected a magnetic needle. In 1766 the Royal Society published, without comment, a report from a surgeon aboard a ship of the Royal Navy to the effect that variation was less at anchor than at sea, "though near the same spot." And although Cook noticed that the magnetic variation appeared least when the sun was on the starboard side and greatest when he was facing the other way and the sun was on the port side, he seems not to have guessed that this anomaly might have been caused by an unequal distribution of iron on his ship.[49]

One of the first to tackle the compass problem was Philippe de La Hire, whose sophisticated observations of terrestrial magnetism were made with a well-marked meridian, well-balanced needles eight inches long, and a site at the Paris Observatory free from ferrous materials. La Hire presented his ideas on magnetic needles to the Académie Royale des Sciences in 1705, and his ideas on azimuth compasses in 1716.[50] A few years later the Académie announced a prize for the best method of observing variation at sea. Pierre Bouguer, then serving as "Hydrographe du Roy au Havre de Grace," captured the 2,000-livre prize with an essay that addressed such instrumental concerns as how to magnetize the needle, as well as such observational concerns as how to determine geographical north.[51] Other claimants included the cartographer Buache (his instrument measured dip as well as variation), the astronomer La Condamine, and M. Meynier, "Ingénieur du Roi pour la Marine." Still other hopefuls came forward after the competition was over. These ranged from the otherwise unknown M. de Mean, who received Académie approval for his complex compass in 1731, to Pierre Le Maire, a Parisian instrument maker trading "au quartier Anglois," who asked the Académie to consider his new azimuth compass in 1747.[52]

Improvements in accuracy were a concern across the Channel as well. In London in 1730 Joseph Harris introduced an azimuth compass that enabled one to read variation "to a single Degree or less," even at sea, whereas the common instrument was said to be reliable to half a point (11½ degrees) at best. In 1738 Christopher Middleton informed the Royal Society that he had devised a very convenient azimuth com-

[48] Guillaume Denys, *L'Art de naviger, perfectionné par la connoissance de la variation de l'aimant* (Dieppe, 1681; 1st ed. 1666).

[49] "Extract of a Letter from Mr. David Ross . . . Relating to the Variation of the Magnetic Needle," *Phil. Trans.*, 1766, *56*:218–223; and James Cook, *The Voyage of the Resolution and Adventure, 1772–1775,* ed. J. C. Beaglehole (Cambridge: Hakluyt Society, 1961), pp. 89, 104.

[50] Philippe de La Hire, "Nouvelles remarques sur l'aiman, et sur les aiguilles aimantées," *Mém. Acad. Roy. Sci., 1705* (1707), pp. 97–109; and La Hire, "De la construction des boussoles dont on se sert pour observer la déclinaison de l'aiguille aimantée," *ibid., 1716,* (1718), pp. 6–11.

[51] Pierre Bouguer, *De la méthode d'observer en mer la déclinaison de la boussole* (Paris, 1731). See also Bouguer, "Description d'un nouveau compas de variation," *Nouveau traité de navigation* (cit. n. 45), pp. 86–88; and "Éloge de M. Bouguer," *Hist. Acad. Roy. Sci., 1758* (1763), p. 130.

[52] Buache, "Construction d'une nouvelle boussole" (cit. n. 45); La Condamine, "Nouvelle manière d'observer en mer la déclinaison de l'aiguille aimantée" (cit. n. 40); and Meynier, *Mémoire sur le sujet du prix* (cit. n. 40). For Mean see *Hist. Acad. Roy. Sci., 1731* (1733), p. 92; for Le Maire see Pierre Le Maire, *Description et usage d'une nouvelle boussole pour observer les amplitudes sur mer* (Paris, 1747). For others see, e.g., on an azimuth compass of de Quercineuf, *Hist. Acad. Roy. Sci., 1734* (1737), p. 105; and Louis Godin, "Méthode d'observer la variation de l'aiguille aimantée en mer," *Mém. Acad. Roy. Sci., 1734* (1736), pp. 590–594.

pass, but he did not provide details; Middleton had earlier been appointed Fellow of the Royal Society for having "communicated to this society several curious observations relating to the variation of the needle in the northern seas."[53]

A major breakthrough occurred in 1744 when a London physician named Gowin Knight showed several particularly strong "artificial magnets" to the Royal Society. Although Knight never divulged his production technique, he was awarded the Copley Medal in 1747. In France, Henri-Louis Duhamel du Monceau, chemist, botanist, and "Inspecteur Générale de la Marine," claimed to have produced similar magnets.[54] Turning his attention to compasses, Knight redesigned the form of the needle, the moment of inertia of the card, and the composition of the cap. John Smeaton made Knight's prototype instruments and then turned Knight's simple boat compass into an azimuth compass. After several trials at sea, the Admiralty authorized the purchase of Knight's compasses for ships of the Royal Navy. These compasses were made by George Adams and tested by Knight himself. After further refining his compass design, Knight obtained a patent in 1766.[55]

These various innovations seem to have improved the performance of compasses, but did not make them terribly reliable. James Cook carried several of Knight's compasses on his second voyage, along with at least one made by Henry Gregory, a London maker who traded "at ye Azimuth Compass." Gregory's compass was similar to Knight's but much larger, and in Cook's view it had some improvements that "in some measure lessen the too quick motion the D^rs compasses are subject to." Still, discrepancies from one moment to the next, or from one instrument to another, often amounted to two degrees or more. Cook routinely dealt with these discrepancies by averaging his several readings and reporting the results to single minutes of arc. In 1776, at the start of his voyage to the northern Pacific, Cook stated again that at sea, and sometimes even on land, the most accurate magnetic observations often differed by two degrees, "without one being able to discover much less remove the cause." Thus, he concluded, since variation cannot be known to within one degree, it cannot be used to determine longitude. Nevertheless, Cook continued to report variation, and dip, to minutes, and sometimes to half minutes.[56]

With adequate instruments for measuring variation seemingly in hand, the Académie Royale des Sciences sponsored a dip circle competition. This contest attracted

[53] Joseph Harris, *A Treatise of Navigation* (London, 1730), advertisement and p. 202; J. K. L., "Christopher Middleton," in *Dictionary of National Biography,* Vol. XIII, pp. 342–343; and Christopher Middleton, "The Use of a new Azimuth Compass for Finding the Variation of the Compass, or Magnetic Needle at Sea, with Greater Ease and Exactness than by any Ever yet Contriv'd for that Purpose," *Phil. Trans.,* 1738, *40:*395–398.

[54] Gowin Knight, "An Account of Some Magnetical Experiments, Shewed Before the Royal Society," *Phil. Trans.,* 1745, *43:*161–166; Henri-Louis Duhamel de Monceau, "Façon singuliere d'aimanter un barreau d'acier," *Mém. Acad. Roy. Sci., 1745* (1749), pp. 181–193; and Duhamel, "Differens moyens pour perfectionner la boussole," *ibid., 1750* (1754), pp. 154–165.

[55] "A Description of a Mariner's Compass Contrived by Gowin Knight," *Phil. Trans.,* 1749–1750, *46:*505–512; and "An Account of some Improvements of the Mariners Compass, in Order to Render the Card and Needle, Proposed by Doctor Knight, of General Use, by John Smeaton," *ibid.,* pp. 513–517. "Dr. Knight's new Azimuth Compass" sold for £5.15.6 in "A Catalogue of Mathematical, Philosophical, and Optical Instruments, Made and Sold by George Adams," in George Adams, *A Treatise on the New Celestial and Terrestrial Globes* (London, 1766).

[56] Cook, *Voyage of the Resolution and Adventure, 1772-1775* (cit. n. 49), pp. 19–20, 76, 78, 525; and James Cook, *The Voyage of the Resolution and Discovery, 1776–1780,* ed. J. C. Beaglehole (Cambridge: Hakluyt Society, 1967), pp. 22–23, 181. For Gregory's trade card see Calvert, *Scientific Trade Cards* (cit. n. 14), p. 24.

such notable mathematicians as Leonard Euler, whose entry came in second, and Daniel Bernoulli, who won the 2,000 franc prize with a sophisticated analysis of the various mechanical factors affecting dip circles. Johann Dietrich, a skilled "ouvrier" in Basel, made a number of dip circles according to Bernoulli's plan. Euler had at least one; Bernoulli had a dozen or so. French artisans, however, were reluctant to adopt these innovations. Bernoulli was particularly dismayed that Alexis Magny, probably the leading mathematical instrument maker in Paris at that time, did not incorporate his ideas in the dip circle that he made for the Académie Royale des Sciences, which Nicolas La Caille used during his trip to South Africa in 1751–1753.[57]

The first to chart dip seems to have been Petrus van Musschenbroek, who included a series of numbers representing the Pound-Cunningham measures of dip in the Indian Ocean on his 1729 reprint of Halley's world chart. Using observations made primarily aboard ships of the Swedish East India Company, Johann Gustav Zegollström in 1755 and Johan Wilcke in 1768 compiled charts of isoclines in the Atlantic and Indian Oceans.[58]

While most interest in terrestrial magnetism focused on its geographical distribution, some investigators sought a detailed picture of its changes over time. From a long series of precise observations conducted in the 1720s, the London horologist George Graham found that the magnetic needles "would not only vary in their Direction upon different Days, but frequently at different times of the same Day." Moreover, these differences seemed "to depend neither upon Heat nor Cold, a dry or moist Air, clear or cloudy, windy or calm Weather, nor the Height of the Barometer." Turning to inclination, Graham found a "very considerable Difference, both in the Quantity of the Dip, and in the Quickness of the Vibrations."[59]

Like Halley's isogonic charts, Graham's observations of the fine structure of terrestrial magnetism provided the spark that ignited countless other investigations. Graham's observation of a strong magnetic disturbance on 5 April 1741, and those of the same event made by Celsius at Uppsala, indicated that such phenomena oc-

[57] Daniel Bernoulli, "Mémoire sur la manière de construire les boussoles d'inclinaison," and Leonard Euler, "De observatione inclinationis magneticae dissertatio," both in *Pièces qui ont Remporté les Prix de l'Académie Royale des Sciences, 1743* (Paris, 1748); and Bernoulli, "Mémoire adressé à messieurs les auteurs du Journal des Sçavans" (cit. n. 33). A photograph of a dip circle "by Daniel Bernoulli, completed by Johann Dietrich of Basel, in 1751," was sent by the Physical Institute in the Bernoullian at Basel for exhibit in London in 1876; see South Kensington Museum, *Catalogue of the Special Loan Collection of Scientific Apparatus* (London, 1877), p. 289; and Balmer, *Geschichte der Erkenntnis des Erdmagnetismus* (cit. n. 23), p. 573.

[58] Petrus van Musschenbroek, *Physicae experimentales, et geometricae* (Leiden, 1729), Fig. 10, "TABULA Totius Orbis TERRARUM Exhibens Declinationes Magneticas, ad Annum 1700, composita ab Edmundo Halleyo simul cum Inclinationibus a Poundio observatis"; see also Musschenbroek, *Essai de physique* (cit. n. 33), Fig. 29. For Zollogström see Gunnar Pipping, *The Chamber of Physics: Instruments in the History of Sciences Collections of the Royal Swedish Academy of Sciences, Stockholm* (Stockholm: Almqvist & Wiksell, 1977), p. 35. For Wilcke see J. C. Wilke, "Forsok til en Magnetisk Inclinations Charta," reproduced in Anita McConnell, *Geomagnetic Instruments before 1900* (London: Harriet Wynter, 1980), p. 33; and "CARTE RÉDUITE QUI INDIQUE les Diverses Inclinaisons de L'AIGUILLE AIMANTÉE par M. Wilcke, Suédois en 1768," in Pierre-Charles Le Monnier, *Loix du magnetisme* (Paris, 1776–1778).

[59] George Graham, "An Account of Observations made of the Variation of the Horizontal Needle at London in the latter Part of the Year 1722, and the Beginning of 1723," *Phil. Trans.*, 1724–1725, *33*:96–107, on pp. 99, 101; and Graham, "Observations of the Dipping Needle, made at London, in the Beginning of the Year 1723," *ibid.*, pp. 332–339, on p. 337.

curred over wide areas of the earth's surface. Celsius also found that variation is "more variable than is commonly supposed," and that the needle was markedly distorted by the aurora borealis. John Canton made some four thousand observations and found a correlation between variation and temperature. Once settled as a professor at Utrecht, Musschenbroek began a research program similar to Graham's. As the Enlightenment got under way, enthusiasm for compiling enormous quantities of data flourished. Jan van Swinden, a Dutch professor of physics, reported that he and his students had observed magnetic variation every hour of the day and night over the course of several years.[60]

Azimuth compasses appropriate for use at sea had been on the market at least since the sixteenth century, but magnetic instruments used for investigations on land tended to be special-purpose creations. George Graham probably built the sophisticated instruments with which he made his very precise observations of variation and dip, and Musschenbroek's beautiful instruments were probably made by his brother Jan. In the 1760s, however, magnetic instruments designed for terrestrial observations began to be produced commercially. Prominent London makers such as Jeremiah Sisson, Edward Nairne, George Adams, and Peter Dollond made particularly fine instruments, as did Georg F. Brander in Augsburg, Canivet in Paris, and Daniel Ekström and later Johan Petter Rosenberg in Stockholm. Some of these instruments measured dip, some measured variation, and some measured both dip and variation.[61]

The rising interest in the fine structure of magnetic variation led to the introduction of the variation compass, an instrument designed for precise measurements in the laboratory and observatory.[62] As these instruments were seldom dated, and as trade literature on this subject is mute, we do not know for certain when variation compasses first came on the market. We do know, however, that Harvard College bought a variation compass from Nairne in 1765. Equipped with a needle 9½ inches

[60] "Memoirs of the Royal Academy of Sciences in Sweden" (on Celsius), *The Gentleman's Magazine,* 1749, *19:*18; John Canton, "An Attempt to Account for the Regular Diurnal Variation of the Horizontal Magnetic Needle; and Also for its Irregular Variation at the Time of an Aurora Borealis," *Phil. Trans.,* 1759–1760, *51:*398–445; and Jan van Swinden, "Recherches sur les aiguilles aimantées, et sur leurs variations régulières," *Mémoires de Mathématique et de Physique, Présentés a l'Académie Royale des Sciences, par Divers Sçavans,* Vol. VIII (Paris, 1780), pp. 1–576.

[61] For an instrument made by Jeremiah Sisson for Lorimer in 1764 see John Lorimer, "Description of a New Dipping Needle," *Phil. Trans.,* 1775, *65:*79–84. For the dip circle Edward Nairne sold to Harvard College in 1765 see David Wheatland, *The Apparatus of Science at Harvard, 1765–1800* (Cambridge, Mass.: Harvard Univ. Press, 1968), pp. 160–161; for that at the Royal Society see Henry Cavendish, "An Account of the Meteorological Instruments used at the Royal Society's House," *Phil. Trans.,* 1776, *66:*395–401. See also "Experiments on two Dipping-Needles, which Dipping-Needles Were Made Agreeable to a Plan of the Reverend Mr. Mitchell, F. R. S. Rector of Thornhill in Yorkshire, and Executed for the Board of Longitude, by Mr. Edward Nairne, of Cornhill, London," *Phil. Trans.,* 1772, *62:*476–480. A dip circle made by Nairne for the Board of Longitude is in the collections of the National Maritime Museum (M.6). For a Nairne & Blunt dip circle marked "Royal Society 38" see McConnell, *Geomagnetic Instruments before 1900* (cit. n. 58), pp. 50–51. For Georg F. Brander see *G. F. Brander, 1713–1783: Wissenschaftliche Instrumente aus seiner Werkstatt* (Munich: Deutsches Museum, 1983). For Canivet see Pierre Charles Le Monnier, "Mémoire sur la variation de l'aimant en 1772 & 1773," *Mémoires, Académie Royale des Sciences, 1773* (1776), pp. 440–444. For Ekström and Rosenberg see Pipping, *Chamber of Physics* (cit. n. 58), pp. 44–47, 146.

[62] In the sixteenth century, the term *variation compass* referred to a navigational instrument. In the eighteenth century the navigation instrument became known as an *azimuth compass,* and the term *variation compass* was used for precision instruments for terrestrial observations of variation.

long, a vernier, and a magnifying lens, it could read positions to one minute of arc; it cost £10.18.0.[63] Thanks to the elaborate description penned by Henry Cavendish, we know that the Nairne variation compass at the Royal Society was also equipped with a telescope for meridian alignment, that it could be read by microscope and vernier to single minutes, and that the discrepancy when the needle was reversed was "so small as to be scarcely sensible." Tiberius Cavallo claimed that his variation compass was as good as any on the market, but "simpler, smaller, and capable of being afforded by the philosophical instrument makers for a very moderate price."[64] Brander in Augsburg mounted his variation compasses on marble bases, equipped them with verniers, and provided them with needles ranging from 8½ to 10 inches long.[65] Charles Le Monnier in Paris had the use of a variation compass, with telescope for meridian alignment, which enabled him to read variation to seconds.[66]

Apparently dissatisfied with these commercial instruments, the Académie Royale des Sciences offered a prize for the best variation compass in 1775. When no adequate entries were forthcoming, the prize was offered again in 1777. This time there were three winners. The instrument maker Alexis Magny was awarded 800 francs for an instrument in which "the suspension of the needle appears ingenious and the compass seems suitable for performing delicate terrestrial observations." The top prize, 2,100 francs each, was shared by a physicist and an engineer who was to become a physicist. One was van Swinden, whose lengthy annotated bibliography of previous observations of variation and discussions of compass design, was soon forgotten. The other was Charles Coulomb, whose sophisticated analysis of a magnetic torsion balance, the first of his important contributions to the study of torsion, electricity, and magnetism, is said to have been "perhaps the most important essay he ever wrote."[67]

III. CONCLUSION

This collection of essays offers a welcome opportunity to consider a number of difficult questions about scientific instruments. The first concerns definitions. What

[63] Wheatland, *Apparatus of Science at Harvard* (cit. n. 61), pp. 158–159. For a variation compass marked "Gowin Knight Inv[t] Made by Geo. Adams Math[l] Instru[t] Maker to His Royal Highness the Prince of Wales in Fleet Street London" see Anthony Turner, *Early Scientific Instruments: Europe, 1400–1800* (London: Sotheby's, 1987), p. 263; for another see McConnell, *Geomagnetic Instruments before 1900* (cit. n. 58), p. 41. For a variation compass with two microscopes made by Peter Dollond see M. V. Brewington, *The Peabody Museum Collection of Navigating Instruments* (Salem, Mass.: Peabody Museum, 1963), pp. 68–69.

[64] Henry Cavendish, "An Account of the Meteorological Instruments used at the Royal Society's House, *Phil. Trans.,* 1776, *66:*385–395; and Tiberius Cavallo, *A Treatise on Magnetism* (London, 1795), Supplement, pp. 26–36.

[65] Georg F. Brander, *Silber-Gewichts-Verhaltnisse, vermittelst* (Augsburg, 1753); Brander, *Beschreibung eines magnetischen Declinatorii und Inclinatorii* (Augsburg, 1779); and *G. F. Brander, 1713–1783* (cit. n. 61), pp. 267–278. An early variation compass on a marble base is illustrated in A. W. Hauch, *Det physiske cabinet* (Copenhagen, 1836); see also Hemming Andersen, *En videnskabsmand af rang: Adam Wilhelm Hauch, 1755–1838* (Arhus, Denmark: Videnskabshistorisk Museum, 1989).

[66] Pierre-Charles Le Monnier, "Construction de la boussole, dont on à commencé a se servir en Août 1777," *Mém. Acad. Roy. Sci., 1778* (1781), pp. 66–68. For a variation compass by Canivet see Le Monnier, "Mémoire sur la variation de l'aimant en 1772 & 1773," *Mém. Acad. Roy. Sci., 1773* (1776), pp. 440–444.

[67] For Magny see Maurice Daumas, *Scientific Instruments of the Seventeenth and Eighteenth Centuries* (New York: Praeger, 1953), quoting from pp. 267, 333. For Coulomb see C. Stewart Gillmor,

is a scientific instrument, what is science, and who is a scientist—and to what extent do these ambiguous, value-laden, and relatively modern terms enhance or inhibit our understanding of the past? A related question concerns the extent to which definitions depend on use. Is an instrument scientific when it is used to investigate nature, and not scientific when it is used for commercial, military, or other obviously practical purposes? A third question concerns the relations between the relatively narrow domain of science and the broader culture in which it is embedded. The widespread production, distribution, and use of instruments is usually seen as evidence of the diffusion of science. One might, however, turn the issue around and consider the extent to which the use of instruments for practical purposes facilitated their use in investigations that we have tended to consider scientific.

Terrestrial magnetism, which does not fit neatly into the realm of science (or of technology), brings these questions into bold relief and forces an appreciation of the distinction between mixed mathematical and natural philosophy. These social and epistemological categories originated in antiquity and played a powerful role in the early modern period. Though their contours may have changed somewhat, their essence remains with us today.

While anyone might have investigated terrestrial magnetism, most investigators in fact came from states desirous of obtaining and maintaining overseas empires. Moreover, responsible statesmen in England and France were well aware that the fate of cargo and crew depended on magnetic compasses that were well designed and constructed, seamen who knew how to take accurate observations, and a knowledge of the ways in which variation varied over time and space. Historians of science may focus their attention on the last of these problems, but many investigators in the early modern period addressed all three.

Some investigators who tackled terrestrial magnetism during the course of the seventeenth and eighteenth centuries would have identified themselves as natural philosophers, but the vast majority were such mathematical practitioners as navigators, mathematical instrument makers, and professors of astronomy. Magnetic instruments were generally considered to be mathematical—not philosophical, and not scientific. And magnetic observations (angular measurements of variation or dip) were remarkably similar to observations in other mathematical sciences, notably the angular measurements made to determine terrestrial and celestial latitude and longitude.

The lack of production and sales records, the infrequent use of serial numbers, and a low survival rate make it difficult to establish the number of magnetic instruments in use in the early modern period. Nonetheless, the numerous discussions of magnetic instruments that appear in navigation and surveying manuals, and the numerous magnetic observations recorded in logbooks, on charts, and in scholarly journals and monographs, suggest a vast number of instruments and of users. These many investigators may attract historians concerned with the advancement of science only insofar as they made significant contributions to knowledge. But however their contributions are rated, their motivations, opportunities, and apparatus do provide important evidence of the culture of science and the cultural milieu in which science has flourished.

Coulomb and the Evolution of Physics and Engineering in Eighteenth-Century France (Princeton: Princeton Univ. Press, 1971), pp. 140–150, quoting from p. 27.

Although details about early magnetic instruments are sketchy at best, we know enough to see that there was little correlation between the purpose of an observation and the instruments employed. The azimuth compass used to steer a ship one day might be used to improve isogonic charts the next. To be sure, access to deep pockets often led to elaborate and expensive instruments, but there is little evidence of investigators relying on cost or glitz to compel assent.

With the quantification of natural philosophy in the late eighteenth century, the term *mathematical* was increasingly restricted to instruments used for practical purposes, while *philosophical* encompassed all instruments used for education or research. Accordingly, azimuth compasses were described as mathematical instruments or as instruments for navigation. Dip circles and variation compasses, on the other hand, were considered to be philosophical instruments or apparatus for experiments on magnetism. In the same vein, investigations of terrestrial magnetism made in observatories and laboratories came to be assigned to the realm of physics, while observations made by navigators and surveyors remained in the realm of the practical.[68]

Some magnetic investigators were attracted by the challenge of understanding nature and some by the challenge of solving a practical problem, but many understood that the widespread and reliable measurements needed to improve navigation would also provide the grounds for philosophical speculations. This dual motivation was not unique to magnetic investigators. Many mathematical practitioners and natural philosophers have long been equally interested in the glory of God and the benefit of mankind.

[68] See, e.g., "A Catalogue of Mathematical, Optical, and Philosophical Instruments, Made and Sold by George Adams," bound with George Adams, *An Essay on Vision* (London, 1792); and "A Catalogue of Optical, Mathematical, and Philosophical Instruments, Made and Sold by W. and S. Jones," bound with George Adams, *An Essay on Electricity* (London, 1799). For a sample of British trade literature see Calvert, *Scientific Trade Cards* (cit. n. 14).

Stanford's Supervoltage X-Ray Tube

*By Bruce Hevly**

IT MAY SEEM INCONGRUOUS to include the history of a failed idea in a volume devoted to the study of instruments. But though never built, the huge X-ray tube planned by Stanford's physics department during 1934 is an intellectual artifact of some importance, and its story serves as a window on the pursuit of experimental physics in America during the interwar years. In effect it was the first large linear electron accelerator planned at Stanford University, and the project helped to lay the intellectual and social patterns for high-energy physics in Palo Alto.

The X-ray tube was the proposal of David Locke Webster, an experimental physicist specializing in X rays, who was then executive head of the Stanford department. Webster had already reorganized a decidedly backward physics program by focusing the department's research on his own specialty. He now planned to use a very large, high-energy X-ray tube to increase the prestige of Stanford physics and to help the department keep up with the other major centers of experimental physics in California: the state-funded University of California at Berkeley and the private California Institute of Technology.

In terms of operating energies, size, and cost, the supervoltage tube represented a large step beyond any of the previous activities of Stanford's physics department. But it was a step directly in line with Webster's path as a researcher and as executive head of the department. He knew that his colleagues at Caltech and Berkeley had successfully found outside support for their research programs by linking them to electrical engineering and to medical research, finding corporate donors and foundations willing to pay for facilities for high-voltage research and for new cancer treatments. Their example provided a sort of "California model" for building up a physics department, and Webster planned similarly to combine his own and his colleagues' research experience in X-ray physics with this well-established model of institutional advancement.[1] He hoped that an impressive experimental edifice, besides providing important facilities for experimentation when built, would attract not only philanthropy but also theoreticians to an experiment-oriented department sorely in need of both. To get a theorist and a first-rate piece of equipment for experimental nuclear physics and to complete Stanford's transition into a first-rank physics

* Department of History, DP-20, University of Washington, Seattle, Washington 98195.

For helpful comments on earlier drafts of this article, I am indebted to Rebecca S. Lowen, to the other authors of this volume, and especially to Thomas D. Cornell, who shared his knowledge of prewar nuclear physics installations and offered important suggestions on the conclusion. Frances Kohler gave invaluable editorial aid.

[1] Two lengthy studies that analyze this "California style" are Robert Wayne Seidel, "Physics Research in California: The Rise of a Leading Sector in American Physics" (Ph.D. diss., Univ. California at Berkeley, 1978); and Rebecca Sue Lowen, "'Exploiting a Wonderful Opportunity': Stanford University, Industry, and the Federal Government, 1937–1965," (Ph.D. diss., Stanford Univ., 1990).

department, Webster was willing to organize and commit his staff to planning and constructing the device, and later to using it—effectively tying the future of physics research at Stanford to its success.

Webster did get his theoretical physicist, Felix Bloch—initially with support from the Rockefeller Foundation program for émigrés from Europe—but his X-ray tube was never funded. Yet though it remained a paper proposal rather than a realized machine, Stanford's imagined device may still be considered in the same terms generally applied to the study of any artifact. In any made device, the physical process of production leaves signs of how the article was actually formed.[2] An artifact's design and intended use, both influenced by a set of social and intellectual values, also preserve evidence of these unstated ideals of its time. The plans for the supervoltage X-ray tube can be read for both types of meaning. They record how small-scale X-ray practice would have been scaled up into a new physical form. More important, they reflect the physics department's judgment of what needed to be included in the design so as to attract outside support. The process of designing this large-scale scientific instrument, conceptualizing its operation, and deciding upon its intellectual goals was carried out in the context of fund raising. If the artifact had actually come into existence, its physical presence might draw attention away from the facts that its original designs mainly took the form of grant proposals, and that the elements of the machine had to be not only physically correct but persuasive to funders as well.

This unrealized instrument had yet another important dimension: work on the supervoltage X-ray tube influenced the Stanford physicists' postwar response to the availability of military support for science. Organizing to fund the tube was one of the last significant corporate experiences they shared before they began their war effort. It had two effects: intellectual and institutional. The process forced them to consider the tube less as a source of X rays and more as a device in which the electron was manipulated in constructed magnetic fields. This conceptual stance proved important when members of the physics faculty collaborated with Russell and Sigurd Varian on a device for the manipulation of electron streams to produce microwaves, the klystron.[3] It also facilitated postwar work on linear electron accelerators. Institutionally, the experience of working under the regimentation imposed by Webster in the department's common efforts before the war made the independence offered by government support to individual physicists seem all the more attractive. Postwar physics at Stanford was redirected by the availability of klystrons as microwave power sources and also by the relationships between physicists and electrical engineers established in the late 1930s, after the failure of the supervoltage project, and cemented by World War II.[4]

[2] David Hounshell, *From the American System to Mass Production* (Baltimore: Johns Hopkins Univ. Press, 1984), App. 2, and Donald Hoke, *Ingenious Yankees* (New York: Columbia Univ. Press, 1990), demonstrate how analysis of an artifact may be attached to a larger historical argument.

[3] E. L. Ginzton, "The $100 Idea," *IEEE Spectrum,* Feb. 1975, *12*(2):30–39; Dorothy Varian, *The Pilot and the Inventor* (Palo Alto, Calif.: Pacific Books, 1983); and Stuart W. Leslie and Bruce Hevly, "Steeple Building at Stanford: Electrical Engineering, Physics, and Microwave Research," *Proceedings of the IEEE,* 1985, *73:*1169–1180.

[4] This broader story is analyzed in Peter Galison, Bruce Hevly, and Rebecca Lowen, "Controlling the Monster: Stanford and the Growth of Physics Research, 1935–1962," in *Big Science: The Growth of Large-Scale Research,* ed. Galison and Hevly (Stanford, Calif.: Stanford Univ. Press, 1992), pp. 47–55.

I. WEBSTER AT STANFORD

In 1933 David Locke Webster retired as vice-president of the American Association for the Advancement of Science and as chairman of its Section B (Physics). Following tradition, he reviewed his field for those assembled at the association's meeting in Boston that year. His talk, "Current Problems in X-Ray Physics," gave an overview of progress in the measurement and interpretation of line and continuous X-ray spectra since H. G. J. Moseley's work in 1913.[5] At the time Webster had been an experimentalist specializing in X-ray physics for almost twenty years, using the radiation emitted by various elements as clues to atomic structures and processes. This work had been the hallmark of his career, and he had made it the central focus of research at Stanford's physics department.

Webster received his training in physics as both an undergraduate and a graduate student at Harvard, where he completed his Ph.D. in 1913 under the direction of the spectroscopist Theodore Lyman. He remained at Harvard as a physics instructor after graduation and extended his research techniques from ultraviolet spectroscopy into the relatively new area of X-ray physics. While the discovery of X rays was almost twenty years old by the time he took up the subject, the recent experimental work of Moseley and the Braggs had made it an important topic of fundamental physics research, one that seemed to encompass some of the central questions of energy, matter, and radiation that animated early work on the quantum theory.[6]

Webster next took a teaching job at the University of Michigan in 1917, then served briefly in World War I as a "scientific pilot" in the Army Signal Corps Air Service. After the war he moved from Michigan to MIT. Dissatisfied by MIT's combination of commercial research and academic commitments, Webster weighed offers that would allow him to experience one or the other in more purified form. He believed that work at General Electric's research laboratories in Schenectady would provide him the most time and resources to pursue research in physics. But the position of executive chairman of Stanford's physics department promised, in addition to a healthier climate for his family, a chance to build a first-rate teaching program. He moved to Stanford in 1921.[7]

As a department head serving an indefinite term, Webster was selected by the

[5] David L. Webster, "Current Progress in X-Ray Physics," *Science,* 1934, *79:*191–197.

[6] On Webster's career see Webster, "Reminiscences of a Rolling Stone," SC 131, addendum, Webster papers, Department of Special Collections, Stanford University Libraries (hereafter **SUA**) . Biographical notes by others include Paul Kirkpatrick, "David Locke Webster II," *National Academy of Sciences Biographical Memoirs* 1982, *53:*367–400; and Bruce Hevly, "Webster, David Locke," *Dictionary of Scientific Biography,* Vol. XVIII: *Supplement II* (New York: Scribners, 1989). Webster's research prior to the supervoltage tube project is summarized in A. H. Compton and S. K. Allison, *X-Rays in Theory and Experiment* (New York: Van Nostrand, 1935). For the relationship of X-ray research to modern physics and Webster's role in the interwar period, see Bruce R. Wheaton, *The Tiger and the Shark: Empirical Roots of Wave-Particle Dualism* (Cambridge: Cambridge Univ. Press, 1983); Roger H. Stuewer, *The Compton Effect: Turning Point in Physics* (New York: Science History Publications, 1975); and Katherine Russell Sopka, *Quantum Physics in America, 1920–1935* (New York: Arno Press, 1980; Los Angeles: Tomash, 1988). In particular, on Webster's use of X-ray experiments to help establish a fundamental relationship in modern physics, see Paul Kirkpatrick, "Confirming the Planck-Einstein Equation $h\nu=(1/2)mv^2$," *American Journal of Physics,* 1980, *48:* 803–806.

[7] Webster, "Reminiscences of a Rolling Stone"; also military records and diaries and early correspondence, Webster papers. On Stanford's choice of Webster, see folder 2, "Physics 1920," SC 64a, box 43, Ray Lyman Wilbur papers, SUA.

president at Stanford, who expected him to take active charge of his department. Given the opportunity to replace retiring professors with younger men, Webster determined to concentrate the department's research efforts on problems of atomic structure, especially investigations of the atom from the outside in, excluding the nucleus itself.[8] He hired Norris Bradbury, Percy Aron Ross, and two physicists trained at Stanford, George R. Harrison and William Hansen. All, like Webster, conducted X-ray research as a method of studying electron orbitals close to the nucleus. When Harrison was lured to MIT, Webster hired another X-ray specialist, Paul Kirkpatrick, as his replacement.[9] Webster felt that in a small and poorly funded department, only this concentration of force would give results.

The Stanford physicists and their graduate students became adept at tabletop X-ray research. Cannibalizing donations of equipment from local physicians for the needed basic components, Webster's group developed a generator and transformer system with an exceptionally smooth and well-controlled electrical output that gave the accelerated electrons a well-defined incident energy when they reached their targets to produce X rays.[10] The group developed the technique of using very thin foils of relatively heavy elements (for example, gold) as targets, in order to produce the K-series of the element's characteristic X-ray spectrum, which resulted from displacing the electrons from inner orbitals. At this point the Stanford physicists could investigate the characteristics of the binding between the nucleus and the inner electrons with relatively low-energy equipment. These experimentalists were not ignorant of the emerging theoretical structures of modern physics: they used them to work on the regions close to the nucleus, where electrons are tightly bound.[11] The experience lent them the knowledge necessary to formulate plans for the supervoltage tube proposal, but these early experiments were scarcely sufficient to compete as other physicists took up nuclear research in the 1930s.

Webster's tabletop approach had delivered its most important results by 1933, and that year also represented a crisis for the dream of a nationally competitive research program at Stanford. Nuclear physics was moving into center stage for academic scientists, propelled by fundamental experimental discoveries and the vigorous applications of modern theory. The means of production for experimental nuclear physics included access to theoreticians and to expensive high-energy equipment.[12]

[8] Webster summarized this approach regularly in his annual reports; see, e.g., *Stanford University Bulletin,* Nov. 1929, pp. 34, 365, and Nov. 1930, pp. 246–247.

[9] Paul Kirkpatrick, "Autobiography" (1971), p. 52, accession no. 83–013, SUA.

[10] Equipment donations are summarized annually in Webster's reports in the *Stanford University Bulletin.* On the 200,000-volt power system see *ibid.,* Nov. 1929, p. 365.

[11] On Webster's ionization function for beta particles approaching K-electrons, see Compton and Allison, *X-Rays in Theory and Experiment* (cit. n. 6), p. 72. See also Gerald L. Pearson, interview by Lillian Hartmann Hoddeson, 23 Aug. 1976, Center for the History of Physics, American Institute of Physics, pp. 6–8.

[12] One of the best-studied sites of such change is Merle Tuve's nuclear physics program at the Carnegie Institution of Washington. See Thomas D. Cornell, "Merle A. Tuve and His Program of Nuclear Studies at the Department of Terrestrial Magnetism: The Early Career of a Modern American Physicist," (Ph.D. diss., Johns Hopkins Univ., 1986); and Michael Aaron Dennis, "A Change of State: The Political Cultures of Technical Practice at the MIT Instrumentation Laboratory and the Johns Hopkins University Applied Physics Laboratory, 1930–1945" (Ph.D. diss., Johns Hopkins Univ., 1991), Ch. 2. See also Finn Aaserud, *Redirecting Science: Niels Bohr, Philanthropy, and the Rise of Nuclear Physics,* (Cambridge: Cambridge Univ. Press, 1990), esp. pp. 17–27 (on integrating theory and experiment), and 231–248 (on the Bohr Institute's "clear turn to nuclear physics" in the context of foundation support).

Although Webster had modernized undergraduate teaching, remade the faculty, and established a small but successful research program, in the early 1930s he found his department falling farther and farther behind its California peers. At Berkeley, E. O. Lawrence had been able to combine reliable state funding with foundation and commercial support to build his "atom smashers."[13] Webster and his World War I commander in the ranks of Army scientists, Robert Millikan, had arrived in California at virtually the same time; Millikan now oversaw a tremendous facility for research, Caltech's Norman Bridge Laboratory of Physics, which drew upon the largess of Southern California industrialists who hoped to be associated with the discovery of the sources of the universe. These donors also expected that science would provide answers for some fundamental technological problems.[14] At Caltech's Kellogg Laboratory of Radiation, Charles Lauritsen oversaw another accelerator facility, this one, like Lawrence's, funded in part by money for medical research.

Institutionally Webster was forced to compare his department to those at Berkeley and Caltech. He could also look beyond Stanford and find examples of experimental apparatus that might help his department's research program compete with theirs. The essential problem of producing highly energetic X rays was that of accelerating electrons, and other institutions had deployed a range of machines to speed up charged particles: Tesla coils, cascade transformers, Van de Graaff machines (built at MIT and also eventually adopted by Merle Tuve at the Carnegie Institution of Washington), and of course E. O. Lawrence's cyclotron. All were systems to accelerate charged particles through electric potential differences. A particular scientist's choice among them had to be the product of experience, resources, local conditions, and even aesthetic sensibilities. Merle Tuve, for example, originally chose Tesla coils over cascade transformers in the mid 1920s in part because he thought them to be more appropriate for a nonindustrial physicist. All these devices required funds that Stanford lacked.[15]

Although it provided an incentive for graduate study, the Depression hit Stanford hard, causing endowment revenue and foundation support to drop. The American scientific community had settled on a policy of "making the peaks higher" in the post–World War I period, but Webster realized he ran the risk of having no peak to build on.[16] He could expect no foundation money, and no eager National Research Council postdoctoral fellows, unless he could find a mechanism to propel Stanford's ascent into modern physics research. Another manifestation of the general threat of poverty and neglect was Webster's inability to attract a first-rate theoretician to join the department. He wrote to many on the East Coast, offering visiting appointments with the potential of permanent jobs. But he could lure no one to Palo Alto for

[13] J. L. Heilbron and Robert W. Seidel, *Lawrence and His Laboratory: A History of the Lawrence-Berkeley Laboratory* (Berkeley/Los Angeles: Univ. California Press, 1989), Vol. I, pp. 207–226.

[14] Robert H. Kargon, "Temple to Science: Cooperative Research and the Birth of the California Institute of Technology," *Historical Studies in the Physical and Biological Sciences* (***HSPS***), 1977, *8*:3–31; and Judith R. Goodstein, *Millikan's School: A History of the California Institute of Technology* (New York: Norton, 1991).

[15] Cornell, "Merle A. Tuve" (cit. n. 12), pp. 198, 225.

[16] Daniel J. Kevles, *The Physicists* (New York: Knopf, 1977), Ch. 13. On Stanford's financial circumstances during the Depression and the university's sense of decline see Rebecca S. Lowen, "Transforming the University: Administrators, Physicists, and Industrial and Federal Patronage at Stanford, 1935–1949," *History of Education Quarterly,* 1991, *31*:365–388, on pp. 367–369.

more than a visit. In contrast, across the bay E. O. Lawrence maneuvered adroitly throughout the Depression, knit together funding from many sources, and had a fine theoretician on the faculty.[17]

In April of the same year that Webster and his colleagues began to seek support to build the supervoltage tube, a leading young European theorist, Felix Bloch, did join the department. But while Bloch spent the rest of his career at Stanford, his original appointment was for only two years, and so he did not seem to represent a permanent solution to Webster's search for a theorist. Bloch came with the aid of funds for displaced scholars; the fortunes of history might have led him to return to Europe after only a brief American stay, or Stanford might have declined to carry the cost of his salary when the Rockefeller subsidy ended.[18]

In the short term, however, Bloch's presence did establish a link to J. Robert Oppenheimer at Berkeley, for the two theorists convened a joint theory seminar. Oppenheimer's interest in the origin of electrons and positrons in P. A. M. Dirac's electron theory soon inspired the Stanford experimentalists and influenced their thinking about the experiments possible with accelerated electrons.[19] They began to think in terms of even more profound experimental goals than they had previously undertaken, such as the creation of new particles from the energy field near the nucleus. It was in this context that Webster turned his department's energies to funding the supervoltage X-ray tube.

II. THE PATH TO THE TUBE

Webster was at the center of the process of seeking support for the supervoltage X-ray tube. When asked by the editors of *Reviews of Modern Physics* to recast his AAAS talk on achievements in X-ray research for a technically trained audience, Webster, himself a member of the editorial board, found that he was too engrossed with work at Stanford. "We are contemplating a general reorganization of the research program of this department, provided funds can be obtained to get into the Super-voltage field," he wrote, explaining his heavy commitment to department business.[20]

Beginning in 1934, the physics department met as a "Supervoltage X-ray Committee" to develop their ideas. Drawing on the department's earlier research practice, they considered the options mentioned above, then proposed a straightforward, albeit huge, device capable of delivering electrons with energies of three to four million electron-volts to a target.[21] As they described it in proposals to the Rockefeller

[17] Heilbron and Seidel, *Lawrence and His Laboratory* (cit. n. 13), pp. 25–29, 260. Webster's best effort was to set up a series of lectures on theory by visitors; it included such figures as George Gamow and Webster's old friend, John Van Vleck, but none was willing to stay on.

[18] *Stanford University Bulletin,* Dec. 1934, p. 10. Bloch's salary came "from funds provided by the Emergency Committee in Aid of Displaced German Scholars and the Rockefeller Foundation."

[19] W. H. Furry and J. R. Oppenheimer, "On the Theory of Electron and Positive," *Physical Review,* 1934, *45:* 245–262; a copy was kept in the Supervoltage X-ray Committee notebooks (see n. 22, below). Joan Bromberg has pointed out that such ideas were common even before Dirac's theory redefined them: Bromberg, "The Concept of Particle Formation Before and After Quantum Mechanics," *HSPS,* 1976, *7:*161–191.

[20] David L. Webster to John T. Tate, 27 Dec. 1934, copy in "Current Progress in X-Ray Physics" notebook, SC 197, box 2, Webster papers.

[21] "Supervoltage X rays" were defined as those created by electrons over 200,000 volts. Surveys of the range of possible equipment for accelerating particles include W. H. Wells, "Production of High-Energy Particles," *Journal of Applied Physics,* 1938, *9:*677–689.

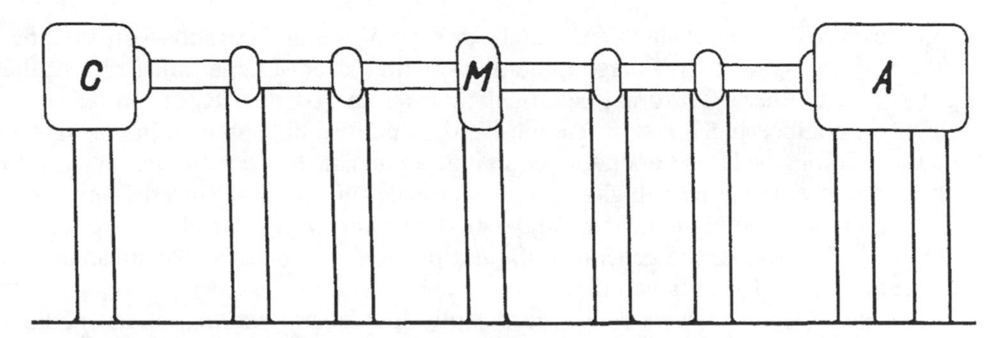

Figure 1. *"Proposed 3-million-volt x-ray tube, showing tubular porcelain sections connecting cathode house (C), steel middle section (M), and anode house (A), all supported by insulating columns," as proposed by Stanford's physicists in March 1935. Courtesy Stanford University Archives, Division of Special Collections.*

Foundation and Carnegie Corporation, this device would consist of a series of ceramic pipes, probably twenty inches in diameter, bolted together in a line by means of metal collars at the end of each section, and standing twenty feet above ground on pyramid-shaped supports (see Figure 1). At either end of the ceramic pipeline, also on stilts, Webster's committee planned to place a "cathode house" and an "anode house," that is, an electron source and a positively charged collector at the end of the tube, which would include a target from which X rays would be produced after its collision with the accelerated electrons.[22]

The ceramic tubes would enclose an evacuated space, within which would be a series of metal pipes. Each of these sizable electrodes would carry a share of the three-million-volt potential difference to be delivered by the system's power supply. With sufficient vacuum and sufficiently precise focusing to keep the electron stream out of contact with stray gases or the solid parts of the tube itself, Webster's group could expect to deliver the electrons to the target and to produce very energetic X rays.

Webster and the members of his department sought more than $100,000 for construction and operation of their device, a vast sum compared to Stanford's Depression-era research budget, and a respectable one even when compared to grants the department received from the federal government soon after World War II. They drew upon three main sources of strength when writing the proposal: their own experience in X-ray research, the prospect of a close link to Stanford's existing facilities for high-voltage electrical engineering investigations, and the promise of a connection to medical research. All three left marks on the design that emerged in the proposals produced in 1935.

Webster's appeal to the department's research experience is discussed in the next section. Besides these departmental resources, Stanford offered Webster two other

[22] "Prospectus on Supervoltage X-ray Research Work at the Ryan Laboratory," 12 March 1935, Supervoltage Project notebook II. The supervoltage committee minutes and related documents, the major sources for this account, are held in the Webster papers, box 1, SC 197, and are henceforth cited as **notebook II** or **III.** Of three loose-leaf binders the first is missing; it was lent to Sperry Gyroscope Company for use in a patent case and is now lost: Webster to Virginia Vorhis, 21 May 1942, notebook II.

significant features relevant to the search for support to build the supervoltage tube. First was the Ryan High-Voltage Laboratory, with its set of huge transformers, the gift of the northern California electrical industry to Harris J. Ryan, professor of electrical engineering.[23] Just off campus in the Stanford hills, in a plain corrugated metal building, the Ryan Laboratory contained a fantastic array of equipment used to investigate the problems of high-voltage electrical transmission. Thus the physicists could propose to build their X-ray tube with the power source already provided, and the tube's size, shape, and general configuration were designed to take advantage of the output of the Ryan transformers.

Stanford's second resource outside the physics department was at its medical school in San Francisco: Dr. R. R. Newell, a specialist in radiation treatments for cancer. The powerful X rays from a three- or four-million-volt tube might prove an effective new resource for therapy, to be used along with natural radioactive sources of energetic gamma rays and the energetic emissions of artificially radioactive sources. In fact, the Stanford physicists had built a 400,000-volt therapy tube, funded by the Stanford Medical School, for Newell—the largest X-ray tube with which they had been associated. It was not used for physics research.[24]

The Stanford physicists' first proposal to the Rockefeller Foundation emphasized both these features, putting it squarely in the tradition of large-scale physics projects developed in California before World War II. As Robert Seidel has pointed out, with the state's relatively abundant hydropower resources located in the mountains and its urban population concentrated along the Pacific coastline, electrical utilities needed to understand how to send alternating current over great distances without suffering tremendous transmission losses. The result was an early, firm connection between the private utility companies and the electrical engineering departments at Berkeley, Stanford, and Caltech, with the companies providing expensive equipment in the form of transformers and generators (or free electricity). The early nuclear physicists at Berkeley and Caltech were able to build on these pre-existing resources,[25] and Webster hoped to emulate them.

The eleemosynary foundations were another important potential source of support, especially for a private institution without access to state money, such as Stanford. It was therefore just as important to appeal to their special interest: the extension of knowledge for the improvement of human life. E. O. Lawrence at Berkeley was a master of connecting his machines to medical research, especially cancer treatment, and he in fact devoted a substantial amount of machine time to producing radioactive substances for therapy and use as biological tracers. Also at Berkeley, David Sloan worked from 1931 to 1935 to build a million-volt X-ray machine, despite dubious commercial and therapeutic prospects. Webster felt the Stanford physicists too would need to exploit the medical connection, although the Caltech example led him to warn his colleagues that "we must consider very carefully where

[23] Thomas Parke Hughes, "The Science-Technology Interaction: The Case of High-Voltage Power Transmission Systems," *Technology and Culture,* 1976, *17:*646–662; and Hughes, *Networks of Power* (Baltimore: Johns Hopkins Univ. Press, 1983), p. 379. See also box 4, folder 4, SC 25, Harris J. Ryan papers, SUA.

[24] *Stanford University Bulletin,* Dec. 1932, p. 287.

[25] Seidel, "Physics Research in California" (cit. n. 1), pp. 28, 81 (on Caltech), 135 (on Carnegie Corporation support), 257, 319–322, 327 (on contributions of California power interests).

we stand in case it looks as though the development ought to go completely over to the therapeutic side as it has in the Kellogg laboratory at Pasadena."[26]

III. THE FIRST PROPOSAL

The tube design went through three basic stages. Its first form was embodied in the proposal submitted by the department to the Rockefeller Foundation early in 1935. This was followed by a period of further research—both on and off campus—and adjustment to the reactions engendered by the first proposal. Finally, toward the end of 1935, the concept of the supervoltage tube was redesigned for an amended proposal directed at the Carnegie Corporation.

The original proposal emphasized the probable and possible benefits of the instrument, set it firmly in position as an extension of research practice at Stanford, and drew on the existence of both the Ryan Laboratory and Newell's cancer therapy research to buttress the application. It preserves a clear image of the instrument the Stanford physicists intended to build and of the way in which they had shaped it to appeal to funders.

The physicists began by arguing that their project would be "a logical extension of the research program of the Department of Physics as it has developed in the past ten or fifteen years." Both their own work, at around 200,000 volts, and their collaboration with Newell on a cancer therapy X-ray tube rated at 400,000 volts gave them experience that would guide their research and justify an investment in it by a donor. As Webster pointed out in the prospectus, sent first to Stanford President Ray Lyman Wilbur, "we have specialized primarily in research on the physics of x-rays, and have acquired not only a wide acquaintance with this field and the theory underlying it, but also a great deal of technical experience which will be useful in any extension of the field."[27] This sounds much like the usual grant-proposal boilerplate. But the project's scientific objectives fell in line with the experience of Stanford's X-ray group and with Webster's policy of concentrating the resources of his small department on an important, well-defined central problem.

Webster's first set of questions contemplated using X rays as probes: he suggested scattering them both from other electrons, to test "Dirac's views of the nature of the electron," and from atomic nuclei. Here the process of writing the proposal produced the beginnings of a change in perspective, as Webster's group began to propose new research questions on nuclear structure that they might begin to answer with a powerful source of X rays.[28] The group did not propose using the accelerated electrons themselves as probes in this first proposal. Rather, they were means to provide a reliable radiation source for experimental nuclear physics, less energetic than the gamma radiation obtained from some radioactive substances, but more intense.

On questions of nuclear structure, Webster wrote, "nothing will touch them but supervoltage rays." High-energy X rays might be used to cause nuclear disintegrations, but more "notable . . . unexpected effects" were possible as well: the

[26] Webster to Kirkpatrick, 15 May 1935, notebook III; Seidel "Physics Research in California" (cit. n. 1), p. 366; and Heilbron and Seidel, *Lawrence and His Laboratory* (cit. n. 13), pp. 214–216, 188–191, 236–239 (on Lawrence), 118–125 (on Sloan).

[27] "Prospectus on Supervoltage X-Ray Research" (cit. n. 22), p. 1.

[28] *Ibid.,* pp. 11–12.

"expulsion of neutrons from the nuclei of some light atoms," "the creation of electron-positron pairs in the strong electric fields of the nuclei of heavy atoms," and the excitation of nuclei, inducing them to emit gamma rays. "Excited nuclei offer special inducements to research, because the excitation of the outer parts of the atoms was the clue to most of our knowledge of their structure, and we can hope that the history of atomic structure research may repeat itself for nuclear structure," Webster concluded.[29]

Thus the proposal's rhetoric appealed to recent history: both the history of atomic physics in the twentieth century, and that of Webster's group at Stanford. Just as physicists had learned to excite the outer electronic structure of atoms, and even the inner electron rings, using smaller X-ray setups—and as Webster's department had in fact tried to focus on understanding the atom from the outside in—so sufficiently powerful equipment might allow them to induce nuclei to behave analogously to excited bound electrons, emitting gamma rays or neutrons, or exhibiting pair production at close range, rather than producing an ultraviolet and X-ray spectrum.

The language of the proposal also appealed to scientific problems more exotic than anything done previously at Stanford. Clearly the spur for building a new device was to open new research areas, but the plans were also shaped by the context of a search for financial support. In that framework a phenomenon such as electron-positron pair production carried great inspirational weight. Donors might sponsor the comprehension of a beginning of existence, as with Robert Millikan's "birth cries of the elements."[30] In this case the search for support to build a new instrument reinforced the tendency, identified by S. S. Schweber, for American theorists to express theoretical concepts in terms amenable to experimental investigation (although here the theorist, Felix Bloch, was only newly an American).[31]

Like his competitors at Berkeley and Caltech, Webster offered a recondite yet inspiring program of nuclear research, organized around a large, advanced apparatus. Yet the proposal failed to win support. Why did the Berkeley-Caltech model fail in Stanford's case?

Stanford labored under a few handicaps. For one, Webster's somewhat abrasive personality did not mark him as likely to cultivate sponsors with the ease of a Millikan or a Lawrence. Stanford had neither Berkeley's financial bedrock of state support, nor the broad support from local industrialists established by George Ellery Hale and Millikan in Pasadena. But more important was the failure of three assumptions that underlay the supervoltage tube proposal: that a link could be made to cancer therapy, that foundation support was available for a large-scale enterprise, especially one related to medical research, and that the Ryan High-Voltage Laboratory was a major asset.

To begin with, Newell at the Stanford Hospital could not be persuaded to lend more than equivocal support, even for the purposes of a grant proposal, to the idea that the X rays from a three-million-volt tube might be useful for therapy. Equally

[29] *Ibid.*, pp. 12–15.

[30] Robert Kargon, *The Rise of Robert Millikan* (Ithaca: Cornell Univ. Press, 1982); and Kargon, "Birth Cries of the Elements: Theory and Experiment along Millikan's Route to Cosmic Rays," in *The Analytic Spirit: Essays in the History of Science in Honor of Henry Guerlac*, ed. Harry Woolf (Ithaca, N.Y.: Cornell Univ. Press, 1981).

[31] Silvan S. Schweber, "The Empiricist Temper Regnant: Theoretical Physics in the United States, 1920–1950," *HSPS*, 1986, *17*:55–98.

energetic radiation, Newell wrote, was available from natural sources, albeit in lower concentrations. While he would not rule out the possibility that research at some future date would show that higher energies were important for therapy, for the present he could only express an interest in the device as "a venture in experimental physics."[32]

This was a sharp blow to the proposal, although the physicists could also point to Newell's admission that there might be long-term benefits from the development of a more powerful tube. Still, the most powerful X-ray tubes Webster knew of—Newell's own 400,000-volt machine, and a GE-built 800,000-volt cancer tube at Seattle's Swedish Hospital—were built for medical uses. Ideally, therapeutic benefit would have scaled up directly with energy, lending a humanitarian justification for ever-larger instruments. Webster later regretfully admitted to Stanford President Wilbur, "If Dr. Newell's statement is 'not very good bait from the cancer standpoint' it is only fair that it should be so," but he continued to search for a medical justification for his projects.[33]

Perhaps because Newell would not certify the medical or biological benefits of the tube, Webster's group got a quick "no" from the Rockefeller Foundation in March 1935. A week after Wilbur wrote a letter of inquiry summarizing the physicists' proposal, Max Mason replied that Rockefeller money was not available to support research in the physical sciences, and so the "proposal would then of necessity lie outside the present program of the Foundation."[34] That left the Carnegie Corporation as a possible source of support. Wilbur reported at the beginning of April that managers of the Carnegie endowment funds would take up Stanford's request after 1 October, as their program was already set for the current fiscal year. Webster found some cause for optimism—"this at least is better than the reply from the Rockefeller Foundation"—and determined to beef up the proposal for submission the following autumn.[35]

Yet the third assumption—that the Ryan Laboratory represented an asset—also failed the Stanford physicists. Upon receiving a copy of the physicists' prospectus for supervoltage X-ray research, which dwelt at length on the benefits to them of access to the Ryan transformers, the electrical engineering professor responsible for their operation weighed in against the use of his equipment for physics research. He objected first on the grounds of operating efficiency. The physicists' plans rested on peak output from the transformers, not their average operating output (root-mean square voltage), but this was only available for the top tenth of each cycle in the transformers' alternating current output. Combined with the fact that the transformers could only be run safely about five minutes out of each hour at their highest voltage, this meant that "three million volts will be applied to the X-ray tube one minute out of every two hours."[36] (That is, the physicists might expect to get bursts

[32] "Prospectus on Supervoltage X-Ray Research" (cit. n. 22), p. 8. Francis Carter Wood, director of Columbia University's cancer institute, ordered a Sloan therapy tube but expressed a similar skepticism concerning the medical efficacy of megavolt X rays; see Heilbron and Seidel, *Lawrence and His Laboratory* (cit. n. 13), p. 121.

[33] Webster to Wilbur, 4 June 1935 (copy), notebook III. The Seattle 800,000-volt tube is illustrated in *Northwest Medicine,* Dec. 1937, *36*(12):1.

[34] Max Mason to Ray Lyman Wilbur, 19 March 1935 (copy), notebook II.

[35] Webster to Wilbur, 15 April 1935 (copy), and supervoltage committee minutes, 14 April 1935 committee meeting, notebook II.

[36] Joseph S. Carroll to Ray Lyman Wilbur, 9 April 1935 (copy), notebook II.

about 1.7 thousandths of a second long, 60 times per second, five minutes out of each hour.)

In addition to what seemed to be a practical objection, the engineer had a territorial one. He feared "a repetition of what our friends down South had with their high-voltage laboratory." Plainly referring to Caltech, he argued that experience had demonstrated that once physicists insinuated themselves into a laboratory, the engineers would be driven away from their lawful occupations. "As nearly as I can make out," he wrote, "the setting up and operation of this tube would more or less completely monopolize the Ryan Laboratory."[37] These objections effectively crippled the physicists' proposal. Without the cooperation of those in control of the transformers, the X-ray tube would have to be redesigned or find its own high-voltage source.

By early spring of 1935, then, Webster's group had formulated their first proposal for the supervoltage X-ray tube and failed to find support. Also, two of the main influences on their design—the assumptions that the tube's potential for cancer therapy and the use of the Ryan transformers as a power supply both naturally strengthened the project—had turned out to be false hopes. Now the physicists had the summer and early fall to revise their design for resubmission to the Carnegie Corporation.

IV. REDESIGNING THE TUBE

The first step in the redesign project was a meeting of the department sitting as the supervoltage committee, at which each member's responsibilities were clearly laid out. According to this division, Webster administered the project, Ross was to design the mechanical supports for the tube and concentrate on other engineering problems, Kirkpatrick was assigned the problem of electron focusing, Bloch took responsibility for developing theoretical ideas, Bradbury would be concerned with the electron paths and electromagnetic fields within the tube, and Hansen took charge of building a model X-ray tube from pyrex pipes and testing its behavior.[38]

Clearly, developing an instrument as complex as this one required some organization; Webster had to discipline his colleagues by setting the tasks they would pursue for the greater good. While the assigned tasks were tailored to each department member's talents and interests, working on the tube led to a certain amount of regimentation. Thus Webster could make inquiries of an electrical utility with experience building the kind of pylons needed to lift and insulate the X-ray tube from the ground, with the confidence that the designated subordinate would take up the engineering details in future conversations: "This question of supports will of course call for careful study which may take some time," he wrote, "and it is in the hands of Professor Ross, who will undoubtedly wish to discuss the matter with you."[39]

While Ross carefully studied legs for the tube and Hansen built a model of it, Webster traveled, looking for more information that might be useful for the project. As a rule Webster traveled each summer. On the way to his sailing vacations around Vancouver Island, for example, he examined the 800,000-volt therapy tube in Seat-

[37] *Ibid.*

[38] Memo to supervoltage group members, 25 March 1935, supervoltage committee minutes, notebook II.

[39] Webster to "Mr. Whisler," Westinghouse, 25 March 1935 (copy), notebook II.

tle. Webster also had family responsibilities on the East Coast and took advantage of his trips there to gather information related to the project of refining the tube design.

The most powerful X-ray tubes Webster knew of were built by GE for the therapy market, under the direction of William Coolidge. Although visits to research centers in the East produced "no sign of any money in any place I have seen yet, including G.E. Co.," they did convince Webster that the Stanford design, which resembled GE's, was essentially on the correct track. Webster decided that Stanford's physicists also needed to learn what they could from the construction techniques used by industrial scientists, especially the commitment to clean mechanical construction, emphasizing positive mechanical connections between parts and eschewing small-scale laboratory measures such as glass blowing, glued joints to hold vacuum seals, and casual "haywire" electrical connections.[40] Physicists elsewhere were coming to the same conclusion. Indeed, after seeing the Stanford group's earlier efforts on Newell's therapy tube during a tour of experimental physics facilities on the West Coast, Merle Tuve wrote, "I think all high-voltage tubes may presently be made in machine shops rather than by glass-blowing." The original proposal had pointed out that "a few supervoltage tubes have been built, and 'built' is the right word, because their construction is not a matter of glassblowing but of engineering." Based upon his tour, Webster asserted, "comparing Lauritsen [at Caltech], Tuve [at the Carnegie Institution of Washington], Sloan & Lawrence [at Berkeley], and Coolidge, my impression is that Coolidge is head and shoulders above the rest, and I am mighty glad that we decided on a tube of his type."[41]

Although he acknowledged the mechanical adeptness of the industrial researchers, Webster believed that the insight of the academic scientists would be required to get the most out of the instrument. For example, he found that Coolidge and others engaged in accelerating particles had little idea of exactly what electrons did while traveling through an X-ray tube. Stanford's physicists, on the other hand, did conceptualize those events: they needed to concentrate on the problems of focusing the electron stream and manipulating it within the length of the supervoltage tube in order to deliver a well-defined beam to the target. Also in the course of working up a revised funding proposal, Webster's colleagues began to think about the accelerated electron itself as an experimental probe, rather than simply as a raw material used for the production of high-energy X rays.

During the rest of the summer the Stanford group concentrated on finishing the details of their design in order to complete a revised proposal for the Carnegie. Focusing the beam engaged the attention of Webster and Hansen upon Webster's return.[42] In addition, the department agreed to give up the idea of using the Ryan

[40] Webster to Paul Kirkpatrick, 8 May 1935, pp. 1 (quotation), 5–6, notebook III. GE regarded some academic efforts with suspicion, as competition in the market for therapy tubes, even accusing MIT of "improper commercial competition" in 1938: Christophe Lecuyer, "The Making of a Science-Based Technological University: Karl Compton, James Killian, and the Reform of MIT, 1930–1957," *HSPS*, 1993, *23*:153–180. The search for foundation support would have forced such competition.

[41] Merle Tuve, "Report on Visit to Pacific Coast," 12 Oct. 1931, Box 15, Tuve papers, Library of Congress Manuscripts Division; "Prospectus on Supervoltage X-Ray Research (cit. n. 22), p. 4; and Webster to Kirkpatrick, 8 May 1935, p. 8. I am indebted to Thomas D. Cornell for providing me with the Tuve material.

[42] Supervoltage committee minutes, 18, 22 March, 14 April 1935, notebook II; W. W. Hansen and D. L. Webster, "Electrostatic Focusing at Relativistic Speeds," *Review of Scientific Instruments*,

Laboratory facilities and added the cost of their own transformer to their proposal. They chose a new site for the instrument: an earthen trench (the earth used for radiation shielding) to be dug in an abandoned quarry on the Stanford campus. Finally, Webster refused to give up the idea of a medical rationale for the X-ray tube. He repeatedly urged Newell to reconsider his opinion, and sought out physicians willing to express contrary views.[43]

The September prospectus declared that the Stanford group was "ready to begin construction as soon as funds are available." Eleventh-hour appeals to Newell having failed, the new proposal briefly outlined the decision to buy a transformer from GE and the results of research on focusing by Kirkpatrick, Webster, and Hansen. The problem of producing electron-positron pairs, which might be tested by injecting energy in the form of fast electrons into the strong field near the target's nuclei, was put forth as the group's "most important research problem,"[44] reflecting a shift in emphasis to using particles rather than radiations as probes of nuclear effects. Still, Carnegie money also was not forthcoming, and despite an attempt at another version of the proposal, by the end of November 1935 the Stanford physicists gave up on the supervoltage X-ray tube and turned to less expensive ways of accelerating electrons.[45]

V. THE INFLUENCE OF THE INSTRUMENT

The most promising option for an economic accelerator was Hansen's idea for an electrically resonant sphere, dubbed the "rhumbatron," which Hansen developed during 1936–1937. The rhumbatron in turn provided part of the artifactual inspiration for the klystron, a versatile microwave tube invented in 1937 by two quasi-independent researchers attached to the department, Sigurd and Russell Varian, with contributions by Hansen and Webster. The klystron quickly attracted commercial support to the department from the Sperry Gyroscope Company, and it found application as part of a series of early radar inventions for blind landing and aircraft detection.[46]

Using klystrons for microwave power, Stanford physicists and electrical engineers developed a series of linear electron accelerators after the war, in projects led by Hansen until his death in 1949, and then by his successors. The electron accelerators were built first in temporary quarters at the Ryan High-Voltage Laboratory—converted after the war to hold a nuclear reactor as well—and then in an earthen trench next to the Stanford academic buildings. Finally, the klystron-powered, two mile-

---

1936, *7:*17–23; and Paul Kirkpatrick and James G. Beckerley, "Ion Optics of Equal Coaxial Cylinders," *Review of Scientific Instruments,* 1936, *7:*24–26.

[43] Supervoltage committee minutes, 28 Aug., 4 Sept. 1935, notebook III. Paul Kirkpatrick was convinced that without the tie to the Ryan transformers, the proposal was doomed: Webster to Kirkpatrick, 29 May 1935, *ibid.* On continuing efforts to tie the tube to therapy see Webster to Kirkpatrick, 15 May 1935, Webster to R. Newell, 15 May 1935, and "Supervoltage Project 3 MVP for Therapy," 11 Sept. 1935, all *ibid.*

[44] "Supervoltage X-Ray Work at Stanford University, Addendum to Prospectus of March 12, 1935," 21 Sept., notebook III.

[45] Robert M. Lester to Ray Lyman Wilbur, 28 Oct. 1935 (copy), notebook III.

[46] Leslie and Hevly, "Steeple Building at Stanford" (cit. n. 3); and Lowen, "Exploiting a Wonderful Opportunity" (cit. n. 1), Ch. 1.

long Stanford Linear Accelerator was built in the hills above the campus, not far from the quarry where Webster's department had hoped to place their supervoltage X-ray tube.[47]

Although never built, the supervoltage tube provided a crucial link in an instrumental tradition at Stanford. That tradition reached back to Webster's early research practices as an X-ray physicist, and to his decision to concentrate the department's efforts in his own field. The tradition's material legacy was Stanford's postwar linear electron accelerators, instruments that persisted at Stanford when most high-energy experimental physicists built machines patterned after E. O. Lawrence's positive-ion accelerating cyclotron, which could be run to higher energies after the war because of the development of phase stability, independently conceived of by Edwin McMillan and V. I. Veksler.

Several elements of Stanford's research practices in high-energy physics can be traced to the effort to build the supervoltage tube. In the process of trying to gather the basic information that would allow the tube to be built, Webster, Hansen, and others had to pay particular attention to the problems of manipulating an electron stream under vacuum in a series of constructed electromagnetic fields. The effort to define the supervoltage tube and its constituent parts engendered the beginnings of a conceptual shift for the experimenters. Members of the group, in particular Hansen, became singularly adept at visualizing the electron physics within conductive chambers. This kind of insight grew out of a changing perspective, as the tube began to be regarded as a region of complex interactions and more than just a source of X rays.

In addition to this intellectual tradition, an important social tradition emerged from the project. In the postwar period Hansen, supported strongly by Felix Bloch, planned to build a microwave laboratory using funds earned from Stanford's share of wartime klystron royalty payments; his major goal was to develop the linear accelerator. Webster, who had disliked the constraints of working under corporate sponsorship during the klystron project, opposed the plan. The outcome was that Webster was deposed as executive chair, to be replaced briefly by Paul Kirkpatrick and permanently by a new arrival, Leonard Schiff; Hansen's microwave laboratory was built to house his work on the electron accelerator.[48]

In the discussion over the construction of the new institution, Hansen and, most vocally, Bloch protested against the "autocracy" of Webster's style of leadership. Their referent must have been the prewar work on the klystron (in which Bloch had no part) and the project to design the supervoltage X-ray tube, both efforts in which the members of the department were organized under Webster's supervision to work on a common project. In contrast, postwar military funding from the Office of Naval Research and the Atomic Energy Commission promised a kind of liberation. After the war, each member of the department had his own patron and his own independent source of funds: Hansen had money for linear accelerators, Bloch for nuclear magnetic resonance studies, and so on. As opposed to the influence wielded by the

[47] Galison, Hevly, and Lowen, "Controlling the Monster" (cit. n. 4), pp. 65–70.

[48] Leslie and Hevly, "Steeple Building at Stanford" (cit. n. 3); and Stuart W. Leslie, "Playing the Education Game to Win: The Military and Interdisciplinary Research at Stanford," *HSPS*, 1987, *18*:55–88, on 61–69.

department chair before the war, after the war individual professors could organize their own research programs, using federal money to purchase autonomy within the department.

This new political economy of research called for a new style of leadership. Webster had used the force of his personality as well as his experience as a leading experimental physicist to remold Stanford's physics department and to try to make it a going concern among the leading institutions in modern physics research. But after the war Schiff was lauded for his skills as a conciliator, a chairman who could broker consensus among colleagues, each of whom enjoyed well-funded independence as a researcher.

VI. CONCLUSION

Intellectually and socially, Stanford's supervoltage X-ray tube formed a bridge between the physics of the 1920s and that of the 1950s. During the 1930s a number of institutions faced the challenge of building big equipment for nuclear physics. Several were more successful than Stanford before the war, especially if they were able to employ veterans of E. O. Lawrence's Berkeley program.[49] But Stanford's physicists began to assemble the conceptual and technological tools they would use for nuclear physics research after World War II. Even planning such a project, however, could transform a physics department, because of the discipline with which its resources and personnel had to be deployed.

What in particular can be learned from Stanford's unbuilt machine? Once built, scientific instruments exert a profound influence on the course of research. A realized instrument contains material kinks and potentials that are explored through the development of tacit knowledge and research practice. Examining an unbuilt machine, of necessity, draws attention to the ways in which its designers are influenced by their scientific, institutional, and economic contexts, and to how the interactions between scientists and influential figures outside their group help to shape an instrumental product before it is constructed. While it was in the design stage, prior experience, new theoretical vistas, the conditions for foundation support, and the expectation that the Ryan Laboratory generators constituted a usable resource all left marks on the design for the supervoltage X-ray tube.

Like Niels Bohr, David Locke Webster knew that success lay partly in being receptive to the changing opportunities presented by foundation support, and this sensitivity in turn influenced the intellectual program of his institution, as it had Bohr's.[50] Webster also observed that GE knew how to "build" equipment. He was commenting on more than shop technique: he recognized that the industrial concern devoted itself to careful planning before the plumbing and glass-blowing began. He hoped to hitch this businesslike facility at machine construction to his academic colleagues' insight into the instrument's research potential for investigations of the nucleus, and so to attract a theoretician to Stanford and establish the department's program in the forefront of research.

[49] Robert Seidel, "The Origins of the Lawrence Berkeley Laboratory," in *Big Science,* ed. Galison and Hevly (cit. n. 4), pp. 21–45, esp. pp. 31–35.

[50] Finn Aaserud, *Redirecting Science* (cit. n. 12), pp. 199–202.

Counting on Invention: Devices and Black Boxes in Very Big Science

By Robert W. Smith and Joseph N. Tatarewicz***

THE HUBBLE SPACE TELESCOPE was launched into space in 1990. One scientific instrument aboard was the "Wide Field/Planetary Camera." Planned, designed, built, and tested in the 1970s and 1980s through the efforts of hundreds of people and tens of organizations at a cost of approximately $125 million, this instrument had at its heart eight thumbnail-sized light detectors—charge-coupled devices, or CCDs. In this article we focus on the highly complex and politically charged process by which, in the face of competition from other detectors, the CCDs came to be used in the wide field–planetary camera, and thus the devices through which most observers—scientific and lay alike—were expected to experience the telescope's views of the universe.

The process involved far more than deciding on scientific goals, taking given devices, inserting them into an existing experimental arrangement, and tinkering with them to produce scientifically acceptable detectors. Instead, the devices and the experimental arrangement had to be developed together. The devices, that is, were part of a wider system. Walter Vincenti, developing an analysis by Rachel Laudan, has distinguished between *devices* and *systems* in that systems are composed of devices combined for a collective purpose, whereas devices are themselves single entities. Here *device* is taken to mean a single piece of technology, but the devices we consider—charge-coupled devices—were not discrete but were linked to other technologies. We also use *system* in the sense elaborated by Thomas Hughes and others. Hence we shall argue that CCDs were part of a "heterogeneous network" that included technologies, institutions, and social networks.

The devices' advocates not only had to produce and shape the devices to certain ends, but to embed them deep within a prospective machine, and so within a hierarchy of prospective hardware and software, and to set them far inside a contested

* Department of Space History, National Air & Space Museum, Smithsonian Institution, Washington, DC 20560.

** 3782 Folly Quarter Road, Ellicott City, Maryland 21042–1412.

Research for this article was supported in part by NASA, under contract NASW-3691, and the NSF, under grant SES-8510336. RWS is grateful for support from the National Humanities Center and the Research Triangle Park Foundation, while writing the article in 1992/93. We are also grateful to those people and institutions who gave interviews or allowed access to active collections of documents, especially C. R. O'Dell, the Goddard Space Flight Center, and the Marshall Space Flight Center; and to Michael Dennis, Paul Forman, Owen Hannaway, Karl Hufbauer, Sharon Kingsland, and Barbara Herrnstein Smith for comments on an earlier draft.

social hierarchy as well. Changes in scientific goals and technology, we shall see, entailed social changes.[1] In our account of the five or so years of detailed design calculations and negotiations on the Space Telescope that led up to the selection in 1977 of the scientific instruments it would carry, we focus on the deliberations of the leading group of astronomers engaged in the planning, the Science Working Group. This group was only one player in a tangled political economy in which the ultimate authority for the choice of instruments and detectors rested with NASA managers. Hence we will see advocates of different instruments and detectors jockeying to influence NASA's choices, and NASA trying to align its choices with various communities.

The black box is another notion crucial to the article. We shall examine contests swirling about the CCDs, contests about the extent to which they as well as other detectors could eventually be made into black boxes. For historians of science and technology, a black box usually means a technical artifact that is "regarded as just performing its function, without any need for, or perhaps any possibility of, awareness of its internal workings on the parts of users."[2] In some recent social studies of science, the term *black box* has acquired an expanded meaning beyond that of well-established fact or unproblematic object. For Trevor Pinch and Bruno Latour the essence of a black box is that in it many potentially disparate elements are made to act as one. Each "time a fact starts to be undisputed it is fed back to the other laboratories as fast as possible. But the only way for new undisputed facts to be fed back, the only way for a whole stable field of science to be mobilised in other fields, is for it to be turned into an automaton, a machine, one more piece of equipment in a lab, another black-box." Latour also stresses that activity is continually needed to keep black boxes closed.[3]

Key in our discussion will be an elaboration on the black box metaphor, the concept of a "provisional black box." The provisional black boxes, however, were not technical artifacts. Rather, they were texts laying out the requirements the artifacts were supposed to satisfy when eventually built. At any stage in the planning process for the Space Telescope, there were large numbers of people and tens of groups and committees and industrial organizations involved throughout the United States and

[1] For Vincenti on *device* and *system* see Walter Vincenti, *What Engineers Know and How They Know It: Analytical Studies from Aeronautical History* (Baltimore: Johns Hopkins Univ. Press, 1990), pp. 200–205. For Hughes on systems see, e.g., Thomas P. Hughes, *Networks of Power: Electrification in Western Society, 1880–1930* (Baltimore: Johns Hopkins Univ. Press, 1983), and Hughes, *American Genesis: A Century of Invention and Technological Enthusiasm, 1870–1970* (New York: Penguin Books, 1990). On heterogeneous networks see John Law, "Technology and Heterogeneous Engineering: The Case of Portugese Expansion," in *The Social Construction of Technological Systems: New Directions in the Sociology and History of Technology,* ed. Wiebe E. Bijker, Thomas P. Hughes, and Trevor J. Pinch (Cambridge, Mass.: MIT Press, 1987), pp. 111–134.

[2] Donald MacKenzie, *Inventing Accuracy: A Historical Sociology of Nuclear Missile Guidance* (Cambridge, Mass.: MIT Press, 1990), p. 26. See also Derek J. de Solla Price, "Gods in Black Boxes," in *Computers in Humanistic Research,* ed. Edmund Bowles (Englewood Cliffs, N.J.: Prentice Hall, 1967), pp. 3–7. We are grateful to James Capshew for bringing this last reference to our attention.

[3] Bruno Latour, *Science in Action: How to Follow Scientists and Engineers through Society* (Cambridge, Mass.: Harvard Univ. Press, 1987), p. 131. See also Trevor Pinch, "Towards an Analysis of Scientific Observation: The Externality and Evidential Significance of Observational Reports in Physics," *Social Studies of Science,* 1985, *15:*3–36. While we seek to exploit some of Latour's telling insights into scientific practice, we do not always follow his account, in particular, his tendency to attribute life to the inanimate. For a critique addressing this point see Simon Schaffer, "The Eighteenth Brumaire of Bruno Latour," *Studies in the History and Philosophy of Science,* 1991, *22:*174–192.

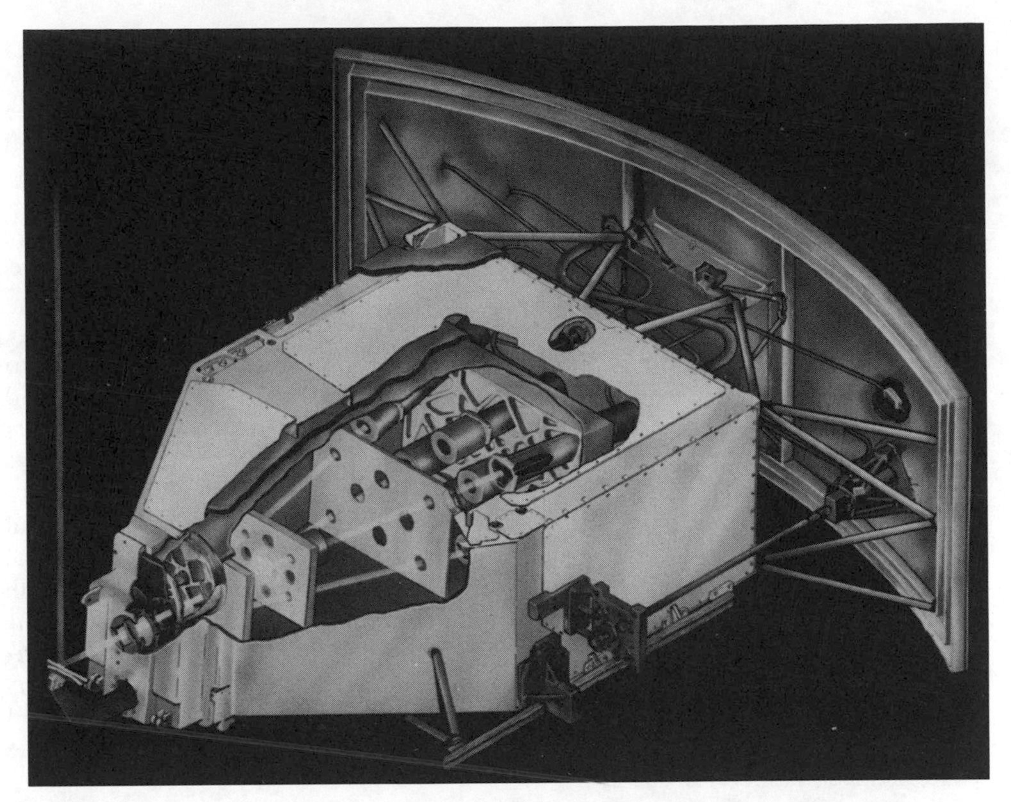

Figure 1. *The wide field–planetary camera for the Space Telescope as envisaged by a Caltech–Jet Propulsion Laboratory Group in 1977, when detailed design and construction of the telescope first got under way. The incoming light beam from an astronomical object is divided into four parts; it is then directed onto the instrument's light detectors, eight thumbnail-sized charge-coupled devices (CCDs): onto one group of four CCDs at one setting and a different four at another setting. Courtesy of NASA.*

Europe. If historians are to make sense of Very Big Science, they need some means of analyzing the myriad of activities that it comprises. The provisional black boxes are one way to bring order to historical accounts out of apparent, and sometimes very real, chaos, as well as to establish links between the microsocial and the macrosocial, the worlds within and outside the laboratory and observatory. Agreed guidelines and requirements toward which everyone engaged on the Space Telescope could work were certainly essential, even though everyone recognized that these guidelines and requirements would necessarily change as conceptions of what the telescope was supposed to be matured. To capture this recognition, we employ the qualifier *provisional.* We shall see provisional black boxes negotiated and defined and used to check at various stages the likely performance of the actual telescope, its instruments, detectors, and associated systems. As the planning developed, so the provisional black boxes could be "opened" and their contents changed. A provisional black box, however, could not be opened by just anyone for any reason, for to do so would have wide ramifications far outside that particular provisional black box. Opening provisional black boxes meant first winning allies and mobilizing resources. Hence much of our account deals with the struggles over the opening and closing of the provisional black boxes of the wide field–planetary camera and its detectors.

Our concerns are somewhat different from those of most recent authors on the history of experiment, who, as Jan Golinski notes in a recent review, have exhibited two complementary tendencies: either narrowing the focus "to a local setting, the laboratory (or, in natural history and the earth sciences, the field trip), where experimental facts first emerge," or investigating the "ways in which such knowledge is transformed in the course of its communication into more public settings (lectures, meetings of scientific societies, published books and journal articles)." By emphasizing procurement, we analyze a set of activities both temporally and ontologically prior to those investigated in most laboratory studies, in which instruments tend to appear (if they appear) as givens. Much had to be done before the detectors could be used in the telescope's "laboratories"—let alone before knowledge and facts could be constructed from that use.[4]

In part we are answering Peter Galison's call for an archeology of instruments to unearth buried theoretical assumptions and experimental practices.[5] To produce an archaeology for the wide field–planetary camera and CCDs, we will excavate not only in many laboratories but at sites far beyond the laboratory where disparate institutions, groups, and individuals intersect: committees advisory to NASA, international conferences on detectors, meetings in the Office of Management and Budget, and so on. We will track the CCD from its first invention at Bell Telephone Laboratories in 1969 through its transformation at many intermediate locations, until we reach the inside of the planned wide field–planetary camera, itself deep inside the planned Space Telescope by 1977.

I. CCDs, 1969–1972

The CCD was not an isolated discovery or invention. It emerged as the electronics industry struggled to produce an alternative to the various image tubes that had been the mainstay of television in all its forms from its very beginnings. The television camera focused the image of the scene onto a target where the incident photons created a facsimile of the image. This facsimile was then "read" off the target by scanning it with an electron beam, measuring the power required to discharge the target at each point in the scan. The target, beam-producing electronics, and powerful electromagnets to control the movement of the beam precisely were all contained within a vacuum tube. Various perceived deficiencies in these image tubes motivated engineers to investigate a variety of solid-state alternatives, one subset of which was known as "bucket brigade devices" because of their method of operation. In these, the image was focused directly onto a solid-state chip producing a facsimile of the scene already broken down into a number of discrete picture elements, or *pixels.* The charges of these pixels were then passed, bucket-brigade fashion, to another area of the chip to be stored and then passed to the associated electronics. CCDs, one type of bucket brigade device, were "self-scanning," as they did not need an

[4] Jan Golinski, "The Theory of Practice and the Practice of Theory: Sociological Approaches in the History of Science," *Isis,* 1990, *81:*492–505, p. 494. See also Timothy Lenoir, "Practice, Reason, Context: The Dialogue Between Theory and Experiment," *Science in Context,* 1988, *2:*3–22.

[5] Peter Galison, *How Experiments End* (Chicago: Univ. Chicago Press, 1987), p. 252.

electron beam. The bulky, power-hungry, and fragile vacuum tubes that went along with electron beams could in principle be replaced by small, rugged, and simple chips. These, one trade press editor declared in 1972, "could prove to be the sleeper of the seventies," and in 1974 Gilbert F. Amelio, who had worked on the device at Bell Labs, declared that the concept underlying the CCD "may someday have an impact on our lives as dramatic as that of the transistor."[6]

Yet as promising as the device was potentially, its significance was only gradually realized. When the Bell engineers responsible for the concept reported their initial results in 1970, there was not an immediate response. According to one observer,

> The charge-coupled device was introduced by Willard Boyle of Bell Labs at a sparsely attended panel discussion in the 1970 [Institute of Electrical and Electronics Engineers] International Convention in New York. Boyle, co-inventor of the technology, along with George Smith, briefly described the CCD structure and sketched out its potential for imaging, memory, and analog delay applications. The audience asked no questions, Boyle volunteered no additional information, and the panel moved on.
>
> But if the IEEE audience was not listening, Boyle's counterparts at major laboratories were, and rightly read the message.[7]

This was a strategic moment. The charge-coupled device in 1970 was a very long way indeed from being a black box—its name described its method of operation but neither its utility nor its import. It was a device that operated *via* charge coupling, but it was not yet a device *for* anything. In order to become a useful, transportable device it had to be developed. The transformation of the CCD would include refining the bare device itself and embedding it within various electronic circuits, themselves embedded within other circuits and systems.[8] It would also include embedding the device within the various social, economic, and political circuits and systems associated with their electronic counterparts.

How the device itself came to be surrounded by these systems associated with astronomical research is the main thread of our story. But it is important at this stage to sketch, however briefly, other applications of the CCD, for these other applications both helped legitimize the CCD as an astronomical detector and provided industry economic incentives for making the devices available at all. Astronomers could not simply procure a CCD "off the shelf" soon after the device's invention at Bell Labs. They as well as the other potential users of the device would have to participate in the entity's development and customize it for their own purposes.

Of all the many commercial applications of television, the largest was in home entertainment and information. The electronics industry decided that some sort of solid-state alternative to the image tube was required not only to make studio cameras more reliable and remote news-gathering cameras more portable and rugged, but to make video cameras available to individual consumers for home use. In this area, one promoter of the CCD declared, a "huge market will exist when the price

[6] Laurence Altman, "Bucket Brigade Devices Pass from Principle to Prototype," *Electronics,* 28 Feb. 1972, pp. 62–70, on p. 62; and Gilbert F. Amelio, "Charge-Coupled Devices," *Scientific American,* 1974, *230:*22–31, on p. 23.

[7] M. F. Tompsett, "Charge Coupling Improves Its Image, Challenging Video Camera Tubes," *Electronics,* 18 Jan. 1973, pp. 162–169, on p. 162. Tompsett, however, was not a disinterested observer; he himself had worked on the device at Bell Laboratories.

[8] See M. J. Howes and D. V. Morgan, eds., *Charge-Coupled Devices and Systems* (New York: Wiley, 1979), preface, p. ix.

is right."[9] Lesser but still significant markets existed for the CCD as an imaging detector, including closed-circuit television security systems, factory automation, and robotics. Bell Labs was actively researching alternatives for their proposed picture-telephone system, although the concept of the CCD seems to have emerged from other lines of research.[10] The potential of the CCD concept was soon grasped by military agencies long interested in low-light-level TV systems, which had been sponsoring research and development on competitors to the CCD. The low weight and power requirements—far lower than those for television-type systems—were attractive to those who wanted, for example, to pursue reconnaissance from air and space.[11] The military in turn represented the largest potential source of funds for the technology. The military services might not buy anywhere near the number of devices the commercial sector would, but they would subsidize the R&D. Thomas Misa has described how in the case of the transistor the military agencies shaped technological change through their entrepreneurial activities and even "bias[ed] the industry" towards certain types of hardware. This process can be seen clearly for CCDs too. Military agencies and, as we shall see, NASA employed a variety of means to shape the development of the charge-coupled concept into devices suited to their particular needs. Yet these specialty applications were still based on, and limited by, the desiderata of the commercial television and home video markets.[12]

Astronomy was one speciality application. Electronic detectors had become widespread in astronomy only after the World War II. Even during the time of their rapid adoption (1950s–1960s), however, they did not displace the photographic plate. These electronic detectors (all tube-based) excelled in their sensitivity and response to light of different wavelengths, but they were limited to measuring the radiation at a single point. In 1972, just as the charge-coupled concept was being translated into provisional operating devices, astronomers were seeking to combine the best features of the photographic plate (especially its coverage of relatively large areas of the sky) with those of the point-source tube into a better "image tube" for astronomy. Image tubes were in fact in widespread use but considered far from satisfactory. They were power-hungry, operated at high voltages, and were temperamental.[13] While based upon commercial television-tube technology and manufacturing infra-

[9] D. F. Barbe and W. D. Baker, eds., *Charge-Coupled Devices* (Berlin: Springer-Verlag, 1980), p. 3.

[10] E. I. Gordon, "A 'Solid-State' Electron Tube for the Picturephone Set," *Bell Laboratories Record,* June 1967, pp. 175–179; and James Janesick and Morley Blouke, "Sky on a Chip: The Fabulous CCD," *Sky and Telescope,* Sept. 1987, pp. 238–242, on p. 238.

[11] For instance, by 1972 Fairchild Research Laboratory was developing CCDs for the Army Electronics Command, in 1973 the Navy began sponsoring an annual conference on CCDs, and by the mid 1970s various public reports contend that CCDs were flying aboard Air Force satellites. See Benjamin F. Elson, "Charge-Coupled Concept Studied for Photo-Sensors," *Aviation Week and Space Technology,* 22 May 1972, p. 73; Barbe and Baker, eds., *Charge-Coupled Devices* (cit. n. 9), preface, p. xx; and William E. Burrows, *Deep Black: Space Espionage and National Security* (New York: Random House, 1986), pp. 244–246.

[12] Thomas J. Misa, "Military Needs, Commercial Realities, and the Development of the Transistor, 1948–1958," in *Military Enterprise and Technological Change,* ed. Merritt Roe Smith (Cambridge, Mass.: MIT Press, 1985), pp. 253–287.

[13] William C. Livingston, "Image-Tube Systems," *Annual Review of Astronomy and Astrophysics,* 1973, *11:*95–114; W. Kent Ford, Jr., "Digital Imaging Techniques," *Annual Review of Astronomy and Astrophysics,* 1979, *17:*189–212; and David H. DeVorkin, "Electronics in Astronomy: Early Applications of the Photoelectric Cell and Photomultiplier for Studies of Point-Source Celestial Phenomena," *Proceedings of the IEEE,* 1985, *73:*1205–1220.

structure, they were very different from their television counterparts. Television tubes had only to produce an aesthetically acceptable scene. Image tubes had to produce the spatial, intensity, and spectral information astronomers desired to high degrees of accuracy. But both astronomical photography and astronomical electronic imaging were piggybacking on large domestic commercial markets. In his analysis of "technoscience," Latour describes a number of strategies of "translation"—that is, strategies of persuading others to further one's own interest—the first of which is a piggyback strategy. As he puts it, "By pushing their explicit interests, you will also further yours."[14] Here we have an example of such a strategy. Just as speciality emulsions were produced at relatively high cost using special handicraft by photographic firms, speciality tubes were produced for astronomers at relatively high cost, in small volume, and with special techniques by electronics firms whose major interests lay more in mainstream commercial markets. By pursuing these speciality applications, however, firms could extend their expertise into new areas at someone else's expense. Without speciality markets, the firms would have had to support such experimentation with their own funds.

As the charge-coupled concept began to be realized in hardware, another issue arose: how to evaluate the CCDs against competing alternatives, such as image tubes. Numerous criteria or parameters were proposed or used. Comparisons were sometimes straightforward, but most often were exceedingly complex because of vast differences in the electronic systems surrounding the several detectors. Hence the CCDs were not simply competing with various image tubes and with various other solid state detectors. The entire respective systems were in competition. Moreover, the operation of the CCD was sufficiently different from that of conventional image tubes to make it difficult for astronomers and engineers even to find and define effective technical criteria with which to compare them.[15]

But it was the *potential* of the CCD that was at stake during this period (1969–1972). Existing prototype devices were decidedly inferior to the established alternatives. Extrapolations and expectations of what the CCD might *become* were compared in the literature with the performance of the various television tubes then available. Similarly, astronomers using ground-based telescopes had at their disposal a variety of proven and familiar technologies which, while not perfect, were serviceable and established. Even astronomers conducting research from telescopes located in space, where photographic emulsions were problematic at best, had been working with several image tubes. CCDs posed major, and to many people very obvious, development problems. For example, in comparison to photographic plates they could cover only a tiny area of sky, their resolving power was poor, and they responded weakly to ultraviolet light. How would the CCD be greeted by astronomers, and most important, how would its proponents present it to them to engage their interest and displace the alternatives?

[14] Latour, *Science in Action* (cit. n. 3), p. 110. For a case study of the links between scientists and industrial suppliers in the development of modern astronomical detectors, see David Edge, "Mosaic Array Cameras in Infrared Astronomy," in *Invisible Connections: Instruments, Institutions, and Science,* ed. R. Bud and S. E. Cozzens (SPIE Institutes for Advanced Optical Technology, 9) (Bellingham, Wash.: SPIE Optical Engineering Press, 1992), pp. 130–167.

[15] James R. Janesick *et al.,* "The Future Scientific CCD," Proceedings of the SPIE Conference, San Diego, California, 21 Aug. 1984, copy provided by the author.

II. THE LARGE SPACE TELESCOPE

As Very Big Science, even in its planning stages the Space Telescope program entailed writing a vast number of formal texts.[16] Such texts bound together the hundreds of people in disparate groups engaged in planning for, and then later the many thousands of people who would actually build, the telescope. One very important subset of these documents contained the different versions of the telescope's technical specifications. In recent years historians of technology have devoted considerable attention to the part played by the writing of technical specifications in the design process. As Eric Schatzberg has argued, "Design . . . is thoroughly embedded in a world of texts [which] . . . embody the criteria that purport to govern the design of the artifact. However, the meaning of these criteria remains indeterminate until interpreted in a particular context. Because of this indeterminacy, technical specifications, standards and data never fully control the form of the designed object." Further, "the design process is influenced by a number of communities whose traditions may not be harmonious. . . . For the designer, conflicting design requirements represent the conflicting interests of the various groups with a stake in the design process."[17]

In support of these claims, Schatzberg draws in part upon the researches of Walter Vincenti, who has examined the role played by technical specifications in aeroplane design. But Vincenti's case studies focus on a highly centralized system in which technical specifications are handed down to a design group. In the case of the Hubble Space Telescope, we instead see various design groups and interest groups—with the astronomers only one relatively small element among many—engaged in activities directed at writing scientific and technical specifications. Also, those involved in writing the specifications were in part those who later proposed to fulfill them. These specifications evolved through both formal and informal negotiations. For the astronomers, the main document to be written was the so-called Announcement of Opportunity. It would stipulate the specifications for the scientific instruments to be flown aboard the telescope. In effect, the Announcement of Opportunity would embody and define a set of provisional black boxes, which were to be set within and alongside the other provisional black boxes that constituted the planned design for the telescope. Instruments proposed for the telescope would have to promise to meet the demands and requirements of the Announcement of Opportunity, but a lot of design work would still be required *after* the Announcement of Opportunity had been issued and potential instruments selected by NASA. The evolving document would serve also as a focus of debate and negotiation for many parties who sought to shape its contents to fit their interests. As such, it would function as what Kathryn

[16] See Robert W. Smith, *The Space Telescope: A Study of NASA, Science, Technology, and Politics* (with contributions by Paul A. Hanle, Robert H. Kargon, and Joseph N. Tatarewicz) (New York: Cambridge Univ. Press, 1989), p. 233.

[17] Eric Schatzberg, "Interpretation and Engineering Design," paper presented to the Conference on Technical Development and Science in the 19th and 20th Centuries, Technische Universität Eindhoven, 7 Nov. 1990, pp. 17, 10. Latour also argues that machines are "drawn, written, argued and calculated, before being built. Going from 'science' to 'technology' is not going from a paper world to a messy, greasy, concrete world. It is going from paperwork to still more paperwork, from one center of calculation to another which gathers and handles more calculations of still more heterogeneous origins": Latour, *Science in Action* (cit. n. 3), p. 253.

Henderson has called a "conscription device." Such devices "enlist group participation and are receptacles for knowledge created and adjusted through group interaction aimed toward a common goal."[18] The process by which the Announcement of Opportunity was written involved a number of steps, and it is these that we now track.

By 1973 serious planning for the Large Space Telescope (LST)—it would be renamed the Space Telescope in 1975 and the Hubble Space Telescope only in 1983—had been conducted for several years. It had chiefly engaged a federal agency, NASA, tens of industrial contractors, and a wide variety of engineers and astronomers based in universities, government laboratories, and industry. At this stage the telescope was slated to be a reflecting telescope with a primary mirror of 120 inches in aperture, to carry seven scientific instruments, and to be launched by NASA's proposed space shuttle. But competing designs were vying for support, and nothing had been finally decided. A variety of conceptual studies and engineering tests using hardware had already been performed, but the telescope was only just about to enter what in NASA parlance is known as "Phase B," that is, a period of feasibility studies in which supposedly detailed cost estimates are also made. Only after Phase B was complete or nearly complete did NASA plan to present the LST for approval by the White House and Congress.[19]

As Phase B got under way in the spring of 1973, NASA managers reckoned that it would last about two years. The final design and construction of the Large Space Telescope were to begin, if all went well, in 1976, with the telescope's launch in 1981. By spring 1973 a project structure was in place characterized by a division of labor and a diffusion of power and responsibility among numerous institutions, teams, individuals, and networks. Even as early as this stage, it makes no sense to talk of an individual acting as *the* system builder for the telescope. Rather, the system builder was itself a complex system.[20]

Management authority for the Large Space Telescope ultimately rested with NASA headquarters in Washington, D.C., but day-to-day management was delegated to two NASA centers, each large organizations employing thousands of civil servants and contractor staff. The main NASA center involved with the project was the Marshall Space Flight Center in Huntsville, Alabama. Charged with planning the telescope's scientific instruments and their detectors was the Goddard Space Flight Center in Greenbelt, Maryland. For the astronomers, the most important group was the "Science Working Group," a committee of about fourteen astronomers that was to meet every two or three months.[21] The Science Working Group had no direct authority: it had no control of budgets and acted in a purely advisory

[18] Kathryn Henderson, "Flexible Sketches and Inflexible Data Bases: Visual Communication, Conscription Devices, and Boundary Objects in Design Engineering," *Science, Technology, & Human Values,* 1991, *16:*448–473, on p. 456.

[19] Smith, *Space Telescope* (cit. n. 16), Chs. 2–6.

[20] In his work on systems (esp. *Networks of Power,* cit. n. 1, pp. 5–6), Hughes identifies certain types of individuals with each phase in the building of a system; single individuals emerge as dominant system builders in each phase. System building in the case of the telescope was a much more diffuse activity than in Hughes's examples: its would-be system builders were subject to numerous checks and immersed in a management structure that severely circumscribed authority and the power to act.

[21] The number varied with time, and often others were invited to attend to discuss specific topics.

fashion. It was nevertheless a major force, comprising prominent astronomers whose support would be essential if the LST was to be sold to the White House and Congress. Several astronomers in the Working Group also led teams of other astronomers who worked with various industrial contractors on developing and analyzing possible designs for scientific instruments. For astronomers, the deliberations of the Science Working Group provided the most influential forum in the project.

During Phase B, NASA managers and engineers planned to conduct a range of technical studies, both in NASA centers and at numerous industrial contractors. Using the information gathered, NASA would then develop and write the requirements it wanted the eventual telescope to meet. The astronomers on the Science Working Group therefore sought to influence not only the requirements themselves, but the way the telescope's requirements would be arrived at, and thus the hardware selected. They were seeking, in short, to have the provisional black boxes designed to their specifications.

The astronomers were particularly concerned to shape the requirements for, and so the eventual design of, the scientific instruments. At the heart of the instruments were the detectors, the means by which light from astronomical bodies would be captured for analysis. For the Science Working Group, the detectors were the most crucial part of the telescope. They thrashed out issues surrounding detectors time and again during the group's early meetings.[22] They debated a range of prospective detectors. None, however, were judged fully satisfactory by NASA or the astronomers to fly on the telescope as they were. It would be necessary to customize the devices chosen, and to do this would take money and resources, and because of the skills and techniques required, necessarily involve industry and mean piggybacking on industry R&D programs. Which detectors, then, deserved some of the project's scarce financial and material resources to develop them further?

In developing detectors NASA was acting as an institutional entrepreneur. As noted earlier, Misa argued that military agencies influenced the development of the transistor through a variety of means: providing research and development funds, sponsoring conferences and publications that helped spread information, subsidizing the construction of facilities, and overseeing the setting of standards, as well as "bias[ing] the industry" toward the development of specific types of transistors.[23] Since its inception, NASA managers had done the same sorts of things in order to advance detectors to be flown aboard spacecraft, not only astronomy satellites but also a wide variety of other applications, including spacecraft to be sent throughout the solar system to pursue planetary science.

One detector NASA had backed for some years was the so-called SEC vidicon, a type of television tube. The SEC vidicon group was based at the Princeton University Observatory and collaborated with Westinghouse. Since 1964 NASA had supported the Princeton group's activities through a series of grants and contracts, and

[22] Smith, *Space Telescope* (cit. n. 16), pp. 104–110. See also, e.g., Margaret Burbidge to C. R. O'Dell, 22 Nov. 1974, File "LST-Misc. Letters and Correspondence," Box 101, Arthur D. Code papers, University of Wisconsin, Madison.

[23] Misa, "Military Needs, Commercial Realities, and the Development of the Transistor" (cit. n. 12), p. 255. On the influence of military funding on physical research, including electronics, between 1940 and 1960 see Paul Forman, "Behind Quantum Electronics: National Security as Basis for Physical Research in the United States, 1940–60," *Historical Studies in the Physical Sciences,* 1987, *18:*149–229.

these funds had brought together what was widely seen in NASA and elsewhere as a good team, well placed to continue developing the SEC vidicon. There was also a strong institutional link between Princeton and the Large Space Telescope program.[24] By late 1973 the generally shared assumption was that the SEC vidicon would be employed in the principal instrument for the telescope—the wide field camera—and for a time even in other instruments too.[25]

In the early planning for the telescope's scientific instruments, the wide field camera dominated the complement, with other instruments filling whatever space remained. Once NASA, the astronomers, and contractors agreed upon a modular design for the instruments, the wide field camera was granted access to the central part of the telescope's light path. The wide field camera's command of the physical space around the telescope's focal plane matched its ranking by the Science Working Group as the telescope's most important instrument.[26]

The Science Working Group had also agreed in late 1973, while shaping the "Preliminary Instrument Definitions"—an important step towards the Announcement of Opportunity—that the wide field camera should include a field of view on the sky of 3-by-3 arc minutes. This size was set by the desire to image in one frame distant clusters of galaxies; it recognized that the telescope's chief scientific goals at this period were cosmological, that is, focused on the faintest and most distant objects in the universe. That size requirement, in turn, determined the size of detector needed for the camera. For the SEC vidicon it translated to a 70-mm photocathode, significantly bigger than the 35-mm SEC vidicon already successfully fabricated. Meeting the size requirement thus meant that a variety of problems had to be solved—and the research funded.[27]

The funds available for detectors comprised only one part of the Large Space Telescope's planning budget, and the Science Working Group continually pressed for additional resources for detector development. The fewer resources available to develop other detectors, the more likely the choice was to be the SEC vidicon for the wide field camera. But long lead times for building the entire telescope translated into an even longer conceptual time for detector development. It strained the ability of astronomers and engineers to foresee the likely characteristics of future detectors, and made them fearful of simply backing known technology almost certain to be seen later as obsolescent. Some of the Science Working Group were uneasy. As one leading member put it in 1974,

[24] See, e.g., Lyman Spitzer, Jr., interview by David DeVorkin, 17 June 1982, p. 45; and John Lowrance, interview by Robert W. Smith, 28 March 1985. All **oral history interviews** cited are from the collections of the National Air and Space Museum, Smithsonian Institution.

[25] See, e.g., C. R. O'Dell, interview by David DeVorkin, 12 April 1982, p. 33; Nancy Roman, interview by Robert W. Smith, 3 Feb. 1984, p. 43; and Large Space Telescope Operations and Management Working Group, minutes of meetings, 11 June, 31 July–1 Aug., 5–6 Nov., and 12–13 Dec. 1973, Project Scientist papers, Marshall Space Flight Center, Huntsville, Alabama (hereafter **Project Scientist papers**). Although known officially as the Operations and Management Working Group, this group was referred to by almost everyone as the Science Working Group, and we shall follow the latter usage in the text proper.

[26] This observation closely parallels a study made of high energy physics detectors and the way to order the "social space" around them. See Marck Bodnarczuk, "The Social Structure of Experimental Strings at Fermilab: A Physics and Detector Driven Model," Fermilab Publication 91/63, March 1990.

[27] See LST Operations and Management Working Group, enclosures with minutes, 13–14 Dec. 1973, Project Scientist papers. The sizes refer to the corner-to-corner length of the photocathode.

> I feel that we are making a grave mistake if we economize on the most important part of the LST—its detectors. To decide now, eight years before the LST can possibly fly, on what is already an out-dated detector, less than state-of-the-art, will be regarded in the future, I think, as a poor choice of the options which are conceivable in other directions for cutting the cost of the LST. It is like deciding to treat a sick patient by cutting out his heart on the grounds he would be saved the energy used by the heart muscles in pumping the blood around his body.[28]

There was ample motivation for astronomers to count on invention.

But at this stage the Working Group saw no good alternative to the SEC vidicon immediately to hand. Studies were under way on so-called intensified charge-coupled devices (ICCDs), that is, CCDs placed inside a television tube, the aim being to use the charge-coupled devices to detect electrons in the television tube rather than to catch photons directly. This strategy would, the ICCD builders hoped, make it far more sensitive to ultraviolet light than a CCD alone. Neither Goddard nor the Science Working Group, however, planned to include this hybrid device in the first set of scientific instruments, slated for launch aboard the LST in 1981. Rather, NASA planned to return the LST to earth every five years or so and exchange old instruments for new ones.[29] At this stage the ICCDs were therefore candidate detectors for a later generation of scientific instruments.

The Science Working Group was dominated by stellar and galactic astronomers. Discussions of the specifications for cameras and detectors—responses to ultraviolet light, field of view, and so on—reflected these interests. Astronomers primarily interested in the study of solar system objects—and so with a different set of concerns—were in a distinct minority. One staunch advocate of planetary science in the group was the Princeton astronomer Robert E. Danielson. Danielson was also team leader of a group of astronomers examining possible high-resolution cameras, a team that included astronomer Gerry Smith of the Jet Propulsion Laboratory (JPL). Smith and JPL, as we shall see in the next section, kept the team abreast of the very latest developments in CCDs.

As early as mid 1973, Danielson's High Resolution Camera Team had proposed to Goddard and the Science Working Group that NASA fund a variety of studies of CCDs, including an analysis of ICCDs.[30] For Danielson, the CCDs were particularly promising because of their response in the red part of the spectrum. This was a very important region for planetary studies—particularly the long-term monitoring of the atmospheric phenomena of Jupiter, Saturn, Uranus, and Neptune—but one where the SEC vidicon, which had a big advantage in the ultraviolet, was not effective. Astronomers and planetary scientists were thus developing different provisional black boxes.

[28] Margaret Burbidge to C. R. O'Dell, 22 Nov. 1974, File "LST—Misc. Letters and Correspondence," Box 101, Code papers.

[29] Smith, *Space Telescope* (cit. n. 16), p. 90.

[30] LST Operations and Management Working Group, minutes, 31 July–1 Aug. 1973, pp. 2–3, Project Scientist papers. For more on Danielson as mediator between astrophysics and planetary science see Robert W. Smith and Joseph N. Tatarewicz, "Replacing a Technology: The Large Space Telescope and CCDs," *Proc. IEEE,* 1985, *73:*1221–1235. On stellar and galactic versus planetary astronomy see also Smith, *Space Telescope,* Chs. 5, 7; and Tatarewicz, *Space Technology and Planetary Astronomy* (Bloomington: Indiana Univ. Press, 1990).

III. THE JET PROPULSION LABORATORY

By the early 1970s, the Jet Propulsion Laboratory in Pasadena, California, had claims to be among the world's leading space installations.[31] Founded as the Guggenheim Aeronautical Laboratory of the California Institute of Technology in 1936, it became the Jet Propulsion Laboratory in 1944. In 1959 it joined the newly established NASA and embarked on a variety of robotic missions to the Moon and planets, including, for example, *Mariner 4,* which secured the first close-up images of Mars. Methods of imaging from spacecraft became a central concern for JPL. It made an early commitment to a vidicon manufactured by the General Electrodynamics Corporation (GEC): slightly modified versions of this vidicon flew on a number of JPL missions, from *Mariner 4* in 1965 to two *Voyager* spacecraft launched in 1977.[32] Yet JPL regarded the GEC vidicons as problematic. The General Electrodynamics Corporation was a small company and might go out of business. Further, like the SEC vidicon, the GEC vidicon had a limited response and was insensitive to red light—an important failing for planetary astronomers, who, as noted, judged there were many major problems to be tackled in this region. The high voltages of vidicon tubes also made engineers and scientists uneasy because of the possibility of a catastrophic failure of the tube. JPL was therefore on the hunt for alternatives to the GEC vidicons.

In 1972 CCDs caught the eye of two members of JPL's space photography section, Kenneth Ando and Gerry Smith (who joined Danielson's High Resolution Camera Team the following year). In December 1972 Ando and Smith made a pitch to NASA headquarters for funds to contract with industry to start work on CCDs tailored to JPL's planetary missions.[33] With funds forthcoming, early in 1973 JPL contracted with Texas Instruments and Fairchild, both already deeply involved with CCDs for commercial—and probably also for military—applications to pursue a variety of studies of the devices.[34]

Some people at JPL were initially skeptical of CCDs. In 1973 the commercially available CCDs had many drawbacks compared to the more mature GEC vidicons, among them their small field of view and their relatively large pixel sizes (and so reduced resolution). But by 1975 JPL had decided that CCDs were making rapid advances. The Texas Instruments program had led to 400-by-400 pixel CCDs that JPL thought promising, although the number of pixels was small by the standards of vidicons, and the device had other problems. In March 1975 JPL even hosted a major international conference on CCDs.[35] By the end of the year CCDs had been

[31] On the history of JPL see Clayton Koppes, *JPL and the American Space Program: A History of the Jet Propulsion Laboratory* (New Haven/London: Yale Univ. Press, 1982).

[32] See Harold Masursky *et al.,* "Planetary Imaging: Past, Present, and Future," *Transactions on Geoscience Electronics,* 1976, *GE-14:*122–134.

[33] Kenneth Ando, interview by Robert W. Smith, 19 Aug. 1991; and "Advanced Imaging Systems Technology: Solid State Sensor Presentation, December 1972" (a set of briefing charts), copy provided by Ando.

[34] See, e.g., Texas Instruments, Inc., "Final Technical Report Charge-Coupled Device Image Sensor Study Contract No. 953673, 3 December 1973," Jet Propulsion Laboratory, Pasadena, California (hereafter **JPL**).

[35] See, e.g., Gault A. Antcliffe (Texas Instruments, Inc.), "Development of CCD Imaging Sensors for Space Applications: Final Report—Phase I, Contract JPL 953788," JPL; and *Proceedings of the*

incorporated into plans for all planetary missions under study at the lab that involved imaging.

IV. A PLANETARY CAMERA

CCDs were also at the heart of a camera Danielson championed for the Space Telescope. Time and again, with the backing of his camera team, he urged the Science Working Group to adopt a "planetary camera" with a longer focal ratio than that planned for the wide field camera. The longer focal ratio would mean higher resolution if a smaller field of view and, Danielson and his team argued, be very suitable for planetary observations. But a camera of such high resolution raised problems for the SEC vidicon. As a NASA staffer put it in mid 1974, the SEC vidicon could not meet these demands, and so another detector system would be required.[36]

In 1975 the Space Telescope—a change of name from the Large Space Telescope and in part a reflection of its reduction to 96 inches from 120 inches following budget battles with the Congress—was slated to be included in the budget President Ford introduced to Congress in January 1976. Time was running out for the participants in the program to influence the requirements NASA would set for the telescope. Throughout 1975 Danielson and his camera team pressed hard for a planetary camera and advocated that it incorporate CCD detectors.[37] He emphasized the importance of such a camera for imaging bright objects such as planets (the CCDs could image bright objects better than vidicons) and for investigating the red end of the spectrum. Danielson cited studies of the atmospheres of Uranus and Neptune that could be pursued with such a camera, providing constraints on theories of the origin and evolution of the solar system. The red response would also be valuable beyond planetary studies, particularly for galaxies so distant that their spectral lines had been redshifted out of the region of sensitivity of the SEC vidicon.

By this time the telescope's budget problems had led Congress to instruct NASA to seek international partners to help share costs, and NASA was negotiating with the European Space Agency (ESA) about a joint venture. One way for the Europeans to become involved was to provide a scientific instrument. ESA showed an early interest in a so-called faint object camera based on a type of television tube with which European astronomers possessed a particular competence.[38] Competition for places in the set of scientific instruments was now more intense than before. A European camera meant one less slot for a U.S.-built instrument, and the budget problems had not only led to downsizing of the telescope, but also to reducing from seven to five the number of scientific instruments. With the European faint object camera very likely to fly, the competition was now for four slots.

The proposed faint object camera would not be very effective in the red or of much use for observing the brighter planets; thus planetary scientists still wanted a planetary camera. As part of a strategy to make it seem less threatening to the wide

---

Symposium on Charge-Coupled Device Technology for Scientific Imaging Applications, March 6–7, 1975 (Pasadena, Calif.: Jet Propulsion Laboratory, 1975).

[36] Stanley Sobieski to Robert E. Danielson, 21 June 1974, copy with LST Operations and Management Working Group, minutes, Project Scientist papers.

[37] See, e.g., LST Operations and Management Working Group, minutes, 18–19 Sept. 1975, *ibid.*

[38] Smith, *Space Telescope* (cit. n. 16), pp. 137–141, 245–246.

field and faint object cameras, Danielson and his team characterized their camera as a *complement* to these two. Danielson also sought allies from outside the Space Telescope program. For example, he kept an influential National Academy of Sciences committee on lunar and planetary exploration informed of developments, and that committee kept pressure on NASA's Space Telescope managers to be sensitive to the needs of planetary science.[39]

As Danielson labored to build support for the planetary camera, he made little headway with the Science Working Group, composed, as we have noted, very largely of stellar and galactic astronomers whose interests it did not strongly engage. This became apparent at a meeting in September 1975, at which the group's main task was to prepare recommendations on the scientific instruments for the Space Telescope's first flight, by then planned for 1982. Other NASA advisory committees would review these recommendations, but this was the group's best opportunity to influence the writing of the forthcoming Announcement of Opportunity, the document laying out the number and type of instruments to be flown, the requirements they were to meet, and the detectors some of them were to employ.

Danielson's team and the Science Working Group endorsed a wide field camera that would employ a SEC vidicon as its detector. This, after all, had been at the heart of the planning for the telescope for some years, and the camera and SEC vidicon were well entrenched in the Space Telescope program. Danielson and his team again also advocated a planetary camera using CCDs. The Science Working Group went through the motions of showing support for the concept of a planetary camera, but in fact passed responsibility to the European Space Agency. It unanimously resolved: "The Working Group believes that the addition of a CCD Planetary Camera . . . greatly enhances the ability of the LST to obtain images of astronomical objects at the highest possible spatial resolution. We, therefore, urge [ESA] to provide both a planetary camera and a Faint Object Camera as its Scientific Instrument contribution." The group did not recommend that NASA expend resources on developing the planetary camera, and the chances that ESA, which most likely had funds for only one instrument and strongly favored a television-type camera system and not CCDs, would sponsor such an instrument were tiny. The Science Working Group also decided that two instruments—the wide field camera and a spectrograph—were so crucial to the Space Telescope that they were essential to its mission, and thus were designated "core" instruments. The Science Working Group analyzed nine instruments, but voted heavily against including a planetary camera in the core complement.[40]

The writing of the Announcement of Opportunity would provide a set of provisional black boxes. But in late 1975 the prospects that the planetary camera with CCDs would be included seemed bleak to its proponents: there was little support for it on the Science Working Group, at Goddard, or at Marshall. In September 1975, however, President Ford called for reducing the proposed federal budget by $28 billion, giving a new twist to the negotiations between NASA and the Office of Management and Budget over NASA's forthcoming budget. In the ensuing horse

[39] R. E. Danielson to G. J. Wasserburg, Handout #1, in LST Operations and Management Working Group, minutes, 18–19 Sept. 1975, pp. 1–3.

[40] LST Operations and Management Working Group, minutes, 18–19 Sept. 1975, pp. 8 (quotation), 6.

trading, NASA agreed to cancel some projects and postpone others, including the Space Telescope.[41] If the telescope project was to be deferred, so too would the writing of the Announcement of Opportunity.

V. OPENING THE PROVISIONAL BLACK BOXES

Our emphasis has been on the process of deciding which detectors could be made in time to meet the demands of—in effect to become—the provisional black boxes already negotiated and defined. In the mid 1970s everyone recognized that production of a detector for the wide field camera that was anything like a true black box was years away. In other words, Space Telescope managers were counting on invention. Once NASA and the White House agreed to postpone the start of the building of the Space Telescope for at least one year, those who had doubts in late 1975 and early 1976 about the collection of provisional black boxes that constituted the telescope's design were able to regroup and attempt to mobilize additional resources to effect changes. Hence the Announcement of Opportunity was again subject to argument. In particular, was it to be open to whatever proposals for instruments and detectors astronomers might wish to make, or was it to follow closely the recommendations—the attempted provisional black-boxing—of the Science Working Group? In the language actually used in the debates, was the Announcement of Opportunity to be "open" or "closed"?

Danielson's High Resolution Camera Team seized this unexpected opportunity. In January 1976 a staffer from JPL addressed the team on the 400-by-400 pixel CCD chips and their improving performance, and the team passed another resolution that a CCD planetary camera be included aboard the Space Telescope. The meeting also returned to the issue of the SEC vidicon for the wide field camera, a subject they had continually debated for the previous three years. Again they judged that given the requirements for the camera of a 3-by-3 arc minute field of view and a response far into the ultraviolet region of the spectrum, then the SEC vidicon "is the only available detector." There was nevertheless a telling shift. The team members also resolved that, as the telescope program was very likely to be delayed, then the choice of detector for the wide field camera should be reexamined once approval for the telescope had been won. They did not want the choice of the SEC vidicon to be taken as given.[42]

At the Science Working Group itself the following month, the members reconsidered the list of priorities for the scientific instruments that they had drawn up a few months before. Some now wanted astronomers to be able to propose instruments not listed, that is, they wanted the Announcement of Opportunity to be open.[43] NASA insiders were nevertheless sure that a wide field camera would fly and that it was highly likely, but not assured, that the SEC vidicon would be the camera's detector.[44]

[41] See Smith, *Space Telescope* (cit. n. 16), pp. 160–163.

[42] High Resolution Camera Team, minutes, 15–16 Jan. 1976, p. 3, copy in Space Telescope (ST) Operations and Management Working Group, minutes, 6 Feb. 1976, Project Scientist papers.

[43] ST Operations and Management Working Group, minutes, 6 Feb. 1976, p. 3.

[44] For example, though protesting a plan to procure the vidicon as government-furnished equipment, the project scientist, C. R. O'Dell, told a Goddard manager, "Almost unquestionably the detector will be the SEC [vidicon] and the place where it is done—Princeton, but let's preserve our options": O'Dell to G. Levin, 5 March 1976, "Reading File, 1976," Project Scientist papers.

NASA's budget problems, however, were changing the political economy of the Space Telescope program by bringing new and powerful forces to bear. The Space Telescope was not the only spacecraft removed from the proposed budget during negotiations in late 1975 between the space agency and the White House. A planned *Mariner* spacecraft for a mission to Jupiter and Uranus had, to the chagrin of JPL and planetary scientists, also been deleted. As we have noted, most of the astronomers closely associated with the Space Telescope saw it as a tool for the study of stars and galaxies, not solar system objects. Nor had the telescope proved attractive in the past to the wider community of planetary scientists. The telescope would orbit the Earth at an altitude of a few hundred miles and so make its observations of planets from essentially the distance of the Earth; planetary scientists generally favored spacecraft that could fly by, or land on, or enter the atmospheres of, planets.[45] But the deletion of the *Mariner* Jupiter-Uranus mission meant that NASA had secured only one new planetary science mission in four years. As the head of the American Astronomical Society's Division of Planetary Sciences told a NASA staffer, there had been "agitation over NASA's reduction of funds for planetary science."[46] One effect was that with future opportunities for such missions seemingly dwindling and the technical performance of detector systems improving, the pursuit of solar-system studies from earth orbit with the Space Telescope began to look much more attractive to planetary scientists than before.

Yet planetary scientists judged that NASA cared little about using the telescope for planetary research. Their anxiety was heightened by the death of Bob Danielson in early 1976. Planetary scientists made known their concerns to NASA in meetings of various advisory groups, as well as by telephone call and letter. NASA headquarters staff, however, were generally sympathetic. They not only thought the telescope capable of high-quality planetary science, but welcomed the additional strength that linking stellar and galactic astronomers to planetary scientists would give the political coalition in the telescope's favor, enhancing its chances of winning approval from the White House and Congress.[47]

Various NASA Space Telescope officials met on 5 August 1976 with four planetary scientists to assess the capabilities of the planned scientific instruments in that area and to explore how the interests of planetary researchers might be accommodated. The four emphasized the poor capabilities of the two planned cameras—the wide field camera with an SEC vidicon and the Europeans' proposed faint object camera—for planetary imaging. They pressed for the type of planetary camera with CCDs that Danielson's camera team had strongly advocated for two years. As a NASA staffer who attended put it, "The planetary astronomers pleaded that the [Announcement of Opportunity] should be sufficiently open to allow a planctary camera to be proposed and to fairly compete with other [scientific instruments]."[48]

[45] Smith, *Space Telescope* (cit. n. 16), p. 183.

[46] William Baum to Nancy Roman, 8 June 1976, file "ST—Faint Object Camera," Program Scientist papers, NASA headquarters, Washington, D.C.

[47] This is discussed at length in Smith, *Space Telescope,* Ch. 5; and Robert W. Smith, "The Biggest Kind of Big Science: Astronomers and the Space Telescope," in *Big Science: The Growth of Large Scale Research,* ed. Peter Galison and Bruce Hevly (Stanford: Stanford Univ. Press, 1992), pp. 184–211.

[48] David S. Leckrone to George F. Pieper, 15 Sept. 1976, file "May–September 1976," Box 3, accession no. 255–81–305, Goddard Space Flight Center, Greenbelt, Maryland. See also C. R. O'Dell, daily notes, 5 Aug. 1976, O'Dell Papers, Rice University; and Nancy Roman, memorandum

At the same meeting, the University of Arizona's Bradford A. Smith showed images taken with a CCD camera at a telescope on Mount Lemmon in Arizona. A small team at JPL had rigged up the camera, which exploited the Texas Instruments 400-by-400 pixel CCDs, in a matter of weeks, with the aim of selling the CCDs to astronomers by demonstrating that they *were* real devices capable of securing astronomical images. To carry the message beyond JPL, two JPL staffers had taken the camera on a tour of a number of observatories, including that on Mount Lemmon. Smith had used it to image a range of astronomical objects. One image was of Uranus in the far red part of the spectrum, and Smith used the facts it purported to convey as a conscription device. Most important, he was able to claim that these "images clearly reveal for the first time cloud structure high in the atmosphere of Uranus."[49]

Certainly after the August meeting, if not before, key NASA headquarters and Goddard staffers were solidly behind the idea of a planetary camera for the Space Telescope. NASA headquarters—where decision-making authority for the selection of instruments ultimately rested—had decided that to win approval for the Space Telescope, the support of planetary scientists was very highly desirable, even essential. The CCDs had also become, in effect, the means for NASA headquarters to mediate the interests of planetary scientists in the Space Telescope. Moreover, just as JPL was reporting that the CCDs were making rapid strides, it seemed to some that the SEC vidicon had not developed as planned. Nancy Roman, a NASA program scientist who had chaired the meeting with planetary scientists on 5 August, contended that the performance of the SEC vidicon justified taking a new look at CCDs.[50] There was unease at Goddard too that, among other development problems, while the SEC vidicon would be effective in the ultraviolet, it still fell well short of the specifications set for it in the red and far red. By September a Goddard staffer was writing, "Because of the growing concern about the potentially poor performance of the [SEC vidicon], Nancy Roman is organizing a full review of candidate detectors for the [wide field camera]" for the next Science Working Group meeting in October.[51] Although a year earlier the Science Working Group had strongly voted down such a move, key NASA staffers wanted to make the planetary camera a core instrument, thereby ensuring that it flew on the Space Telescope. But the group's chairman disagreed. Instead, a deal was cut in which two of the planetary camera's advocates at the 5 August meeting—Bradford Smith and James Westphal—were invited to the Science Working Group's October meeting to make a pitch for it.[52]

Much of the last, like the very first, Phase B Science Working Group meeting was

---

(file copy) "Meeting . . . Planetary Use 8/6/76," file "Meeting of Planetary Users of the ST," 19 Aug. 1976, Project Scientist papers.

[49] Bradford A. Smith, "Astronomical Imaging Applications for CCDs," in *Proceedings of the Conference of Charge-Coupled Device Technology and Applications, Nov. 30–Dec. 2, 1976* (JPL SP 43–40) (Pasadena, Calif.: Jet Propulsion Laboratory, 1976), pp. 135–137, on p. 136; and B. A. Smith, "Uranus Rings: An Optical Search" (letter), *Nature,* 1977, *268:* 32. On the tour to different observatories and JPL's preparations see James Janesick, interview by Robert W. Smith, 12 Aug. 1991. For conscription devices see Henderson, "Flexible Sketches and Inflexible Data Bases" (cit. n. 18).

[50] O'Dell, daily notes, 11 Aug. 1976, O'Dell papers.

[51] David S. Leckrone to George F. Pieper, 12 Aug. 1976, T. Kelsall to Leckrone, 24 Aug. 1976, and Leckrone to Pieper, 15 Sept. 1976, file "May–September 1976," Goddard (cit. n. 48).

[52] O'Dell, daily notes, 8, 13 Sept. 1976, O'Dell papers.

devoted to the issue of detectors. Roman again expressed her opinion that although selected as the detector for the wide field camera, "the SEC vidicon had not progressed as fast as expected and there is now doubt that it can meet the minimum performance specifications" established before. There were presentations and discussions on a range of detectors, including the SEC vidicon, the ICCD being worked on at Goddard, and the CCD developed for JPL (where the speaker noted that JPL expected to have an 800-by-800 pixel CCD in eighteen months for the planned Jupiter Orbiter Probe mission, later renamed Galileo). As many new pieces of testimony were introduced about the SEC vidicon and CCDs, assessments about their relative merits began to shift in favor of CCDs. Westphal, as agreed, argued for the advantages to planetary science of a camera using them.

The issue of whether to preselect the SEC vidicon was debated explicitly. The specifications set for the wide field camera had been fashioned to balance the ideal and the anticipated SEC vidicon performance. But, Roman judged, given the state of progress on the vidicon and the other detectors, it might not be possible to achieve some of the specifications negotiated earlier, such as red (or, for the CCDs, ultraviolet) response. There was a vigorous debate, but the Working Group passed a resolution that the detector for the wide field camera should be selected through open competition in the Announcement of Opportunity. That is, astronomers could propose whatever detectors they wished, and the SEC vidicon would not necessarily be chosen. A NASA staffer now noted that "a proposed Wide Field Camera which has a planetary capability is more likely to be selected." NASA had delegated the planetary capability to the CCDs, and the clear meaning of this announcement was that a proposal without CCDs would stand little chance. The provisional black box—a preselected wide field camera using the SEC vidicon—had been pulled open.[53]

What had happened? The third strategy of translation in winning support for a position that Latour has identified is a "detour" strategy. "In this new rendering of others' goals," Latour writes, "the contenders do not try to shift them away from their goals. They simply offer to guide them through a short cut."[54] For such a strategy to be effective, three conditions have to be met: the old route is clearly cut off, the new detour is well signposted, and the detour appears short. If the SEC vidicon was the old route, then by late 1976, though not clearly cut off, for many astronomers it was chancier than before. By then the CCD advocates had pointed to a detour, and the new route, the CCDs, had become well signposted. Images of astronomical objects had been secured with a CCD camera, and there was now substantial testimony to their improving performance. Hence the detour appeared relatively short, and NASA decided it would not be going too far out of its way to procure CCDs. Another potential benefit of selecting CCDs was more support from planetary scientists. In fact, as we shall soon see, to NASA the CCDs had become indispensable.

But the new route still possessed obstacles, in particular the lack of sensitivity of CCDs to ultraviolet light and the small area of sky that they could cover. How those who proposed CCDs for the wide field camera sought to map ways around these obstacles is the subject of the next section.

[53] ST Operations and Management Working Group, minutes, 18–19 Oct. 1976, pp. 7–8, Project Scientist papers.

[54] Latour, *Science in Action* (cit. n. 3), p. 111.

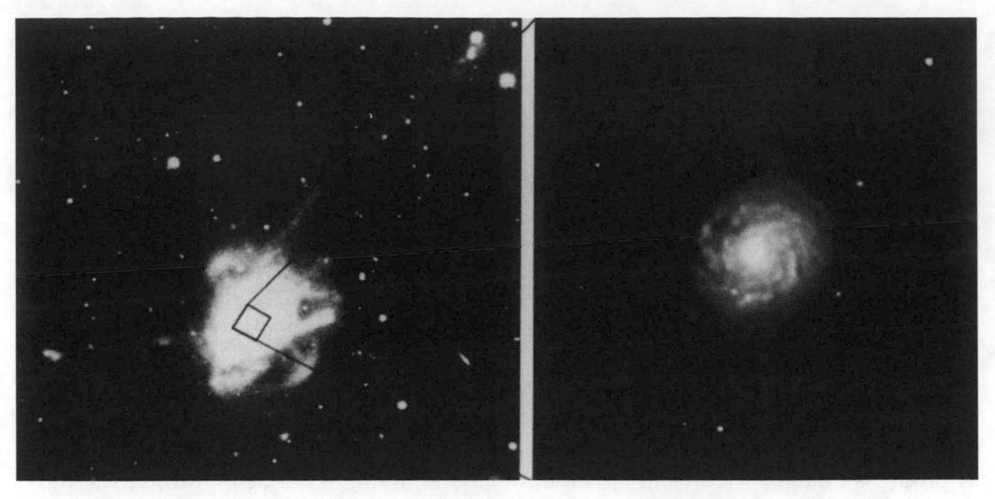

Figure 2. *An image of the center of a galaxy as secured with the wide field–planetary camera (*right*), compared with that secured with a large ground-based telescope (the small square in the center,* left*). The wide field–planetary camera image does in fact reveal far more detail, despite spherical aberration. Courtesy of NASA.*

VI. RESPONSE TO THE ANNOUNCEMENT OF OPPORTUNITY

The Announcement of Opportunity issued in March of 1977 defined a set of provisional black boxes. Among its pages were those specifying the limits within which one could design the wide field camera. Since the camera had now been opened to competition, its design and especially its detector enjoyed a certain latitude. The three resulting proposals exhibited the design traditions established by the different competing groups, but all included CCDs. Here we have an example of another of Latour's translation strategies, like the detour strategy but more forceful: a required, not an optional, detour. The strategy involves becoming indispensable, and it "means that whatever you do, and wherever you go, you *have* to pass through the contenders' position and to help them further their interests."[55] By the fall of 1976 the advocates of the CCDs had so altered the political economy of the Space Telescope Project that NASA had in effect decided that anybody proposing a wide field camera had to pass through the CCDs, even the group at Princeton that had for over a decade worked on the SEC vidicon.

After over a decade of receiving NASA funding to develop the detector, Princeton felt obligated to propose the SEC vidicon.[56] But while chiefly based on the SEC vidicon, the Princeton proposal incorporated a CCD to enhance the camera's response at the far red end of the spectrum. The Goddard proposal used a Texas Instruments 490-by-327 pixel CCD, modified from one developed for military television applications, for the visible and far red; it employed another identical CCD incorporated into an image-intensifier tube (itself originally developed for the Army Night Vision Laboratory) for the ultraviolet. Using the image intensifier tube to translate ultraviolet photons into electrons aimed at the CCD, the Goddard team planned to circumvent the CCD's poor ultraviolet response. Hence their proposal too was

[55] *Ibid.*, p. 120.
[56] See, e.g., Lyman Spitzer, Jr., interview by David DeVorkin, 17 June 1982, pp. 72–74.

strongly shaped by the programs Goddard had conducted for three years with Texas Instruments and the Night Vision Laboratory, programs that had tied them to the ICCDs.[57] The instrument's optical arrangement allowed it to function as both a wide field camera and as a planetary camera.

The final, and winning, proposal came from Caltech and JPL. It incorporated eight 800-by-800 pixel CCDs—though only 400-by-400 pixel devices had yet been built—from the imaging system that Texas Instruments and JPL were developing for the *Galileo* spacecraft. The Caltech-JPL proposal thus directly linked the interests of the Space Telescope and of *Galileo,* making them allies through the CCDs.[58] The leader of the proposal was James Westphal of Caltech. Westphal, one of the planetary scientists who had met with NASA staff in August 1976 to discuss the Space Telescope's capabilities for planetary science, had also pitched the planetary camera to the Science Working Group in October 1976. Although he had been aware of CCDs for some years, he had at first not been much impressed by their performance. But in mid 1976, after news of the latest JPL developments, he became enthusiastic. The question now was, "How do you lay your hands on one?" In choosing to propose a camera with nearby JPL, Westphal also had an enormous advantage: he had worked on astronomical detectors in the 1960s with Bruce Murray. Murray, the newly appointed director of JPL, thought Westphal was a "miracle worker" and was more than happy to make JPL resources available to him. In constructing the proposal, Westphal's team's aim was "to sell the device, to sell the CCD, to sell the Wide Field Camera. We did that on purpose recognizing that if we were to compete with whoever else might be out there competing for the Wide Field Camera, including Princeton with the SEC Vidicon, that we had to have a really overwhelmingly unbelievable case."[59] To that end they generated as much evidence as possible to show that they could be counted on for invention, while also addressing the two major scientific drawbacks to CCDs for the Space Telescope: the poor ultraviolet response and the small field of view.

To enhance the CCDs' ultraviolet response, the Caltech-JPL team coated them with an organic phosphor sensitive to ultraviolet light. The idea had been examined five years before at Goddard, but rejected. To increase the field of view the Caltech-JPL team devised a scheme of optically "mosaicking" together four 800-by-800-pixel CCDs, thereby in effect turning them into 1,600-by-1,600-pixel chips. The mosaicking was to be done with a pyramid with four faces that would split the single beam of light entering the camera into four. But, the team realized, they could have the pyramid itself rotate to two positions, illuminating two sets of CCDs, and thereby combine a planetary camera *and* a wide field camera in the same instrument.[60] Thus was born the wide field–planetary camera. Through an established commitment to

[57] For the proposal see Goddard Space Flight Center, "A Wide Field and Planetary Camera for the Space Telescope" (July 1977), Part I: "Summary," Part II: "Technical," and Part III: "Cost Management." We are most grateful to the principal investigator for the proposal, Stanley Sobieski, for a copy.

[58] On the importance of this alliance in winning support for the telescope see Smith, "The Biggest Kind of Big Science" (cit. n. 47), pp. 184–211.

[59] James A. Westphal, interview by David DeVorkin, 14 Sept. 1982, pp. 182, 199; and Bruce Murray, interview by Robert W. Smith, 14 Aug. 1991.

[60] For the proposed method see "Technical Proposal, Investigation Definition Team, Wide Field/Planetary Camera" (1977), pp. 28–29, JPL. On the preparation of the proposal see Smith, *Space Telescope* (cit. n. 16), pp. 248–253.

CCDs and some fortuitous ingenuity the Caltech team succeeded in proposing what were widely seen as elegant ways to meet all of the instrument's main desiderata.

After a lengthy and complex review by its advisory committees, NASA finally selected the Caltech-JPL proposal. At the prompting of a National Academy of Sciences committee, NASA also first assembled a special team of experts to assess the strengths and weaknesses of the proposed detectors. In their opinion, the CCDs were a very promising and attractive technology.[61] Even though the CCDs the team planned to use did not even exist, the proposal, consisting of a host of elements, had been assessed and accepted as potentially a black box. At the time of selection, the wide field–planetary camera therefore consisted of many heterogeneous elements: actual hardware known to exist, hardware promised to exist eventually, personnel with substantial reputations as clever inventors and able managers, testimony of reviewers both formal and informal, and a massive package (many hundreds of pages) of testimony by the designers and potential designers. But in the eyes of the judging committees, the provisional black box contained plausible accommodations to the conflicting desiderata of the competing groups. By selecting this potential black box, and not those proposed by Goddard and Princeton, NASA managers and officials were in fact counting on invention.

VII. CONCLUSION

In the introduction we referred to our goal of producing an archaeology of CCDs. The case of the CCDs and the wide field–planetary camera demonstrates how many factors such an archaeology entails. If we ask *why* there was a camera that exploits CCDs aboard the Hubble Space Telescope, then we soon confront a complex process permeated by social, institutional, political, and economic issues, as well as scientific and technical ones. If in pursuing our analysis we had simply stayed inside Westphal's lab, many of the elements of this process would not have been evident. We would have seen CCDs entering and leaving his lab, but had no real grasp of why.

We pointed to the importance for a Very Big Science project such as the Hubble Space Telescope of defining and negotiating provisional black boxes, especially in the early stages. We argued that the pressure to open such provisional black boxes could come from different groups for many different reasons and in a variety of circumstances, and thus the provisional black boxes were by nature transient and unstable. We have argued further that technologies embody interests, and that what gets built is a negotiated accommodation of various interests. Hence to understand the wide field–planetary camera and the Space Telescope, it is essential to examine the early design period, when interests were negotiated; as we have seen, these negotiations were fundamental in shaping the technology. Although the technology had not become black boxed, even when the instruments were selected following the Announcement of Opportunity, the basic designs had been established in the years between 1972 and 1977, and the room to maneuver during the detailed design and construction of the telescope (that is, the period after 1977) was severely circumscribed. As the telescope program developed after 1977, so the interests changed,

[61] J. D. Rosendhal, active papers, files "Status of Detectors Proposed for the Space Telescope Wide Field Camera and Faint Object Spectrograph 9/77" and "Space Telescope Detectors—Working Papers", NASA headquarters. On the NAS intervention see O'Dell, daily notes, 4 May 1977, O'Dell papers. On the review see Smith, *Space Telescope* (cit. n. 16), pp. 254–257.

raising continual tensions between the existing design, which manifested the old set of interests, and proposed changes to the design, which different groups—principally the science team for the wide field–planetary camera—argued better suited their interests.

The actual implementation of the CCDs in the wide field–planetary camera provides a coda to this story. The final camera that was launched into space in 1990 was different in several ways from that accepted for development in 1977. That provisional black box was opened many times during the overall project. The camera's actual ultraviolet performance, for example, was, for a variety of reasons, much worse than its advocates expected it would be in 1977. The camera's status as one element deep inside a complex system was also dramatically underlined by the spherical aberration of the telescope's primary mirror, discovered soon after it went into orbit, and this optical flaw significantly diminished the camera's performance.

NASA had already decided in 1985 that the wide field–planetary camera was so essential, for scientific and political reasons—it was, for example, to provide the bulk of the images that would justify the telescope to its patrons in the White House and Congress and to the public—that it would fund the building of a "clone" wide field–planetary camera, WF/PC II, to replace the original camera after some time in orbit. WF/PC II, NASA mandated, would use CCDs from the batch built by Texas Instruments for the original camera, rather than later generation CCDs.[62] Time and again NASA managers resisted the scientists' wishes to employ different CCDs. But once one of the Texas Instruments devices failed, causing a loss of confidence in the others, the provisional black box that was the WF/PC II was opened up, and NASA chose to fly a new kind of CCD. By 1992 planning groups were even considering refurbishing the original wide field–planetary camera if the scheme went ahead for the space shuttle astronauts to replace it with WF/PC II. After return to Earth, the original wide field–planetary camera might become WF/PC III and eventually replace WF/PC II.

Hence to judge from the story of the CCDs and the wide field–planetary camera, those who wish to see research scientific instruments in a very large scale scientific enterprise as true black boxes for any length of time are chasing a chimera.

[62] See Smith, *Space Telescope* (cit. n. 16), pp. 254–257.

INSTRUMENTS & CULTURE

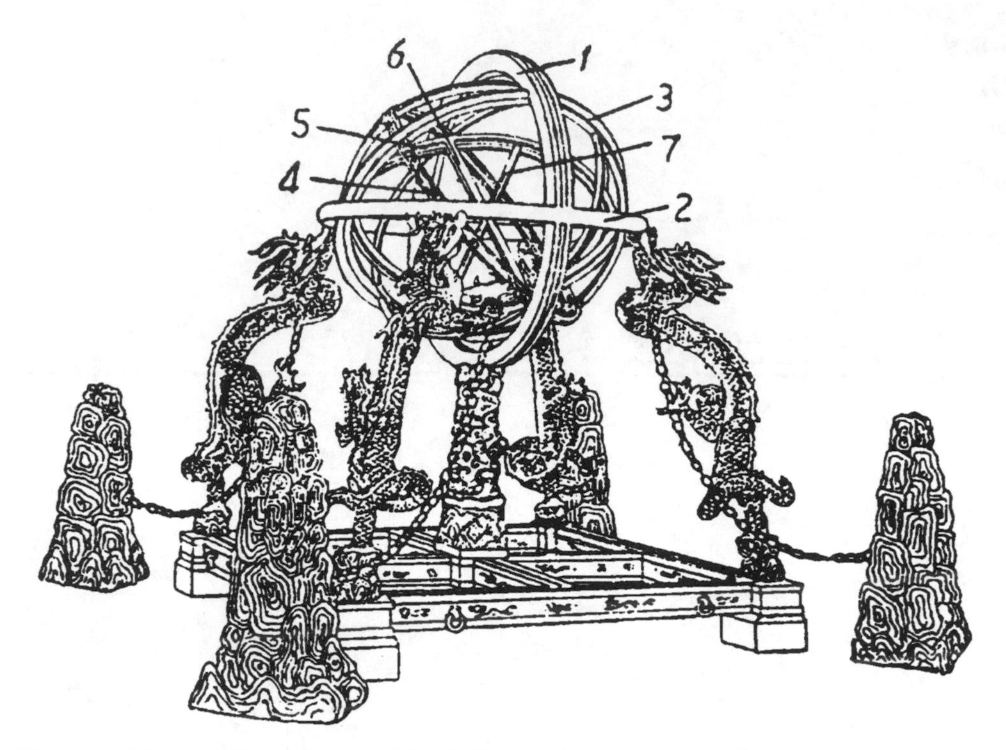

Figure 1. *The armillary sphere used by the Imperial Astronomical Bureau in the Ming dynasty: (1) meridian ring; (2) horizon ring; (3) fixed equatorial ring; (4) ecliptic, colure, and movable equatorial rings; (5) movable right ascension ring; (6) polar axis; (7) sighting tube. From Chen Zungui,* Qingchao tianwen yiqi jieshuo *(see note 17), p. 53.*

Instruments and Observation at the Imperial Astronomical Bureau during the Ming Dynasty

By Thatcher E. Deane*

THROUGHOUT THE IMPERIAL PERIOD in China, the activity that a modern astronomer or historian of science would identify as astronomy was carried out within an imperial Chinese astronomical bureau. From the first articulated bureaucratic government in the Chinese culture area, in the third century B.C., to the fall of the Qing dynasty twenty-one centuries later, each polity in that region sustained a special organ within its bureaucracy whose task it was to observe, interpret, and where possible predict celestial phenomena. The source of this governmental concern with the celestial realm was a political cosmology that identified the ruler of China as the Son of Heaven (*tianzi* 天子), a mortal being empowered by Heaven with the rule of man. The emperor had a sacred obligation to monitor the celestial events through which Heaven conveyed its evaluation of his behavior as Son of Heaven and to respond with oaths of gratitude or contrition as the circumstances dictated. As the astronomical bureau of each dynasty fulfilled this obligation throughout the imperial period of Chinese history, its personnel, usually of low rank, used observational instruments of various sorts supplied and authorized only by the emperor's edicts. These were the basic tools upon which an astronomical bureau relied as it engaged in its routine activities: observing celestial events, interpreting those events to the emperor, preparing the official calendar, and developing and occasionally examining the accuracy of the calendric systems used to generate calendars.

The purpose of this study is to examine the use of the observational instrumentation of the Imperial Astronomical Bureau of the Ming dynasty (1368–1644) within the context of the mission and actual activity of this organ of government. The Ming period is the earliest for which there exist sufficient historical records and physical evidence to explore in a systematic fashion the routine operation and instrumentation of an astronomical bureau. True, the Ming dynasty has the reputation of being a period in which material development stagnated and intellectual discourse narrowed. Chinese historians of science ignore this period because this stability came just before and provided considerable resistance to the scientific reforms of the seventeenth century that were inspired by contact with European Jesuits. Western scholars, for their part, have generally been more interested in earlier, brighter eras or in the Jesuit period itself. This disdain on the one hand and neglect on the other

* 1532 N.E. 106th Street, Seattle, Washington 98195–6514.

combine to prove that the Ming is a period of normal science, the best to begin with when looking at a tradition not one's own.[1]

Our examination will illustrate how the important position of astronomical activity in China led one dynasty to maintain an astronomical bureau and equip it with large astronomical instruments, but could not ensure that those instruments were appropriate for more than the bureau's routine work. The astronomical instruments provided to the Imperial Astronomical Bureau in the Ming appear to have been poorly adjusted from the start and poorly maintained thereafter. This situation limited the bureau's ability to judge competing calendric systems but did not prevent it from conducting noteworthy systematic and consistent observations.

I. THE ASSIGNMENT: OBSERVATION AND INTERPRETATION OF ANOMALIES

According to the *Collected Statutes of the Ming Dynasty* (*Daming huidian* 大明會典), the Astronomical Bureau was required to watch for any strangeness or anomaly (*bian yi* 變異) with respect to the sun, moon, stars, planets, conjunctions, or wind, clouds, fog, and dew.[2] Because celestial anomalies were seen as important signs of heavenly criticism of the ruler's actions, the task of observing anomalies was taken very seriously. The various staff members (who in the Ming were selected from a hereditary pool) and students at the bureau's observatory were required to keep watch day and night. In the 1389 version of the Ming legal code, a punishment of sixty blows of the heavy bamboo was mandated for any failure on their part to make anomalies known to the emperor. Upon observing an anomaly, the bureau personnel were to draw up an open memorial of interpretation (*baiben zhanzou* 白本占奏) indicating the circumstances of the event and its significance.[3]

Early in the dynasty, memorials of this sort were signed by the individual official who wrote them and were submitted to the throne via the regular channels. Beginning with the first reign of Yingzong (1436–1449), the interpretive memorials were no longer signed by individuals but rather bore the seal of the Astronomical Bureau as a whole, thus implying that these reports were written jointly by the upper echelon

[1] For more on the subject of this article see Thatcher Deane, "The Chinese Imperial Astronomical Bureau: Form and Function of the Ming Dynasty *Qintianjian* from 1365 to 1627" (Ph.D. diss., Univ. Washington, 1989), esp. Ch. 5. On the Jesuit-inspired reforms see Henri Bernard, "L'Encyclopedie astronomique du Père Schall (*Tch'ong-tcheng li-chou,* 1629, et *Si-yang sin-fa li-chou,* 1645): La réform du calendrier chinois sous l'influence de Clavius, de Galilée et de Kepler," *Monumenta Serica,* 1938, *3*:35–77, 441–527; Joseph Needham, *Science and Civilisation in China,* 5 vols. (Cambridge: Cambridge Univ. Press, 1954–), Vol. III, *Mathematics and the Sciences of the Heavens and the Earth* (1959) (hereafter **Needham, *Science and Civilisation,* Vol. III**), pp. 437–458; Nathan Sivin, "Copernicus in China" in *Colloquia Copernicana, II: Etudes sur l'audience de la théorie héliocentrique* (Warsaw: Ossolineum, 1973), p. 103; and Hashimoto Keizo, "*Sūtei rekisho* ni miru kagaku kakumei no ichi katei" (On the introduction of European astronomy into late Ming China) in *Tōyō no kagaku to gijutsu: Yabuuchi Kiyoshi sensei shōju kinen rombunshū* (Science and skills in Asia: A festschrift for the 77th birthday of Professor Yabuuchi Kiyoshi) (Tokyo: Domeisha, 1982), pp. 370–390.

[2] *Collected Statues of the Ming Dynasty* (*Daming huidian;* hereafter ***Huidian***), 222, 1b–2a, pp. 2955b–c. This work, the major source for administrative regulations for the Ming dynasty, was published in two editions. References here are to the 1963 reprint of the Wanli edition (Taibei: Dongnan shubaoshe), and append (as do those to other reprints of early sources) the pagination added to the reprint.

[3] *Daminglü zhijie* (Direct explanations of the great Ming code) (1395), 30 *juan* (Seoul: Chōsen sōtokufu chūsūin, 1936), 12, 6b, p. 284. The astronomical bureau at the auxiliary capital in Nanjing was also required to report anomalies.

officials in the bureau.[4] By shifting the authorship of anomaly interpretation memorials from individuals to the bureau as a corporate entity, Yingzong probably hoped to reduce the potential for manipulating them. With the entire upper echelon of the bureau responsible for their contents, no given interpretation would be likely to contain overly pointed criticisms. Indeed, a request by the bureau later in the dynasty to restore individual anomaly interpretations was rejected.

The reports of anomalies by the Ming Astronomical Bureau are preserved thirdhand in the *Veritable Records of the Ming Dynasty* (*Damingshilu* 大明實錄), an ongoing chronological compendium of government documents compiled after the conclusion of each emperor's reign.[5] The criteria for inclusion are given in the "Compilation Principles" (*xiuzuan fanli* 修纂凡例) section of the *Veritable Records* for the Jiajing reign (1522–1566): "[The compilers] write in all instances when the Astronomical Bureau memorialized celestial phenomena, atmospheric observations (*qihou* 氣候), all eclipses of the sun and moon, and intrusions and close conjunctions among the Seven Governors [i.e., sun, moon, and five planets]. Also included shall be instances of capital or provincial memorials on auspicious or weird events and [the respective] congratulations or [imperial] prayers [made on the occasion of such reports.]"[6] In 1575 the famed statesman Zhang Juzheng (1525–1582) noted in a memorial on the preparation under way for the *Veritable Records* for the reigns of Shizong and Muzong that Astronomical Bureau astrologic notices as well as auspicious and inauspicious (*xiangyi* 祥異) notices were to be included chronologically in the compilation. Zhang also explained that the Astronomical Bureau was not required to send the material to the compilers in its original form as memorials but rather could simply send copies on common white paper in clear handwriting.[7] Although it is difficult to know what percentage of the bureau's internally recorded observations were recorded in the completed *Veritable Records,* the "Compilation Principles" and Zhang's directive suggest that most were. A preliminary analysis of entries for the planets Mercury (see below, Section IV) and Saturn, as well as entries on lunar and solar eclipses, for example, reveals a high proportion of the observations one would expect the bureau to make, given its mission.[8] Although the interpretive portions of memorials on celestial anomalies submitted to the emperor were not routinely included in the *Veritable Records,* those cited indicate that the interpretations were clearly based upon established tradition and classical codifications.

[4] *Huidian,* 222, 1b–2a, pp. 295b–c.

[5] The *Veritable Records of the Ming Dynasty* (*Damingshilu*) are by far the single most important annalistic source on central government action for the dynasty. Citations are to the 183-volume reprint published by the Academica Sinica Institute of History and Philology in Nangang, Taiwan, between 1962 and 1967, and use the form ——— ***shilu.*** (*Taizu shilu,* i.e., refers to the reign of Taizu, the first emperor of the Ming dynasty.) On how the *Veritable Records* were compiled see Wolfgang Franke, *An Introduction to the Sources of Ming History* (Kuala Lumpur: Univ. Malaya Press, 1968), pp. 8–23, esp. pp. 11–15.

[6] *Shizong shilu,* "Xiuzuan fanli," 8b.

[7] *Shenzong shilu,* 35, 12a–15a, pp. 825–831.

[8] A study of statistically significant variations in the quantity of these records and their distribution within individual and between the several separate editions of the *Veritable Records* might settle the question of political influences on their inclusion. To have value such a study would require a baseline of astronomical phenomena we could reasonably expect the bureau to have observed, as well as an intimate knowledge of the political context and procedural details of compilation of each edition of the *Veritable Records.*

Interpretation of celestial anomalies in the Ming no doubt relied strongly on distillations in such works as *Great Ming Manual of Astrology and Field Allocation Clearly Classified,* attributed to Liu Ji, first director of the Ming Astronomical Bureau and close advisor to the Ming founder.[9]

Anomaly reports generated by the Astronomical Bureau were available at least in part to the general bureaucracy. We know this because such reports were sometimes cited by other memorials for rhetorical effect. In 1500, for example, the Five Military Commissions and Six Ministries began a memorial by recounting recent anomalies and disasters in the empire, as a more or less standard prelude to presenting the substance of their particular agenda. A report of a comet by the Astronomical Bureau appears verbatim at the head of the list.[10] Predictions of eclipses by the Ming Astronomical Bureau were also made available to the bureaucracy through formal channels.[11]

II. THE EQUIPMENT

To facilitate their watch for celestial anomalies, astronomical bureaus were provided with an imperial observatory or "observing platform." The observing platform of the Ming dynasty's original bureau, apparently first located south of Nanjing, was moved to Jiming Mountain northwest of the city in 1385, at which time the old location was given over to the Islamic Astronomical Bureau, first created under the Mongol Yuan dynasty and continued under the Ming as a second source for eclipse predictions.[12] The instruments used at both observatories were collected from the Yuan Astronomical Bureau near Beijing and transported south, as ordered in a directive of late 1368.[13]

When Beijing became the main capital in 1402, an astronomical bureau was again created there, followed shortly by a new observatory. This "star-observing platform" (*guanxingtai* 觀星台) was built on top of the southeast corner of the Beijing city wall above what is now the Jianguo Gate. The facility was used by the Astronomical Bureau throughout both the Ming and Qing dynasties; recently it was restored as it was in the Qing and, in 1983, opened to the public as a museum.[14] Before the new observatory was constructed, observations were probably performed at the site of the old Yuan observatory, and for a time without significant instruments.[15] Once operational, the Beijing observatory used large bronze instruments cast in 1439 from

[9] *Daming qinglei tianwen fenye shu,* microfilm copy, East Asian Library, University of British Columbia. For early examples of anomaly interpretation see Fang Hsuan-ling, *The Astronomical Chapters of the Chin-shu,* ed. and trans. Ho Peng Yoke (Le monde d'outremer passé et présent, 2e série: Documents, 9) (Paris/The Hague: Mouton: 1966), pp. 121ff, 149ff.

[10] *Xiaozong shilu,* 162, 4a–8a, pp. 2917–2925. On this comet see Herman Mucke, *Helle Kometen −86 bis +1950: Ephemeriden und Kurzbeschreibungen* (Vienna: Astronomisches Büro, 1972), pp. 36–37.

[11] *Huidian,* 222, 2b, p. 2955d.

[12] Jiao Lian, *Daming yitong zhusi yamen guanzhi* (1541), 16 *juan* (Taipei: Xuesheng, 1970), 2, 3a, p. 87; and *Taizu shilu,* 176, 1b, p. 2666. The computational component of the Islamic Bureau was incorporated into the main Astronomical Bureau in 1398.

[13] *Taizu shilu,* 35, 3b–4a, pp. 632–634. On the question of latitude adjustments for proper use of these instruments at the new location see below.

[14] G. E. W. Beekman, "New Glory for the Ancient Beijing Observatory," *New Scientist* (London), 20 Sep. 1984, p. 54. On its history see Yi Shitong, "Beijing guguanxiangtai de kaocha yu yanjiu" (The old astronomical observing platform in Beijing), *Wenwu,* 1983, *8:*44–51.

[15] Yi, "Beijing guguanxiangtai" (cit. n. 14), p. 45.

wooden replicas of the Yuan instruments (which were themselves left in Nanjing), a process initiated in 1437 from a memorial by Huangfu Zhonghe, then vice-director of the Astronomical Bureau.[16] To the dynasty's fall, Ming emperors maintained duplicates of most central government offices in Nanjing, among them an auxiliary astronomical bureau. Since the successor Qing dynasty did not use Nanjing as an auxiliary capital, the Nanjing auxiliary astronomical bureaus and their observatories were abolished and their major instruments were brought north in 1670. Late in the reign of the Kangxi emperor (1662–1722) these original Yuan instruments were melted down for scrap to make new ones, apparently on the order of European Jesuit directors of the Qing Astronomical Bureau, an act for which the Chinese have never quite forgiven them.[17]

The two primary instruments employed at the observatory of the Ming Astronomical Bureau, the armillary sphere (*hunyi* 渾儀) and the simplified instrument (*jianyi* 簡儀) were both copies made in 1439 of instruments constructed under the direction of Guo Shoujing (1231–1316) at the beginning of the Yuan dynasty in the late thirteenth century. Unlike the originals from which they were copied, these Ming instruments are still extant and can be seen at the Purple Mountain Observatory in Nanjing.[18] The armillary sphere consists of three nests of rings representing the fundamental celestial circles.[19] The outermost nest consists of fixed meridian, horizon, and equatorial rings (see Figure 1). The inner diameter of these rings is about two meters. The middle nest of rings turns as a unit about the equatorial pole. In addition to a second equatorial ring, this nest has ecliptic, solstitial, and equinoctial colure rings. The innermost nest consists of a single meridian ring movable about the celestial pole. A sighting tube in the plane of this ring pivots about its center.

The single degree of freedom of the middle nest of rings permits its rings as a unit to be aligned with the celestial circles they represent. The two degrees of freedom of the innermost meridian ring and its sighting tube allow the operator to align on a particular celestial object. Once set in this way, the coordinates of the observed object are determined from the graduated scales of the various rings.

The simplified instrument serves the same purpose as the armillary sphere but has fewer parts in a different arrangement (see Figure 2).[20] This instrument consists of two sets of two rings, one set in the equatorial coordinate system and the other in the horizon coordinate system. In each set the two rings are perpendicular to each other, but instead of their centers coinciding, the edge of one ring is mounted on the

[16] *Yingzong shilu,* 60, 3b, p. 1146, and 27, 5b–6a, pp. 540–541 (the memorial).

[17] See, e.g., Chen Zungui, *Qingchao tianwen yiqi jieshuo* (Explanation of the astronomical instruments of the Qing dynasty) (Beijing: Zhonghua quanguo kexue jishu puji xichui, 1956), p. 50.

[18] See Pan Nai, "Nanjing de liang tai gudai cetian yiqi—Mingzhi hunyi he jianyi" (Two ancient astronomical instruments—the Ming armillary sphere and the simplified instrument), *Wenwu,* 1975, *230*(7):84–89, Pl. 9. Pan refuted the idea that the armillary sphere was based on a Song rather than Yuan original, in "Xiancun Mingzhi fangyi hunyi yuanliu kao" (An examination of the origin of the extant armillary sphere copied in the Ming), *Zirankexue shi yanjiu,* 1983, *2*(3):234–245. See also Needham, *Science and Civilisation,* Vol. III, pp. 372ff, as well as Figs. 156, 163–166.

[19] On Chinese armillaries see Needham, *Science and Civilisation,* Vol. III, pp. 339–382. The most complete description of this instrument is that of Chen, *Qingchao tianwen yiqi jieshuo* (cit. n. 17), pp. 52–55; see also the studies by Pan Nai cited in note 18.

[20] The earliest extant description of the simplified instrument was included with accounts of several instruments recorded in the "Monograph on Astrologics," *Yuanshi* (History of the Yuan dynasty), 48, 990–991; see the translation by Alexander Wylie in "The Mongol Astronomical Instruments in Peking," *Chinese Researches* (Shanghai: 1897; rpt. Taipei: Chengwen, 1966), Pt. II, pp. 1–27. See also Needham, *Science and Civilisation,* Vol. III, pp. 371ff.

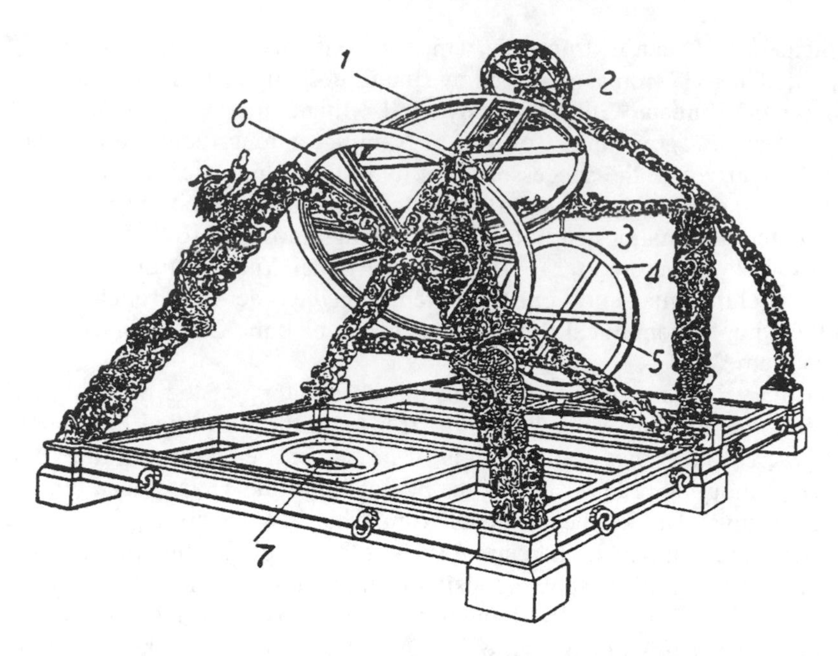

Figure 2. The Ming simplified instrument: (1) right ascension ring; (2) north pole; (3) zenith point; (4) azimuth ring; (5) horizon line; (6) equatorial ring; (7) sundial. From Chen, Qingchao tianwen yiqi, *p. 55.*

center of the other ring. As in the armillary sphere, the diameter of these rings is about two meters. The fixed ring of the larger of the two sets is parallel to the celestial equator, making its companion ring movable about the polar axis parallel to great circles of right ascension. A sighting tube is fixed to the center and rotates in the plane of this right ascension circle to indicate declination. This freedom of motion, along with the motion of the right ascension circle, allows the sighting tube to be aligned with any celestial object. The coordinates of that object in the equivalents of right ascension and declination are then read off the scales of the two circles. The second set of circles, identically arranged but oriented in the horizon system of coordinates, is located under the north end of the larger equatorial set. This instrument was deemed "simplified" because these two sets of coordinate circles were separated. Because the centers of the rings no longer shared the same point in space and there were fewer of them, the simplified instrument achieved a greater range of vision than some of the more complicated armillary spheres. After its adaptation in the Yuan, the simplified instrument became the instrument which most exemplified Chinese astronomy in East Asia. The instrument was copied and used in the royal court of Korea in the Yi dynasty (1392–1910), for example.[21]

Although not included in a list of the instruments completed in 1439, a copy of a Yuan gnomon (*guibiao* 圭表), or shadow-casting instrument, was probably also

[21] See Joseph Needham, Lu Gwei-djen, John Combridge, and John Major, *The Hall of Heavenly Records: Korean Astronomical Instruments and Clocks* (Cambridge: Cambridge Univ. Press, 1986), passim.

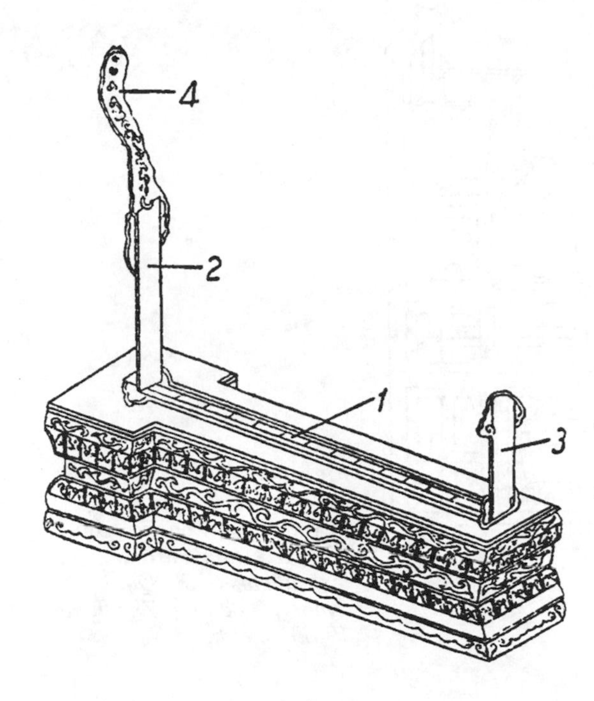

Figure 3. *The Ming gnomon: (1) horizontal scale; (2) gnomon; (3) vertical scale extension; (4) aperture. From Chen,* Qingchao tianwen yiqi, *p. 57.*

made about this time.[22] Huangfu Zhonghe had included the Yuan gnomon in his 1437 memorial requesting that copies of instruments in Nanjing be made. Since there is no indication that approval of his request excluded the gnomon, we may conclude that it too was duplicated at this time. The relative simplicity of the gnomon's construction compared to that of the armillary sphere and the simplified instrument might explain why it was not mentioned together with them in the 1439 notice. The gnomon consists of a vertical member about 2.5 meters tall and a horizontal bench or beam upon which its shadows fall (see Figure 3). The gnomon is oriented in the north-south meridian line so that shadows at local noon fall on the horizontal scale. Although a time-honored instrument in China, where it was employed with great ingenuity, gnomons are in practice restricted to measuring the position of the sun and even in this task are not very accurate.[23]

A time-measuring device is of paramount importance to quantitative observation. Although other forms of timekeeping were used in China, astronomical and official timekeeping was done by means of a water clock or clepsydra (*louhu* 漏壺) (see Figure 4). The water clock used by the observatory of the Astronomical Bureau was probably a three-chamber inflow-type clepsydra.[24] No water clock was among the instruments at the Nanjing Observatory copied for the station in Beijing in 1437–

[22] On Chinese gnomons and the Yuan gnomon in particular see Yi Shitong, "Guibiao zhi yanjiu–Yuandai guibiao fuyuan shexiang" (Study of the gnomon: Restoration of the Yuan era gnomon), paper presented at the People's Republic of China National Conference for History of Science, 1980.

[23] For a mathematical analysis of the accuracy of gnomon observations compared to those using an armillary sphere see Nakayama Shigeru, *A History of Japanese Astronomy: Chinese Background and Western Impact* (Cambridge, Mass.: Harvard Univ. Press, 1969), App. 4, pp. 242–244.

[24] The clepsydra now seen in the Palace Museum in Beijing might well resemble in basic form the one used at the Astronomical Bureau's observatory. See the photograph reproduced in Needham,

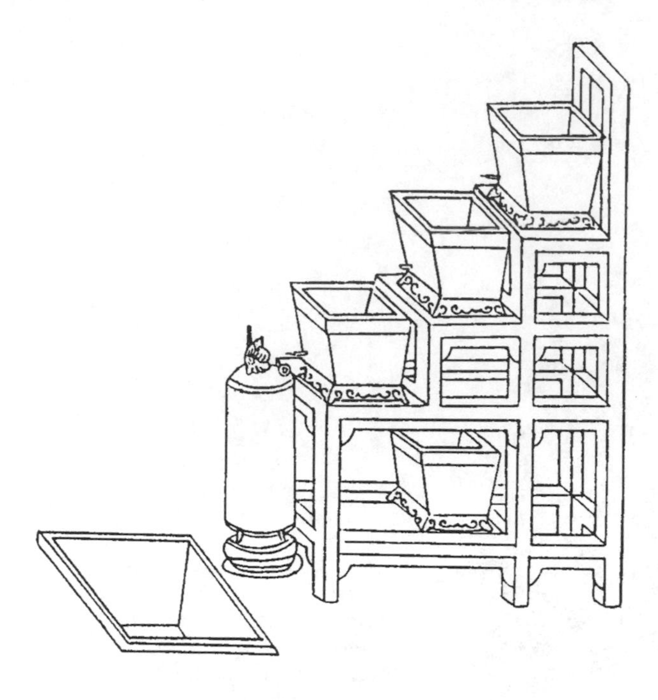

Figure 4. *A clepsydra, or water clock. From Chang Fuyuan,* Tianwen yiqi zhilue *(On the extant astronomical instruments of the Qing dynasty) (Beijing: Zhenhuage, 1921), p. 39b.*

1439, but the Astronomical Bureau did maintain a hall for a water clock at the southern capital and after the move to Beijing.[25] That there was a water clock at the observatory of the Astronomical Bureau in Beijing is clear from passing references,[26] but its construction and accuracy, unlike those of the armillary sphere and the simplified instrument, were not specifically discussed.

III. PRECISION OF THE OBSERVING INSTRUMENTS

It is likely that neither the armillary sphere nor the simplified instrument in Beijing was ever capable of precise measurements. The record is not clear on whether the two instruments were even properly constructed for the latitude of Beijing. Huangfu Zhonghe's 1437 memorial had called for adjusting the altitude of the celestial pole of the instruments before bronze copies were to be made from wooden copies of the originals at Nanjing.[27] Since the Nanjing originals were first constructed and used in Beijing, there should have been no need for an adjustment in copies intended for use in Beijing unless the originals had somehow been adjusted when brought to Nanjing, something for which there is no other evidence.[28] Between Beijing and

Science and Civilisation, Vol. III, Fig. 143 (Pl. XLVI). On Chinese water clocks generally see *ibid.*, pp. 313–329.

[25] *Yingzong shilu,* 239, 3a, pp. 5205.

[26] E.g., *Shenzong shilu,* 477, 6b–7b, pp. 9012–9014.

[27] *Xianzong shilu,* 27, 5b–6a, pp. 540–541.

[28] The pioneering Jesuit missionary Matteo Ricci left an account of astronomical instruments he saw at the Nanjing Observatory in 1600, which included the original armillary sphere and simplified instrument. He stated that their construction indicated that they were made for a latitude of 36 de-

Nanjing there is a latitude difference of about seven degrees; thus the major axes of rotation of the armillary sphere and simplified instrument should have been inclined seven degrees less in Nanjing than in Beijing. The armillary sphere could have been modified appropriately if the points of connection of the equatorial nest were shifted inside the outer framework, but there is no evidence that this was done. A latitude adjustment would have been more difficult for the simplified instrument, since the framework of the instrument was built around the equatorial ring itself. If its axis were lowered by placing wedges under the feet of the instrument to raise the south end of the base relative to the north end, this would have rendered useless both the secondary rings in the horizon coordinate system and the water channels set into the base of the instrument to monitor its leveling. Alternately, the six supporting members of the instrument could have been adjusted. If any latitude adjustment was made in the simplified instrument for its use in Nanjing, it probably was this latter modification, since the former and less satisfying modification would have left obvious traces. In 1502 Zhang Shen, then bureau director, wrote an account of problems with the Beijing copies. He explicitly stated that the latitude setting of the simplified instrument was too low, but he does not say by how much.[29] Without a quantitative measure, it is impossible to know whether to attribute this latitude error to the original instrument, to an adjustment made to it, to a failure to copy it properly, or to a failure to adjust the copy correctly or to some combination of these possibilities.

The precession of the equinoxes also affected the accuracy of ecliptic longitudes and latitudes measured with the armillary sphere. Since this armillary was constructed near the beginning of the Yuan dynasty in the last quarter of the thirteenth century, if we assume precise construction for that epoch, over one degree of precession error would already have been evident in the original instrument by the beginning of the Ming dynasty. Considering the problems recounted above concerning the instruments' latitude calibration, one may also justifiably conclude that the copies were not corrected for precession. On this assumption, in 1502, when Zhang Shen first referred to precession with respect to the armillary, there would have been 3.1 degrees of precession error in the instrument.[30]

The problems caused by precession were probably no larger than inaccuracies in the basic construction of the instruments. Already in 1478 a memorial from the Astronomical Bureau had noted that the observing instruments were all in disrepair.

---

grees. Needham thinks this indicates that they were used in Pingyuan (modern Linde), Shanxi, but there is no indication that the Yuan instruments were ever anywhere but Beijing before their appropriation by the Ming. More likely possibilities are (1) that there was a scribal error of 36 for 39, the latitude of Beijing; (2) that an unsuccessful attempt was made to adjust the instruments for the latitude of Nanjing; or (3) that Ricci's original determination was simply incorrect. After all, his description of the size of the instruments is only roughly quantitative, and he does not explain how he determined the latitude setting of the instruments. See Needham, *Science and Civilisation,* Vol. III, pp. 367–368, and note *f,* pp. 368–369.

[29] *Xiaozong shilu,* 182, 1b–2a, pp. 3346–3347.

[30] *Ibid.* How this error would have affected observations depends on how the armillary was aligned before an observation. If initialized by aligning a degree mark on the equatorial ring with a known star, as would have been the normal procedure, the equinoctial and solstitial points on the instrument would have been displaced from their correct positions by an amount which depended on the star used in the alignment. Once the instrument was misaligned in this way, all ecliptic measurements would have had periodic errors of around 3.1 degrees in longitude and smaller periodic errors in latitude. Equatorial measurements, however, would not have been affected.

When asked to address this question, the Ministry of Works offered the explanation that the artisans who constructed the copies of the original armillary sphere and simplified instrument were not familiar with such devices.[31] The Ministry of Works was then ordered to make repairs with the aid of another simplified instrument from the palace observatory, a small office attached directly to the imperial palace that had some instruments used in an irregular fashion by various imperial attendants. What these repairs consisted of we do not know, but they must have been insufficient, for just eleven years later, in 1489, the bureau director Wu Hao requested that the simplified instrument and the armillary be completely rebuilt. Permission was granted at that time only for Zhang Shen, then bureau vice-director, to oversee the construction of wooden prototypes while the question of casting new instruments was deferred.[32] Thirteen years later the wooden models were reported completed. That report also cited an earlier memorial by Zhang (who had become bureau director in the interim), charging specifically that the armillary sphere had never been aligned properly and that the latitude setting of the simplified instrument was too low.[33] At this time the Ministry of Rites recommended that the wooden models Zhang had constructed should serve the bureau in actual use for a long time. Two decades later, in 1523, the record shows that repairs were made to some observatory instruments, including the simplified instrument, but does not specify what repairs or whether they were made to the bronze instrument or to Zhang's wooden substitutes.[34] The next report of the condition of the bureau's observatory and its instruments comes nearly a century later in a memorial advocating calendric system reform from the Ministry of Rites. This 1612 memorial paints a scene of advanced decay at the observatory, describing the instruments as out of alignment because the very structure on which they were mounted had become tilted and decrepit.[35] As with so many other memorials of the later Wanli reign, this report was kept within the palace, and no imperial response was issued and no action taken.

In the context of these complaints and this evidence of deficiencies, it hardly seems to matter how the instruments were calibrated. In a 1956 study Chen Zungui indicated that the rings of both the armillary sphere and the simplified instrument were marked in Chinese degrees but not whether the scales were further subdivided.[36] As a practical matter, it is unlikely that these instruments could have had scales divided into units much less than six minutes of arc. The practice in traditional Chinese astronomy was to divide celestial circles into 365¼ degrees (*du* 度), rather than 360 degrees; Chinese degrees are thus about 1.4 percent smaller than modern degrees. The size of these instruments (circles two meters in diameter) makes it reasonable to assume that their scales were divided down to tenths of a Chinese degree (*fen* 分; six modern minutes of arc) or an arc length of 3.44 mm at the circumference. This is between the scale accuracy of 10 minutes available to Copernicus and the accuracy of 1 minute attained by Tycho Brahe.[37]

[31] *Xianzong shilu,* 175, 3a, p. 3157.
[32] *Xiaozong shilu,* 31, 2b–3a, pp. 684–685.
[33] *Ibid.,* 182, 1b–2a, pp. 3346–3347.
[34] *Shizong shilu,* 31, 5a, p. 819.
[35] *Shenzong shilu,* 490, 1a–b, pp. 9219–9220.
[36] See Chen, *Qingchao tianwen yiqi jieshuo* (cit. n. 17), pp. 52–55.
[37] Allan Chapman, "The Accuracy of Angular Measuring Instruments Used in Astronomy between 1500 and 1850," *Journal of the History of Astronomy,* 1983, *14:*133–137, p. 134, Fig. 1.

This sketch of the Astronomical Bureau instruments suggests that it was never capable of particularly precise observations. Despite the imperial sanction under which it operated, the bureau labored under the burden of inaccurate, uncorrected instruments that were probably incapable of reliable absolute observations to better than a degree or so. However, more important than the bureau's failure to achieve a Tychonic level of observational accuracy is the issue of how well it was able to accomplish its assigned task with the resources at its disposal. Its mandated responsibilities in the area of anomaly observation did not require particularly quantitative observation: in the tradition and theory of anomaly observation and interpretation, locating a given event with a high degree of precision was in fact completely irrelevant. Relative and nonprecise measures were the normal expectation for observations of such phenomena. It was entirely adequate to say that a planet had intruded upon (*fan* 犯) a star or another planet, or that it had entered (*ru* 入) a lunar lodge. The notable exception entailed solar and lunar eclipses. Even here, precision was expected not so much in predicting or observing the location of the event in the heavens as in predicting its time, duration, and magnitude. Indeed, the relationship between successful prediction of eclipses and precise observations was not generally or consistently understood even by some bureau officials. Even as eclipse predictions began to fail more severely in mid dynasty, calls for reform of the calendric system used to make such predictions were slow to include the need for precise observations, and hence precise instruments. The political ideology that sanctioned an astronomical bureau also allowed the instruments provided it in the emperor's name to become marginal to the bureau's activity, since as imperial artifacts these instruments could not be modified without imperial authority.

IV. OBSERVATION OR FABRICATION

The significant ideological and rhetorical role of celestial anomalies in the politics of the Chinese state and the dubious quality of the instruments used in observation raise the question of whether the observations were manipulated or fabricated. The possibility that imperial Chinese astronomers fabricated observations was first raised by Wolfram Eberhard in 1933. Eberhard noted that a list of solar eclipses recorded in Han sources included many whose visibility in China was not verified by modern computation and lacked others whose visibility was. From this disparity he concluded that eclipses were reported or suppressed in order to make or withhold indirect criticism of the emperor.[38] In a later study, based on a loosely statistical analysis by Hans Beilenstein of celestial and other portents recorded in the *History of the Former Han Dynasty* and on his own examination of the Chinese calendar in the Han, Eberhard concluded that all "astronomy" and "astronomers" in Han China were controlled by purely political motives. This theory was largely adopted by Homer Dubs in the notes to his translations from the *History of the Former Han Dynasty.*[39] Although concerned with a much earlier time than dealt with in this

[38] Wolfram Eberhard, *Beiträge zur kosmologishen Spekulation der Chinesen der Han-Zeit* (Berlin, 1933), pp. 93–94.

[39] Wolfram Eberhard, "The Political Function of Astronomy and Astronomers in Han China," in *Chinese Thought and Institutions,* ed. John K. Fairbank (Comparative Studies of Cultures and Civilizations) (Chicago: Univ. Chicago Press, 1957), pp. 33–70, 345–352, on p. 70. See also Hans Bielenstein, "An Interpretation of the Portents in the Ts'ien-Han-shu, *Bulletin of the Museum of Far Eastern Antiquities,* 1950, *22:*127–143; and *History of the Former Han Dynasty by Pan Ku: A Critical*

article, and well refuted by Nathan Sivin in 1969,[40] Eberhard's characterization of Chinese astronomy lingers and calls into question the basic nature of observations reported by the astronomical bureau in all periods.

To assess whether the Ming Astronomical Bureau falsified its astronomical observations requires selecting an appropriate test set of celestial events. Eclipse records are not appropriate, since analyzing them for accuracy and completeness is complicated, in the Ming as in any other epoch of Chinese history, by the bureau's varying ability to predict eclipses as well as by the ideological basis that called for them to do so. More routine records of putative observations produced by the Astronomical Bureau can be easily tested by modern calculations to determine whether they were the result of actual observation. The notices of the planet Mercury contained in the *Veritable Records* seem ideal for several reasons. First, the criterion of testability rules out unpredictable, erratic celestial phenomena such as meteors and sunspots and leaves the more or less regular motions of the Sun, Moon, planets, and comets. Comets have several disadvantages: they are few in number, often of an unpredictable brightness, and difficult to verify with modern computations.[41] Of the Sun, Moon, and five planets visible with the naked eye, the planet Mercury exhibits behavior with several relevant properties. Since it is the closest planet to the Sun, it can be observed from Earth only with difficulty and for very limited times. Moreover, the orbit of Mercury, of all the visible planets, has the greatest inclination to the plane of the ecliptic, which means its possible path through the stars is relatively wide. These factors make an observation of Mercury a good one to fake both because such a faked report is extremely unlikely to be refuted by an untrained observer and because the putative fabricator has a fair range to choose from in designating a plausible star with which to associate the planet. Thus if the Astronomical Bureau habitually invented observations of the planets, an examination of its Mercury notices would reveal such fabrications. Mercury's elusive behavior has several other advantages for our examination. Unlike the situation with lunar and solar eclipses, the Chinese methods for predicting Mercury's position were not good enough for them to create consistent confusion between observed and predicted positions. The results of bureau computations for the location of Mercury could not be expected to be very close to its actual position much of the time for two reasons.[42] First, the Chinese

---

Translation with Annotations, ed. and trans. Homer H. Dubs with Phan Lo-Chi and Jen Thai, 3 vols. (Baltimore: Waverly, 1938), Vol. I, p. 90.

[40] Two problems with Eberhard's argument are worth brief mention. First, our sources for Han history are simply inadequate for deciding whether the data they give was manipulated. Second, the Ming Astronomical Bureau and its predecessors had good reasons, independent of making criticisms of the ruler, to overpredict and report solar eclipses. Already in the Han, lunar eclipse prediction methods generously overpredicted that phenomenon. See Nathan Sivin, *Cosmos and Computation in Early Chinese Mathematical Astronomy* (Leiden: E. J. Brill, 1969), also published as an article in *T'oung Pao,* 1969, *55:*1–73; see pp. 25–27.

[41] For Chinese comet records see the old, but still useful, list compiled by John Williams (1797–1874), *Observations of Comets from 611 B.C. to A.D. 1640 Extracted from the Chinese Annals* (London: Strangeways & Walden, 1871; rpt. Hornchurch: Science and Technology Publishers, 1987). For correlations with other observations and technical data see Mucke, *Helle Kometen −86 bis +1950* (cit. n. 10).

[42] For an overview of planetary position theories in Chinese mathematical astronomy see Yabuuchi Kiyoshi, "Chūgoku temmon ni okeru gosei undōron" (Theories of motion of the five planets in Chinese astronomy), *Tōhō gakuhō,* 1962, *26:*90–103. On the earliest theories see the excellent monograph of Michel Teboul, *Les premières théories planétaires chinoises* (Paris: Collège de France, Institute des Hautes Études Chinoises, 1983). For later advances, discussed in modern analytic terms,

were never compelled to develop a method for predicting planetary latitudes and so could only make informed guesses where Mercury was in a range of over four degrees on each side of the ecliptic. Second, of all visible planets, Mercury has by far the largest eccentricity in its orbit, making the prediction of its longitude a very tough problem, especially as it is difficult to observe with the naked eye.[43] Thus, so long as the recorded observations of Mercury made by the Astronomical Bureau can be shown to be in close accord with the planet's known position based on modern calculations, there will be no question that the basic notices preserved in the *Veritable Records* resulted from actual observations, not simple political fabrications.

Of fifty-two notices of bureau observations of Mercury in the *Veritable Records,* the large majority were, by modern calculation, well within a degree, and many within half a degree (or the diameter of the Moon), from the cited reference star or planet.[44] Thus, although the vast bulk of the planetary observations cited in the *Veritable Records* are more qualitative than quantitative, examination leaves no doubt, at least in the case of Mercury, that these notices reflect anything but actual observations. In only one case does the possibility that a Mercury notice was manipulated seem plausible: the record of 10 June 1404. The notice for this date is typical and reads as follows: "Evening, the Water Star (Mercury) intruded on the third star in the west of the Princes (*zhuwang* 諸王)."[45] The Chinese constellation of the Princes is six faint stars in a line just north of the ecliptic in Taurus. In modern nomenclature, these six are, from east to west, 136, 125, 118, 103, 99, and Tau Tauri. The "third star from the west" would then be 103 Tauri. However, modern computations show that Mercury at this time was south of the ecliptic by nearly two degrees and was much closer—about three-fourths of a degree of arc southwest—to Iota Tauri, the

see Li Changhao, ed., *Zhongguo tianwenxue shi* (The history of Chinese astronomy) (Beijing: Kexue, 1981), pp. 147–160.

[43] It was these difficulties which led even Ptolemy to construct an inaccurate if distinctive theory of Mercury's motion. See Otto Neugebauer, *A History of Ancient Mathematical Astronomy,* 3 pts. (Berlin/Heidelberg/New York: Springer-Verlag, 1975), Vol. I, pp. 158–169. See also Olaf Pedersen, *A Survey of the Almagest* (Odense: Odense Univ. Press, 1974), pp. 309–328; and, reading critically, to avoid the polemics, Robert R. Newton, *The Crime of Claudius Ptolemy,* (Baltimore/London: Johns Hopkins Univ. Press, 1977), pp. 257–299.

[44] There are probably more than the fifty-two notices of Mercury in the *Veritable Records,* since the Chinese graphs for Mercury (*shuixing* 水星) and those for Jupiter (*muxing* 木星) were on occasion confused by copyists. Five notices (not included in the fifty-two) in the Academia Sinica edition of the *Veritable Records,* for example, have Mercury written in error for Jupiter, as modern computations that show Jupiter, not Mercury, at the indicated location make clear. (For two instances—only—another edition had the correct reading.) Presumably some notices of Jupiter in the *Veritable Records* are in turn actually observations of Mercury, but the large number overall precluded identifying the erroneous ones at this time.

My computations of Mercury's position were made by third-order interpolation from the values in Bryant Tuckerman, *Planetary, Lunar, and Solar Positions A.D. 2 to A.D. 1649 at Five-day and Ten-day Intervals* (Philadelphia: American Philosophical Society, 1964). The conversion to equatorial coordinates and precession calculation to bring the values to the epoch 2000.0 were carried out with programs written by Andrew J. P. Maclean for the Hewlett-Packard HP-41C programmable calculator. To identify the modern designations of traditional Chinese stars and asterisms the following works were used: the excellent charts in Yi Shitong ed., *Quantian xingtu 2000.0* (Chinese-modern star charts for the epoch 2000.0) (Beijing: Tudi chubanshe, 1984), a refinement of his earlier work; Alexander Wylie's work, based on Qian Jiyue, "List of Fixed Stars," in Wylie, *Chinese Researches* (cit. n. 20), pp. 110–139; and the still valuable work of Gustave Schlegel, *Uranographie Chinoise* (Leiden: E. J. Brill, 1875; rpt. Taipei: Ch'eng-wen, 1967). I hope at a later date to make separate studies of the planetary and comet observations in the *Veritable Records.*

[45] *Taizong shilu,* 31, 1b, p. 556.

first star of the Chinese asterism Celestial Elevation (*tiangao* 天高), than to any of the stars of the Princes. No other naked-eye planet was closer to the reported location. Tau and Iota Tauri, which have magnitudes of 4.3 and 4.7 respectively, are the brightest stars in the Princes and Celestial Elevation. With such faint background stars the bureau observers may simply have misidentified the star which Mercury had "encroached" upon. Yet at the time the Yongle emperor was still in the process of legitimizing his insurrection and usurpation of the throne, and someone at some point in the observation-reporting procedure or historiographic process may conceivably have deliberately modified a correct record with the aim of creating heavenly support of the efforts to contain the power and influence of the other sons of Taizu, the emperor's competitors. However, there is no additional evidence that the record was created in this way.

This brief examination into the authenticity of the extant observational records of the Ming Astronomical Bureau therefore leads to the conclusion that the bureau's observations, at least in this period and in the case of Mercury, were conducted in a consistently objective manner.

V. CONCLUSION

The ideological position of astronomical activity in China, although sanctioning the Astronomical Bureau itself and equipping it with large astronomical instruments, did not ensure that the instruments were appropriate to more than the bureau's routine observational work. From poor initial implementation to minimal maintenance, the instruments provided to the Astronomical Bureau were well understood to be seriously deficient for observations more accurate than a degree of arc. As such, the large Ming astronomical instruments were more akin to the presentation instruments that grace museum collections than to the instruments of Tycho Brahe to which they have been compared. As with the development of the calendric systems used to predict eclipses, for astronomical instruments also imperial resources and willingness to engage in reform were most evident at the beginning of a dynasty, less so at the beginning of an individual emperor's reign, and almost never at other times, when such expenditures were not direct investments in legitimizing state and ruler.

It should not surprise us, however, that this situation did not prevent the bureau from conducting systematically consistent observations of various celestial phenomena. The form and content of Chinese astronomical activity was conditioned strongly by its context within the Ming Astronomical Bureau. However symbolic it had become, the bureau's mission was still fundamentally to legitimize the Emperor's position as Son of Heaven. This meant that grand instruments were required, that observations must be made and reported, and that the calendar should go forth as it always had. It meant also that despite their ultimate reliance on their instrumentation, the bureau's routine activities were not a basis from which successful action for improved instrumentation could be made. The ideological context of Chinese astronomy in the Ming limited the quality of instrumentation and its maintenance at the outset, yet provided a basis for sustained consistently accurate observational activity.

The Ocular Harpsichord of Louis-Bertrand Castel; or, The Instrument That Wasn't

*By Thomas L. Hankins**

IN HIS ACCOUNT OF THE GREAT CAT MASSACRE Robert Darnton brings to history a lesson learned from anthropology, that one can enter an unfamiliar culture most easily by studying those aspects that are most incomprehensible. From a bizarre massacre of cats by printers' apprentices in Paris during the 1730s Darnton explains the apprentices' life, their ceremonies, their behavior, their hatred for their master, and the peculiar significance of cats in their rituals. The apprentices found the torture of cats hilariously funny, while we, reading about it in the twentieth century, "don't get the joke." Precisely because we don't get the joke means that we have something to learn.[1]

Historians of science have traditionally ignored that which they do not "get." If an idea, book, organization, or instrument does not make sense from the perspective of twentieth-century science, it is ignored, and if it is found in the writings of someone we have learned to revere, it is regarded as downright embarrassing. The last fifteen years have seen a great change in this regard, and historians of science have learned that they cannot study what used to be called the "progressive element" of science in isolation without doing violence to history as a whole.

One problem with studying the unfamiliar in science is that we dissolve the disciplinary boundaries of our subject. We have no objective criterion by which we can say whether an instrument or idea is "scientific" or not. This is not altogether bad. By dissolving our own disciplinary boundaries, we can then ask the more important historical question of how the instrument or idea was regarded by its creator and by those who used it and how it fit *their* disciplinary boundaries.

"Philosophical" instruments like the telescope, microscope, and air pump were new in the seventeenth century and still carried the flavor of natural magic. As a result they were suspect and their value had to be demonstrated. The process of determining what was acceptable practice in natural philosophy also required a decision about what were acceptable instruments. And since the new instruments were radically different from the old ones and so important for the new experimental philosophy, the choice of instruments helped to define the philosophy.

* Department of History, DP-20, University of Washington, Seattle, Washington 98195.
Research for this project was supported by a grant from the National Endowment for the Humanities.

[1] Robert Darnton, *The Great Cat Massacre and Other Episodes in French Cultural History,* New York, Basic Books, 1984, pp. 75–104.

Figure 1. *The cat piano. From* La Nature, *1883, 2:519–520. Courtesy of the University of Washington Libraries.*

Not all instruments were accepted, of course. If they had been, we would be hard pressed to say what we mean by "natural science." The telescope, microscope, and barometer were big winners. The speaking tubes, magic glasses, and hydraulic fountains were losers. Of most interest to us as historians are those instruments that were, so to speak, "on the margin"—those instruments that caused confusion as to whether they were truly philosophical or not.

I. FROM CAT PIANO TO OCULAR HARPSICHORD

In keeping with Darnton's methodology and subject matter, we might want to look at the cat piano. Unfortunately the cat piano never got anywhere near the margins of acceptable science. As far as I can tell, this is the third mention of it in print since Athanasius Kircher wrote about it in his great *Musurgia universalis* of 1650, and one must earnestly hope that it was never actually built (see Figure 1). In order to raise the spirits of an Italian prince burdened by the cares of his position, a musician created for him a cat piano. The musician selected cats whose natural voices were at different pitches and arranged them in cages side by side, so that when a key on the piano was depressed, a mechanism drove a sharp spike into the appropriate cat's tail. The result was a melody of meows that became more vigorous as the cats became more desperate. Who could not help but laugh at such music?—and so was the prince raised from his melancholy.[2] The cat piano confirms Darnton's discovery

[2] According to several authors (see below), Athanasius Kircher described the cat piano in his *Musurgia universalis,* 2 vols, (Rome, 1650), facs. repr. (Hildesheim: Olms, 1970), but I have not been able to find it there. His pupil Caspar Schott described it in his *Magia universalis naturae et artis, sive recondita naturalium et artificialium rerum scientia,* 4 vols. (Würzburg, 1657–1659), Vol. II, pp. 372–373, attributing it to Kircher. It appeared again in the popular French journal *La Nature,* 1883, 2:519–520, described by a Dr. Z——, from which I have taken the account of its invention. The cat piano was not unique. Caspar Schott proposed a donkey chorus, and Pierre Bayle tells us that the Abbé de Beigne built a pig piano at the order of Louis XI. In every case the animal instrument was

that most early modern Europeans found the torture of cats funny. It also illustrates Kircher's fascination with the relationship between the art of music and the natural production of animal sounds. But for us it is an instrument that has mercifully been forgotten.

However, the cat piano did appear once during the eighteenth century in a place prominent enough to attract notice. Louis Bertrand Castel described it in 1725 in an article announcing his famous *clavecin oculaire* or ocular harpsichord. The ocular harpsichord was like a standard harpsichord except that it played colors instead of sounds. The possibility of such an instrument depended on the analogy between the seven spectral colors and the seven tones of the musical scale. He used the example of the cat piano to show that sound was not beautiful by itself and that the beauty of music lay only in the sequence and harmony of the notes. The cat piano might conceivably have produced a recognizable tune, but the effect would certainly not have been one of harmony. It was only a joke to illustrate Castel's important discovery. The optical harpsichord was a different matter. It would produce beautiful harmonies for the eye. According to Castel, it would be the "universal instrument of the senses."[3]

Whether the optical harpsichord was a scientific instrument or not depends on one's point of view. Castel claimed in his announcement that his harpsichord would not merely give a simple impressionistic idea of sound in color, but would really paint sounds by a precise and natural correspondence between color and pitch, so that a deaf listener could enjoy music that was originally written for the ear. He would demonstrate this correspondence following reasons of fact and geometrical analysis. He would accept only that which was proven.[4]

Reaction to Castel's announcement of the ocular harpsichord was not generally favorable, but it did cause considerable excitement, enough so that Castel could reasonably ask why his opponents were willing to spend so much time combating what they claimed was a worthless idea.[5] Part of the problem was Castel's indepen-

---

created to entertain a noble patron. See Pierre Bayle, *The Dictionary Historical and Critical,* 5 vols. (London, 1736), facs. repr. (New York: Garland, 1984), Vol. III, p. 803; and Isaac Nathan, *Musurgia vocalis,* 2nd ed. (London, 1836), p. 160. The cat piano occasioned a recent debate in *Experimental Musical Instruments,* 1989/90, *5*(5):6, and 1990/91 *6*(1):4; *6*(2):3, and *6*(5) 2.

[3] Louis-Bertrand Castel, "Clavecin pour les yeux, avec l'art de peindre les sons, et toutes sortes de pieces de musique, Lettre écrite de Paris le 20 Fevrier 1725 par le R. P. Castel, Jesuite, à M. Decourt, à Amiens," *Mercure de France,* Nov. 1725, pp. 2552–2577. The best study of Castel is Donald S. Schier, *Louis Bertrand Castel, Anti-Newtonian Scientist* (Cedar Rapids, Iowa: Torch Press, 1941). On Castel's ocular harpsichord see Anne-Marie Chouillet-Roche, "Le clavecin oculaire du P. Castel," *Dix-huitième siècle,* 1976, *8:*141–166; Albert Wellek, "Farbenharmonie und Farbenklavier: Ihre Entstehungsgeschichte im 18. Jahrhundert," *Archiv für die Gesamte Psychologie,* Aug./Dec. 1935, *94:*347–375; and, the most recent and most detailed, Maarten Franssen, "The Ocular Harpsichord of Louis-Bertrand Castel: The Science and Aesthetics of an Eighteenth-Century *Cause célèbre,*" *Tractrix,* 1991, *3:*15–77. Franssen discusses the importance of the ocular harpsichord for theories of aesthetics in the eighteenth century. I am grateful to him for sending me a preprint of his article.

[4] Castel credited Kircher, a fellow Jesuit and the author of the cat piano, with the idea of the ocular harpsichord. It is true that Kircher had not made a harpsichord nor had he found the exact correspondence between musical pitches and colors, but he had provided the "seed," the key analogy, from which the theory of the ocular harpsichord could be constructed: Castel, "Clavecin pour les yeux" (cit. n. 3), pp. 2553–2560. Kircher describes the analogy in his *Ars magna lucis et umbrae* (Rome, 1646), pp. 131–132. See also Kircher, *Musurgia* (cit. n. 2), Vol. II, pp. 567–568; and Kircher, *Phonurgia nova* (Kempten, 1673), preamble 1, fol. 6.

[5] Louis-Bertrand Castel, "Suite et sixième partie des nouvelles expériences d'optique et d'acoustique: Addressées à M. le Président de Montesquieu, par le Père Castel Jesuite," *Journal de Trevoux,*

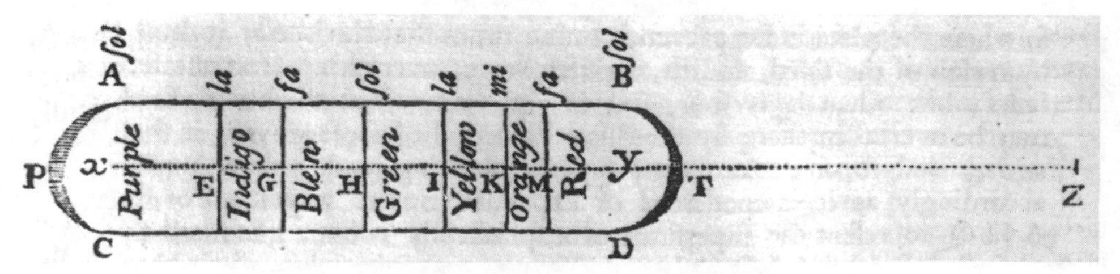

Figure 2. *Isaac Newton's illustration of the color-tone analogy. From Thomas Birch,* The History of the Royal Society of London *(London, 1757), Vol. III, p. 263. Courtesy of Special Collections, University of Washington Libraries.*

dence of mind, which led him to argue with everyone. Voltaire called him the "Dom-Guichotte des mathématiques" because of his tendency to attack the giants, including Newton, Leibniz, Réaumur, and Maupertuis.[6] Voltaire could have included Rameau, Rousseau, Dortous de Mairan, and Voltaire himself. That Castel should have warranted the attention of such illustrious foes is in itself remarkable.

Castel had joined the Jesuits as a novice in 1703 at age fifteen. In 1720 he came to the notice of Bernard de Fontenelle, who was instrumental in having him transferred from Toulouse, where he had been teaching rhetoric, to Paris, where his teaching expanded to include physics, infinitesimal calculus, mechanics, pyrotechnics, and architecture. In Paris he became the unofficial science editor for the Jesuit *Journal de Trevoux* and in this capacity wrote on every conceivable subject from the Northwest Passage to the squaring of the circle. In this he followed the tradition of the great Jesuit polymaths like Kircher, who admitted no limits to their breadth of knowledge.

He announced his ocular harpsichord in 1725 at the urging of the composer Jean-Philippe Rameau, who had been organist at Clermont when Castel taught there. The analogy between color and musical tone was by no means original with Castel. Newton had stated it very prominently, as had Kircher. Newton had studied musical harmony in 1664–1666 and throughout his life retained a belief in the *musica mundana,* or universal harmony of the world. His attention was called to the analogy between color and tone by Robert Hooke, who mentioned it in his criticism of Newton's first optical paper of 1672; Newton, in his second optical paper of 1675, did Hooke one better by showing that the seven bands of color in the spectrum have widths in the same harmonic ratios as the string lengths on the monochord that produced the musical scale (see Figure 2).[7] Because Newton also read Kircher, it is possible that Kircher was the source for Newton's analogy as Voltaire claimed, but it is also certain that Newton's supposed discovery of a new harmonic relation between the colors in

Dec. 1735, pp. 2642–2768, on p. 2654. (The formal name of this journal is *Mémoires pour l'histoire des sciences et des beaux arts,* but in the eighteenth century it was almost always referred to as the *Journal de Trevoux.*)

[6] Voltaire, "Lettre à Mr. Rameau, Mars 1738," *Correspondence,* ed. Theodore Besterman, 135 vols. (Genève: Institut et Musée Voltaire, 1953–1977), Vol. VII, app. 29, pp. 477–480, on p. 480.

[7] Penelope Gouk, "The Harmonic Roots of Newtonian Science," in *Let Newton Be!* ed. John Fauvel *et al.* (Oxford: Oxford Univ. Press, 1988), pp. 101–126. See also *The Optical Papers of Isaac Newton,* ed. Alan E. Shapiro, Vol. I, *The Optical Lectures, 1670–1672* (Cambridge: Cambridge Univ. Press, 1984), pp. 542–545.

the spectrum brought the color-tone analogy into prominence. Newton wrote: "As the harmony and discord of sounds proceed from the properties of the aerial vibrations, so may the harmony of certain colours . . . and the discord of others . . . proceed from the properties of the aetherial. And possibly color may be distinguished into its principal degrees, Red, Orange, Green, Blew, Indigo and deep Violet on the same ground, that sound within an eighth is graduated into tones."[8]

The most immediate stimulus for Castel was probably Nicolas de Malebranche, who in the sixteenth elucidation to his *Recherche de la Verité* referred specifically to the analogy between light and sound. Malebranche used Newton's experiments as evidence for his theory that both light and sound were caused by vibrations propagated in media composed of small vortices, and Castel adopted the same analogy of similar vibrations, although he repudiated Malebranche's little vortices.[9]

Castel's most important patron was Charles de Secondat Montesquieu, with whom he began correspondence soon after his arrival in Paris. For a while he had Montesquieu's son as a pupil at the Collège Louis le Grand and hoped through that contact to persuade Montesquieu to publish in the *Journal de Trevoux.* In 1735 he wrote an extremely long and verbose account of "new experiments on optics and acoustics" in the form of letters addressed to Montesquieu and published in the *Journal de Trevoux.*[10] Montesquieu's friendship was valuable to Castel, but it did not include any great enthusiasm for the ocular harpsichord.

The greatest boost for the ocular harpsichord came from Voltaire, who devoted Chapter 14 of his *Eléments de la philosophie de Newton* (1738) to the color-tone analogy and to Castel's instrument. Voltaire wrote that he believed Kircher to be the source for Newton's analogy between light and sound, and he praised Kircher as "one of the greatest mathematicians and most learned men of his times." Kircher had argued entirely by analogy, and Voltaire favored instead Newton's experimental method. Yet even Voltaire was willing to admit that "this secret analogy between light and sound leads one to suspect that all things in nature have hidden connections, that perhaps will be discovered some day."[11] In spite of his sympathy (limited, to be sure) for Castel's ideas, Voltaire quarreled with him and took revenge by attacking him in the public "Letter to Rameau," in which he also ridiculed the ocular harpsichord.[12] What disturbed Voltaire was not the idea of an ocular harpsichord (after all, Newton had given serious attention to the color-tone analogy) so much as Castel's style of inquiry: unlike Newton, he employed analogy in place of

[8] *The Correspondence of Isaac Newton,* ed. H. W. Turnbull *et al.,* 7 vols. (Cambridge: Cambridge Univ. Press, 1959–1977), Vol. I, p. 376; quoted from Gouk, "Harmonic Roots," p. 118. Newton's statement of the analogy between the spectrum and the octave caused some confusion because he employed an old system of solmization rather than the more recent system of equal temperament. Note on the illustration that Newton repeats "Sol" at the fifth and at the octave. See Wellek, "Farbenharmonie und Farbenklavier" (cit. n. 3), pp. 351–353.

[9] Nicolas de Malebranche, *The Search after Truth,* trans. Thomas M. Lennon and Paul J. Olscamp, and *Elucidations of the Search after Truth,* trans. Lennon (Columbus: Ohio State Univ. Press, 1980), pp. 686–718.

[10] Louis-Bertrand Castel, "Nouvelle expériences d'optique et d'acoustique," *J. Trevoux,* 1735, Aug., pp. 1444–1482; the various "suites" appeared *ibid.,* 1619–1666; Sept., pp. 1807–1839; Oct., pp. 2018–2053; Nov., pp. 2335–2372; and Dec., pp. 2642–2768 (this last cit. n. 5).

[11] Voltaire, *Oeuvres complètes,* ed. Louis Moland, 52 vols. (Paris: Garnier, 1877–1885), Vol. XXII, pp. 503–507, on pp. 503, 505.

[12] Voltaire to Maupertuis, 15 June 1738, *Correspondence* (cit. n. 6), Vol. VII, letter 1454; and Voltaire,"Lettre à Mr. Rameau" (cit. n. 6). See also Chouillet-Roche, "Le clavecin oculaire" (cit. n. 3), p. 162.

experiment. And even though he may not have been able to follow all of Newton's mathematical arguments, Voltaire understood Newton's style and method as well as anyone in France. He concluded that Castel's style did not sufficiently grasp "the spirit of this century."[13] Castel was no child of the Enlightenment.

Others examined directly the analogy between color and tone. Jean-Jacques Dortous de Mairan criticized Castel's ideas in 1737. In 1739 the composer Georg Philipp Telemann wrote *Beschreibung der Augen-orgel, oder des Augen-clavicimbels,* based on his observations of the instrument during his visit to Paris in 1737–1738. In 1742 the Saint Petersburg Academy also devoted a seance to the ocular harpsichord, at which Georg Krafft expressed his doubts about the usefulness of the analogy.[14] Even Jean-Jacques Rousseau, who befriended Castel in 1741, had no use for the instrument.[15] Thus one can conclude that Castel's ocular harpsichord received plenty of attention, but only limited acceptance.

The *philosophe* most willing to give serious consideration to Castel's invention was Denis Diderot, who found in it a natural theme for his *Lettre sur les sourds et muets* (1751). When Diderot's imagined deaf mute sees Castel's machine, he thinks the colors are a form of speech and concludes that the inventor must have been a deaf mute, too. Diderot's interest in the formation of the senses meant that he would take the color-tone analogy seriously, but in the *Encyclopédie* he joined the chorus of those urging Castel to make the instrument and demonstrate the harmony of colors directly rather than talking about it interminably.[16]

One would expect that having conceived of an instrument to exploit the analogy between color and tone, Castel would have been eager to make the instrument or have it made. This was not the case, however, and there is reason to doubt that a working ocular harpsichord was ever made during Castel's lifetime—by him or by anyone else.

Part of the problem was the technical difficulty of making such an instrument in the eighteenth century. In 1730 Castel had exhibited some kind of device, but appar-

[13] Voltaire to Thieriot, 7 Aug. 1738, *Correspondence* (cit. n. 6), Vol. VII, letter 1509.

[14] Jean-Jacques Dortous de Mairan, "Sur la propagation du son dans les différents tons qui le modifient, IV: En quoi l'analogie du son et de la lumière, des tons et des couleurs, de la musique et de la peinture, est imparfaite, ou nulle," *Mémoires de l'Académie Royale des Sciences, Paris,* 1737, pp. 34–45; Georg Philipp Telemann, *Beschreibung der Augen-orgel oder des Augen-clavicimbels* (Hamburg, 1739), repr. in Lorenz Christoph Mitzler von Kolof, *Musikalische Bibliothek, oder gründliche Nachricht von alten und neuen musikalischen Schrifften und Büchern . . . ,* 4 vols. (Leipzig, 1739–1754), Vol. II, pp. 262–266.; and *Sermones in solemni academiae scientiarum imperialis . . .* (St. Petersburg, 1742). Many of Castel's critics had reasons other than the ocular harpsichord for opposing him. Dortous de Mairan was protecting the Academy of Sciences, some members of which Castel had criticized.

[15] Rousseau called the color-sound analogy false in his *Essai sur l'origine des langues* (written 1749, published 1781); and Castel responded by attacking Rousseau's music theory in his *Lettres d'un académicien de Bordeaux sur le fonds de la musique à l'occasion de la lettre de M. R*** contre la musique françoise* (1754), and *L'homme moral opposé à l'homme physique de M. R.**** (1756). In his *Confessions* Rousseau called Castel "fou mais bonhomme au demeurant"; see Chouillet-Roche, "Le clavecin oculaire" (cit. n. 3), p. 165.

[16] Denis Diderot, "Lettre sur les sourds et muets," *Oeuvres complètes,* ed. Herbert Dieckmann, Jean Fabre, Jacques Proust, and Jean Varloot (Paris: Hermann, 1975–), Vol. IV, p. 145; and Diderot, "Clavecin oculaire," *Encyclopédie, ou Dictionnaire raisonné des sciences,* 35 vols. (Paris, 1751–1780), Vol. III, pp. 511a–512a. "Le facture de cet instrument est si extraordinaire, qu'il n'y a que le public peu éclairé qui puisse se plaindre qu'il se fasse toujours et qu'il ne s'achève point." Diderot also included Castel and the ocular harpsichord in his *Bijoux indiscrets* (1747). Castel was "un certaine brame noir, fort original, moitié sensé, moitié fou" whose writings were a "tissu de rêveries." See Chouillet-Roche, "Le clavecin oculaire" (cit. n. 3), p. 164.

ently all it did was raise colored slips of paper into view.[17] Supposedly this modest instrument created so much excitement in Paris that Castel was obliged to close his rooms to visitors and postpone his efforts. On 21 December 1734, with much fanfare, he demonstrated a more advanced instrument, but admitted that it was "only a model and therefore very imperfect."[18] His anonymous English assistant later made an instrument which he demonstrated in London after Castel's death. This harpsichord contained five hundred lamps (probably candles) and must have given off a prodigious quantity of heat. That is probably why a manuscript note attached to the description of the English ocular harpsichord says that it was there to be observed in Soho, but was never played.[19] All descriptions of the instrument during Castel's lifetime are distressingly vague. He had no problem obtaining support for his invention. The Prince de Conti offered his support, and Castel actually accepted two thousand livres from Comte Maillebois and a thousand crowns from the Duke of Huescar, the Spanish ambassador.[20] With this money Castel was able to employ workmen to help with the construction, but their efforts came to naught.

Yet even aside from the technical difficulty of building an ocular harpsichord, Castel seemed to have had no desire to build the instrument in the first place. His response to critics after he announced his harpsichord in 1725 was, "I am a mathematician, a philosopher . . . and I have no desire to make myself into a bricklayer in order to create examples of architecture."[21] For Castel the idea and not the artifact was what counted. It apparently did not occur to him that one might construct an instrument for the purpose of testing a theory. We are confronted here with a thoroughly unfamiliar approach to the natural world, one that we could easily dismiss as unfruitful and therefore unimportant. But Castel's disinclination to make the ocular harpsichord demands an explanation, and it is our task to try to understand it.

II. THE HARPSICHORD AS THOUGHT EXPERIMENT

The ocular harpsichord was a kind of "thought experiment," a realization of an idea in an imagined instrument. Castel claimed that even if he did actually construct an instrument, it would not and could not settle whether there was a real analogy between light and sound. As he explained to Montesquieu, the public clamor to see the ocular harpsichord was misguided. Montesquieu would understand that it was nobler and more scientific to approach the problem through the mind than through the senses. And it would not be possible to judge color harmony immediately from the ocular harpsichord in any case. Castel insisted that one had to become accustomed to any kind of music to appreciate it. "One has to learn to appreciate even Homer."[22]

The cat piano can assist us again in understanding Castel's argument. In his letters

[17] *Explanation of the Ocular Harpsichord upon Shew to the Public* (London, 1757), pp. 2–3; and Chouillet-Roche, "Le clavecin oculaire," pp. 156, 166.

[18] Castel, "Suite et sixième partie des nouvelles expériences" (cit. n. 5), p. 2722.

[19] Chouillet-Roche, "Le clavecin oculaire" (cit. n. 3), p. 158.

[20] Schier, *Louis-Bertrand Castel* (cit. n. 3), pp. 179, 183.

[21] Louis-Bertrand Castel, "Difficultez sur le clavecin oculaire, avec leurs reponses," *Mercure de France,* March 1726, p. 455.

[22] Louis-Bertrand Castel, "Suite et séconde partie des nouvelles expériences d'optique et d'acoustique adressées à M. le Président de Montesquieu," *J. Trevoux,* Aug. 1735, pp. 1619–1666, on pp. 1620, 1624.

to Montesquieu he claims that animals cannot create or appreciate music; the sounds they make are only cries. Therefore the cat piano is a product of human art, not cat art, and it produces music only to the extent that the sounds are controlled by the human playing it. Animals cannot make music at all. Music can only be created and appreciated by the human *mind;* it will not "make sense" to the senses alone.[23] Nor will just any mind appreciate just any kind of music. Only a mind prepared by previous experience can respond to a new kind of music or instrument. Castel uses the quarrel over the relative superiority of French and Italian music to illustrate this last argument, asserting that French music portrayed the French character in a unique way, and that a Frenchman could not immediately appreciate Italian music.[24]

This characteristic of music leads Castel to argue that the ocular harpsichord is artificial, even though it is based on a real analogy in nature. In fact, he argues that just as the best fruits and flowers are the product of the art of agriculture, so is the best music the most artificial. The less natural the ocular harpsichord is the better:

> All of which leads me to say: 1. That the more color-music is refined, artificial, scientific even, that is, nonhabitual, the more beautiful and agreeable it will be, not at first, but *col balsamo di costume;* and thus 2. I must attempt to make it known to the taste, to the mind, to the reason, to the internal sense in order to make it felt by the external sense, the eye.[25]

Of course the reality of the color-tone analogy must not be denied. It exists in nature, but it must be revealed to the mind before it can be appreciated by the senses.

Moreover, the purpose of an instrument like the ocular harpsichord is not to test a theory or to produce a new idea. Physics is the subject of our everyday experience: "Everyone is a bit of a physicist to the extent that he has an attentive mind capable of natural reasoning." Castel bases his physics "not on arbitrary hypothesis or particular and personal experience, but uniquely on history and on the general observation of nature and art."[26] Therefore an instrument in physics has the purpose of confirming what we already know to be true from reason acting on our general experience. It cannot by itself be the basis for constructing a theory that generalizes beyond the single phenomenon that it produces.[27]

This conception of the role of an instrument explains in part Castel's hostility to Newton. Newton's prism experiments are entirely different from Castel's ocular harpsichord. They rest on an *experimentum crucis,* a single test of a single idea. The ocular harpsichord on the other hand, illustrates an analogy understood from general experience. It is not surprising, then, that Castel dislikes Newton's prism: "I distrust the prism and its fantastic spectrum. I regard it as an art of enchantment, as an unfaithful mirror of nature, more proper by its brilliance to create flights of imagination and to serve error than to nourish minds solidly and to draw obscure truth from

[23] Castel, "Suite et cinquième partie des nouvelles expériences . . . ," *J. Trevoux,* Nov. 1735, pp. 2335–2372, on pp. 2350–2351.

[24] Castel, "Suite et séconde partie des nouvelles expériences" (cit. n. 22), p. 1622.

[25] *Ibid.,* p. 1625.

[26] Louis-Bertrand Castel, *Le vrai système de physique générale de M. Isaac Newton . . .* (Paris, 1743), p. 6; and Castel, *L'optique des couleurs, fondée sur les simples observations, et tournée surtout à la pratique de la peinture, de la teinture et des autres arts coloristes* (Paris, 1740), p. 375.

[27] Castel illustrates the methods described by Peter Dear, "Jesuit Mathematical Science and the Reconstitution of Experience in the Early Seventeenth Century," *Studies in History and Philosophy of Science,* 1987, *18:*133–175.

deep wells." The prism is the apparatus of the imposter and the instrument of the "spectre magique."[28]

Castel asks what right the prism has to credence. Does its geometrical shape prove that the colors coming from it are primitive? Why are the colors produced by the prism any more fundamental than the colors of the tricolor flower? Besides, the prism provided Newton with only a single unique fact, that is, the dispersion of the colored rays, from which Newton constructed his entire theory. But a unique fact is a monstrosity, a single event, from which no general conclusion or universal theory can be drawn. "My philosophy . . . considers only facts, but facts that are natural, daily occurring, constant, and a thousand times repeated, habitual facts rather than facts of the moment, facts of humanity rather than facts of one man. A unique fact is a monstrous fact."[29] Castel uses the word *fact* here partly in its original Latin meaning of something made or done. Thus the validity of a "fact" depends on testimony of observers and on the veracity of the person claiming the fact. Because he depends on facts, Newton makes the error of turning effects into causes, phenomena into principles, and experiments into explications. While Castel does not doubt the experiments that Newton describes, he dislikes his tendency to claim as "fact"—that is, as a deed—what is only an interpretation of a phenomenon. Moreover, Newton's jargon is meaningless. His notion of a "ray" of light makes no more sense than his notions of "attraction" and "gravitation."[30]

Newton is imperious and his followers are far more dogmatic than the Cartesians. They have to accept his arguments without question, because Newton's arguments demand complete assent or complete denial. At least Descartes was modest enough to realize that his system of the world was a hypothesis that could be modified by subsequent reasoning and experience. Descartes's hypotheses have flexibility. They are intelligible and his followers can reason with him.[31] But not the Newtonians. They are not allowed to question. Newton transforms his readers into spectators, not participants. The Newtonians claim that they present "facts," not hypotheses, and argue that the facts cannot be denied. This makes them totally unyielding. Reasoning does not force consent, but facts do, and only God can claim facts. The method of facts is emphatic and disdainful. It leads only to occult qualities and error.[32] It is a mistake to claim that a system contains absolute truth.

Newton's system also is difficult and inaccessible. His experiments require that he remove himself from the natural world and enter an artificial world of prisms and rays. There is no need to shut oneself up in a camera obscura in order to understand light. Nature is everywhere and reveals itself constantly to our senses. The rainbow appears in the presence of the sun.[33]

[28] Castel, *L'optique des couleurs* (cit. n. 26), pp. 376, 393.

[29] "Projet d'impression," MS 15747, Bibliotheca Hulthemiana, Royal Library, Brussels, p. 10, quoted from Schier (cit. n. 3), p. 107; and Castel, *L'optique des couleurs,* p. 403.

[30] Castel, *Le vrai système de Newton* (cit. n. 26), pp. 450, 6, 456.

[31] *Ibid.,* pp. 10–13. D'Alembert argued just the opposite—that the lack of flexibility in Newton's system was its greatest strength: Jean d'Alembert, "Elémens de philosophie," *Mélanges de littérature, d'histoire et de philosophie,* 5 vols. (Amsterdam, 1770), Vol. IV, p. 231.

[32] Castel, *Le vrai système de Newton,* 442, 476; and Castel, *L'optique des couleurs* (cit. n. 26), pp. 410–414. Castel here registers his particular dislike of the British form of scientific rhetoric that Shapin and Schaffer call "virtual witnessing." See Steven Shapin and Simon Schaffer, *Leviathan and the Air-Pump: Hobbes, Boyle, and the Experimental Life* (Princeton: Princeton Univ. Press, 1985), Ch. 5.

[33] Castel, *L'optique des couleurs* (cit. n. 26), p. 488.

In these criticisms we recognize an attitude towards instruments that preceded what we call the Scientific Revolution. Experiment had value only to the extent that it confirmed experience, and reason naturally preceded experiment, so that the necessity of an experimental test could only be regarded as a sign of defeat. There could be no crucial experiment, because a crucial experiment was only a single instance, a monstrous event.[34] In fact, experiment should be the last resort of the natural philosopher, not the first step of an investigation, as Newton had argued.

III. THE HARPSICHORD AS RHETORIC

The rhetorical form of argument was also important for Castel. He followed his 1725 announcement of the ocular harpsichord by a "geometrical demonstration" of it. This geometrical demonstration, however, did not contain any geometry as such. It was a set of propositions, followed by demonstrations with an occasional scholium thrown in for good measure, and it could be called geometrical only because it discussed musical harmony, a subject that was traditionally part of mathematics. As Peter Dear has shown, it was characteristic of Jesuit scientists after Christoph Clavius to use the form of a geometrical proof in order to give universality to experiential statements, and this was obviously Castel's purpose.[35] He wrote that in his first publication he merely stated the question and proposed the possibility of an ocular harpsichord. In his second article he wanted to extend his demonstration to all the senses, because "a discovery which is fecund ought always to move forward into a new order." An important discovery cannot exist alone, because it will always lead to more discoveries "as one harvest provides the seeds for a new harvest." In fact, Castel claimed that the ocular harpsichord, or at least the *idea* of the ocular harpsichord, would become the universal instrument of the senses, and poets would discover in it a complete *musurgie* that would account *a priori* for "all sounds, tones, accords, dissonances and, what has never yet been attempted, for the pleasure of all things." By casting his argument in geometrical form, Castel generalized it and extended it to all the senses.[36]

Castel's use of geometry was obviously different from Newton's use of it. Castel called Newton an excellent geometer, but a poor physicist (a criticism that he probably borrowed from Malebranche), and while he praised the geometrical method, he criticized Newton for overreliance on mathematics in his optics. In applying mathematics to physics, Castel ascribed the greatest value not to theorems and calculations, but to the logical form and the generalizing power of geometry.

Castel claims that analogy is the basis for discovery in natural philosophy and that analogy reveals important connections between science, art, and literature. While there may be many arts and sciences, there is only one truth, which the arts and sciences express from different points of view. In particular, philosophy and poetry have the same object, the same nature and the same truth—a sublime thought in poetry is equivalent to a discovery in natural philosophy. Therefore analogy is crucial for making the transition from one expression of truth to another: "Now it is

[34] Dear, "Jesuit Mathematical Science" (cit. n. 27), pp. 145–146.

[35] *Ibid.,* p. 142.

[36] Louis-Bertrand Castel, "Démonstration géométrique du clavecin pour les yeux et pour tous les sens, avec l'éclaircissement de quelques difficultez, et deux nouvelles observations," *Mercure de France,* Feb. 1726, pp. 277–292, quoting from pp. 277, 287, 291.

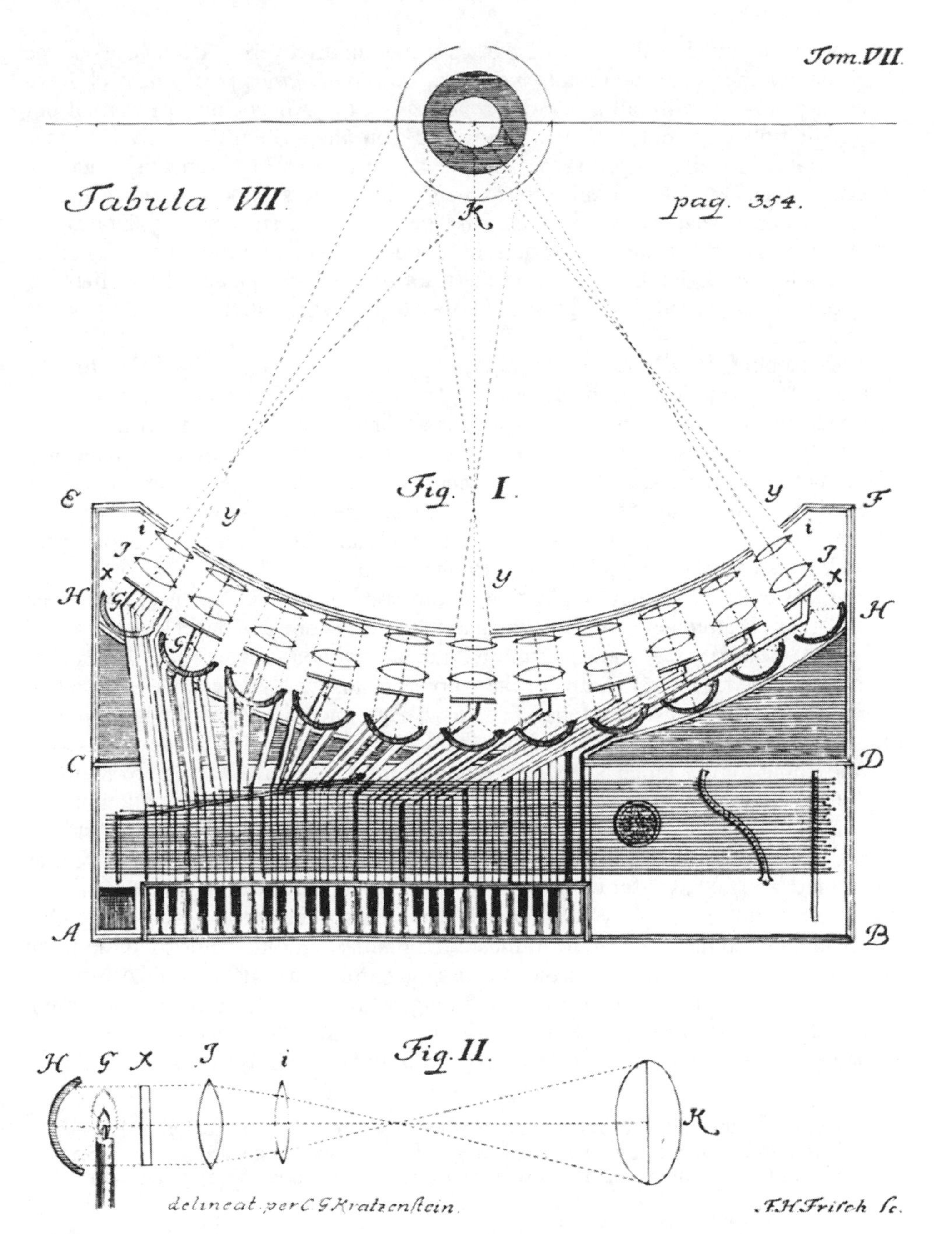

Figure 3. *Castel left no illustration of his ocular harpsichord. The instrument depicted here is a variant proposed by Johannes Gottlob Kruger in his "De novo musices quo oculi delectantur genere,"* Miscellanea Berolinensia, *1743, 7:354. Courtesy of Special Collections, University of Washington Libraries.*

analogy which renders these poetic flashes fecund in discoveries. Because what one calls among the poets and orators *metaphor, similitude, allegory, figure;* a philosopher, a geometer will call *analogy, proportion, ratio.* All our discoveries, all our scientific truths, are only truths of ratio. And from there often the figurative sense degenerates into the proper sense and the figure into reality" (emphasis added). Castel gives as his rule the following. When he encounters a poetic or other literary statement about nature that is especially beautiful and sublime, he applies the method of geometrical analysis, that is, he assumes it to be true and sees what consequences he can derive from it. From the truth of the consequences he verifies the original statement, and if he is persuaded of its truth, he then attempts to demonstrate it to others.[37]

For example, Virgil "paints the night" when he writes *rebus nox abstulit atra colores* ("black night took the color away from things"). The sublimity of this expression lies not in tropes, figures, allegories, or metaphors, but in its truth. It is Descartes who has shown that because colors are only modifications of light, they cannot exist in the dark, and therefore when the night chases the light it also chases the colors. "This thought of Virgil has all the character of the sublime, of the grand, of the beautiful, being in the first place true, and in addition new, marvellous, profound, paradoxical even, and contrary to our presumption."[38]

For Castel the aesthetically pleasing and the rational are the same. It is also the basis of his disagreement with Newton about the color-tone analogy. Newton associated the seven colors of the spectrum with the seven notes of the musical scale by comparing the measured widths of the colored bands with the lengths of vibrating strings that sound consonant tones, but Castel's argument is very different: "Among the colors, violet is a sad color and one that takes much from black, being the color of mourning for our kings and for the Church. . . . violet is the passage from affliction to joy; the rainbow is a sign of joy, but of a joy which follows an affliction, and to which the affliction serves as a contrast and as a base."[39] Therefore violet should serve as the base for the color scale. Later, however, Castel decided that blue was the "fundamental bass" for color harmony, because the study of dyes and pigments convinced him that there were three primary colors—red, yellow, and blue—that corresponded to the major triad in music. Beginning with his color triad, he filled in the rest of the colors to create the twelve-note chromatic scale. Nor did he hesitate to point out that these twelve tones had long been called "chromatic," indicating that musicians had recognized the analogy between color and tone long before the seventeenth century—and that the analogy could not be purely verbal:

> But why is this scientific system of half-tones called chromatic and colored? It is doubtless a metaphor, a comparison, an analogy of discourse, and consequently, it seems to me, of thought, of reasoning, of science. Because in the arts above all, and in the sci-

[37] Louis-Bertrand Castel, "Reflexions sur la nature et la source du sublime dans le discours: Sur le vrai philosophique de discours poétique, et sur l'analogie qui est la clef des découvertes," *Mercure de France,* June 1733, pp. 1309–1322, on pp. 1320 (quotation), 1321.

[38] *Ibid.,* pp. 1311–1312. Castel raises a significant problem that had challenged natural philosophers since Aristotle. See Henry Guerlac, "Can There Be Colors in the Dark? Physical Color Theory before Newton," *Journal of the History of Ideas,* 1986, *47:*3–20.

[39] "Journal . . . de la pratique et exécution du clavecin des couleurs," MS. 15746, Royal Library, Brussels, p. 50, quoted from Schier, *Louis-Bertrand Castel* (cit. n. 3), p. 100.

> ences, there is no affected term, [no conceit] which does not express an idea, and is often the result of several truths and an implied theory.[40]

Not only does analogy serve as a means of scientific discovery; it is also a valuable rhetorical tool. This is because any new truth, and especially a scientific discovery, is shocking and revolts the reader. It should be enclosed in a rhetorical "envelope" that conceals the full harshness of the new truth, piques the curiosity of the reader, and provides only analogies to the new idea. This was Descartes's error. He should have presented his ideas in poetry and allowed the commentators to reveal his principles in full light.[41] While Descartes's style was too direct, Newton's was even worse, and the plain declarative style of writing so favored by the British philosophers was, for Castel, a detriment to the proper pursuit of science.

We can now understand why it was a Castel and not a Newton who came up with the idea of an ocular harpsichord, why Athanasius Kircher was Castel's hero, why Voltaire changed his mind so completely when he learned what was behind the instrument, and why the ocular harpsichord was destined to remain a "marginal" scientific instrument. It was an instrument perfectly suited to Castel's way of studying the natural world. It was based on analogy, the analogy between color and tone, and it connected the aesthetic with the rational. Castel argued that man inhabited an artificial world intermediate between the supernatural and the natural and that as an artificer he was a mediary between God and nature.[42] Instruments like the ocular harpsichord are one way that man can illustrate the hidden analogies that rule nature.

IV. THE HARPSICHORD AFTER CASTEL

Castel's optical harpsichord had much in common with other instruments in the natural magic tradition that combined aesthetics, entertainment, and natural philosophy in a single apparatus. Electrical instruments before 1780 had much the same character. They did not measure anything and were designed to elicit wonder in the spectator. During the last quarter of the eighteenth century, when instruments became much more quantitative, the ocular harpsichord became increasingly irrelevant to most natural philosophers.

One natural philosopher, however, did advance arguments on color similar to those of Castel: Johann Wolfgang Goethe. Most striking is Goethe's attack on Newton's color theory. Both he and Castel insist that color is a modification of white light caused by the interaction of light and dark. Both argue that the spectrum observed by Newton does not occur at all distances from the prism and therefore that Newton was looking at a special case. Both claim that Newton's prismatic colors were produced by modification of the edges of a beam of white light, and that green is not a primary color, but a mixture of blue and yellow rays coming from the edges of the white beam. Castel compares the white light "shattered" by the action of the prism—splintered into colors when bent by it—to a wooden rod that splinters when it is bent.[43]

[40] Castel, "Nouvelles expériences d'optique et d'acoustique" (cit. n. 10), pp. 1458–1459. Castel attributes this method to Kircher.

[41] Castel, "Réflexions sur la nature" (cit. n. 37), pp. 1318–1319.

[42] Louis-Bertrand Castel, "Lettre à M. C*** (sur l'existence d'un milieu entre le naturel et le surnaturel, qu'il appelle artificiel)," *J. Trevoux,* Dec. 1722, pp. 2072–2097.

[43] Castel, *Le vrai système de Newton* (cit. n. 26), pp. 447–448.

Even more striking is the similarity in Castel and Goethe's criticisms of Newton's method. Both locate the error of Newton's method in his rhetorical style. Both argue against the authority of fact as Newton uses it, Goethe accusing Newton of "insufferable arrogance," and both claim that Newton's arguments assume what they set out to prove. Both deny the validity of a single experiment and both argue that only a collection of observations will lead to an understanding of the phenomena.[44]

Both Castel and Goethe insist on the subjective nature of experiment, Goethe going so far as to argue that "insofar as he makes use of his healthy senses, man himself is the best and most exact scientific instrument possible" and that artificial instruments that set nature apart from man are a great misfortune for physics. Not surprisingly, both Goethe and Castel criticize Newton's abstract concept of a"ray," and both approach the phenomenon of color through the study of pigments and dyes, not the "adventitious" colors produced by the prism. Goethe does not like the prism any more than Castel and insists that nature can never be understood by subjecting her to torture.[45]

But Goethe did not share Castel's enthusiasm for the ocular harpsichord. One might expect that Castel's desire to find a truth that transcends both science and poetry and gives validity to both would appeal to Goethe as well.[46] While he sympathized with much of Castel's theory of color, Goethe criticized Castel's excessive use of analogy, and since analogy was at the root of the entire concept of an ocular harpsichord, Goethe could not accept it.[47] Because he employed analogy willingly in his own natural philosophy, we can only conclude that he was not opposed to analogy as such, but only to the kind of analogies employed by Castel. Goethe had of course read the criticisms of the color-tone analogy by Voltaire, Dortous de Mairan, and Krafft and knew that it had few supporters, but it is likely that Goethe's criticism reflected less the opinions of others than a feeling that Castel's method represented an outdated, naive, and undisciplined search for cosmic harmony that ignored any close study of natural phenomena.[48]

Castel had frankly admitted that he did not like bothering with details and that he preferred to grasp the truth by generalizing from daily experience. Goethe, on the other hand, described his own method in natural philosophy as "concrete thinking" (*gegenstandliches Denken*). "My thinking does not separate itself from concrete objects; . . . the elements of the objects or rather my perception of them, enter into my thinking and are most intimately penetrated by it; and . . . my perception itself is thinking, my thinking perception."[49] Goethe was a close observer who worried

[44] Johann Wolfgang von Goethe, *Scientific Studies,* ed. and trans. Douglas Miller (New York: Suhrkamp, 1988), pp. xvi, 14. See Neil M. Ribe, "Goethe's Critique of Newton: A Reconsideration," *Studies in History and Philosophy of Science,* 1985, *16:*315–335, on p. 324.

[45] Goethe, *Scientific Studies,* pp. xvi, 167, 200, 307.

[46] *Ibid.,* p. 276.

[47] Johann Wolfgang von Goethe, *Die Schriften zur Naturwissenschaft,* hrsg. im Auftrage der Deutschen Akademie der Naturforscher zu Halle (Weimar: H. Böhlaus, 1947–), Part I, Vol. IV, p. 329. On the ocular harpsichord see Part II, Vol. VI, pp. 199–204.

[48] Goethe was prepared to describe Castel as an "ingenious man" who, though not one of the first figures of his time, was at least "one of the most distinguished minds of his nation" in spite of the fact that his writing style was "long-winded, nit-picking, and prolix." Goethe, *Schriften zur Naturwissenschaft,* Part I, Vol. VI, pp. 328, 333.

[49] Castel, *Le vrai système de Newton* (cit. n. 26), p. 499; and Goethe, quoted from Walter D. Wetzels, "Art and Science: Organicism and Goethe's Classical Aesthetics," in *Approaches to Organic Form: Permutations in Science and Culture,* ed. Frederick Burwick (Dordrecht: Reidel, 1987), p. 76.

very much about the details. In fact, he described subjective color phenomena like "colored shadows" and afterimages better than anyone before him.[50] He did not use complex apparatus in his experiments, and he denied the possibility of an *experimentum crucis,* but his hostility to Newton did not mean that he neglected experiment. Castel's method of grasping at analogies without worrying about "the details" could only have exasperated Goethe. The ocular harpsichord was a product of this unsatisfactory method. It was also an artifice, an artificial way to create an analogy, which, if it truly existed in nature, should be evident without a complex mechanism.

Although Goethe repudiated the color-tone analogy, it became an important theme in romanticism as an example of synesthesia, the substitution of one sense for another. Poetry, music, and painting all employed the analogy during the nineteenth century, but in a very different way from that used by Newton and Castel. Castel's ocular harpsichord depended on a precise correspondence between color and tone. A particular color corresponded to a particular musical pitch, not to a mood or emotion. While the precise correspondence claimed by Castel still held for those individuals who were "synaesthetic," that is, who could find a given pitch by associating it with a given color, the color-tone analogy as it was used by the romantics usually associated color with the mood of the music and not its pitch.[51]

The ocular harpsichord did continue to suggest itself to inventors after Castel, most of whom reinvented the instrument and discovered afterwards that it had been suggested long before. On 6 June 1895 Alexander Wallace Rimington performed on his great color organ for the first time at St. James Hall in London, and Alexander Scriabin's symphony *Prometheus* (1911), which has a part written especially for a color organ, continues to be performed.[52] Thomas Wilfred toured the United States and Europe in the 1920s with his clavilux, a modern ocular harpsichord. Performances of this sort led Albert Michelson to exclaim in *Light Waves and Their Uses:*

> Indeed, so strongly do these color phenomena appeal to me that I venture to predict that in the not very distant future there may be a color art analogous to the art of sound—a "color-music"—in which the performer seated before a literally chromatic scale, can play the colors of the spectrum in any succession or combination, flashing on a screen all possible gradations of color, simultaneously or in any desired succession, producing at will the most delicate and subtle modulations of light and color, or the most gorgeous and startling contrast and color chords![53]

[50] Goethe, *Die Schriften zur Naturwissenschaft* (cit. n. 47), Part 1, Vol. III, p. 66. See also Dennis L. Sepper, *Goethe contra Newton: Polemics and the Project for a New Science of Color* (Cambridge: Cambridge Univ. Press, 1988), pp. 88–90; and Michael J. Duck, "Newton and Goethe on Colour: Physical and Physiological Considerations," *Annals of Science,* 1988, *45:*512–515.

[51] On the phenomenon of synesthesia see Lawrence E. Marks, "On Colored-Hearing Synesthesia: Cross-modal Translations of Sensory Dimensions," *Psychological Bulletin,* 1975, *82:*303–331; which includes an extensive bibliography. Marks (p. 304) claims that the first "scientific" reference to synesthesia was by John Thomas Woolhouse, an English opthalmologist who lived in Paris, was a friend of Castel, and supported his work on the ocular harpsichord. See also Schier, *Louis-Bertrand Castel* (cit. n. 3), pp. 10, 20–22, 155. Also valuable are Albert Wellek, "Das Doppelempfinden im abendländischen Altertum und Mittelalter," *Arch. Gesamte Psych.,* 1931, *80:*120–166; Wellek, "Renaissance- und Barock-Synästhesie," *Deutsche Vierteljahrsschrift für Literaturwissenschaft und Geistesgeschichte,* 1931, *9:*534–584; and Wellek, "Zur Geschichte und Kritik der Synästhesie-Forschung," *Arch. Gesamte Psych.,* 1931, *79:*325–384.

[52] For a description of a Carnegie Hall performance of *Prometheus* on 20 March 1915 see H. C. Plummer, "Color Music—A New Art Created with the Aid of Science," *Scientific American,* 1915, *112:*343, 350–351.

[53] Albert Michelson, *Light Waves and Their Uses* (Chicago, 1903), quoted from Adrian Bernard Klein, *Colour-Music: The Art of Light,* 2nd ed. (London: Crosby Lockwood & Son, 1930), p. 223.

As this quotation shows, the ocular harpsichord is too attractive an idea to disappear completely, and we can expect it to reappear in one form or another, although perhaps not as the instrument that Castel envisioned. So far painting and photography appear to have been the most important media for exploiting the color-tone analogy.[54]

The ocular harpsichord was one of those marginal instruments that served science for a while and then disappeared, only to pop up again occasionally in subsequent history. One cannot really say that the analogy upon which it was based was proven false, just that it did not lead anywhere in the form that Castel proposed, nor in the direction that natural science subsequently took. From the way that Castel looked at the world it made perfect sense. From the way that we look at the world it belongs in the same category as the cat piano. In the eighteenth century it was not obvious where it belonged.

[54] Ernst H. Gombrich, "Epilogue: Some Musical Analogies," *The Sense of Order: A Study in the Psychology of Decorative Art* (Oxford: Phaidon, 1979), pp. 285–305; and Judith Zilczer, "'Color Music': Synaesthesia and Nineteenth-Century Sources for Abstract Art," *Artibus et historiae: An Art Anthology,* 1987, *16:*1101–1126. Most important for disseminating the concept of the musical analogy in abstract painting was Arthur Wesley Dow, whose ideas were picked up by Alfred Stieglitz, Eduard Steichen, Georgia O'Keeffe, Max Weber, and Arthur Dove.

Machine Philosophy: Demonstration Devices in Georgian Mechanics

*By Simon Schaffer**

For the Skrew, Axle and Wheel, Pulleys, the Lever and inclined Plane are known in the Schools.
For the Centre is not known but by the application of the members to matter.
For I have shown the Vis Inertiae to be false, and such is all nonsense.
For the Centre is the hold of the Spirit upon the matter in hand.
For FRICTION is inevitable because the Universe is FULL of God's works.
—Christopher Smart, "Jubilate Agno" (ca. 1760)[1]

DEMONSTRATION DEVICES, ranging from orreries and astronomical globes to the philosophical tables with which mechanical principles were taught, provided the stock-in-trade for makers of natural-philosophical instruments in Georgian Britain. These devices were deployed in public lectures to recruit audiences by using artifice to display a doctrine about nature. For example, the pulleys and falling weights involved in the instruments of Newtonian mechanics were designed ingeniously to minimize friction and distinguish inertia from gravity. The aim of demonstration was to make a specific doctrinal interpretation of these devices' performance seem inevitable and authoritative. Such an aim required hard work from the demonstrator. But this work needed to be carefully managed lest the audience's attention be drawn towards the artifice involved in the demonstration. So demonstration demanded a means of rendering tacit the gestures of the demonstrator and thus demanded rigorous training in the use of these devices. Different means of instrumental instruction, training, and performance were proper to different social settings. This article explores the different settings in which demonstration devices were used and the different forms of authority invested in work with these machines.

Whether conducted in coffee-houses, salons, or universities, lectures were both performances in front of audiences and also collaborative productions by interested practitioners. The term *demonstration* captures this range of uses rather well, because in Georgian vocabulary it might refer either to geometrical synthesis—the extraction of a proof from undeniable axioms—or else to theatrical showmanship—the display of dramatic phenomena. The relationship between mathematical certainty and public shows was always troublesome. The trouble focused on the

* Department of History and Philosophy of Science, Cambridge University, Free School Lane, Cambridge, England CB2 3RH.

[1] Christopher Smart, "Jubilate Agno," fragment B, lines 180–185, *Selected Poems,* ed. Karina Williamson and Marcus Walsh (Harmondsworth: Penguin Books, 1990), p. 70. Smart, resident poet in Cambridge for 1739–1749, composed these verses in St. Luke's lunatic asylum between 1759 and 1763.

authority and role of the performer. Some natural philosophers, such as Joseph Priestley, well-experienced in teaching chemistry, electricity, mechanics, and optics, contrasted machines for demonstration with those of philosophy. Only the latter could properly be said to transmit the messages of creation.

> Philosophical instruments are an endless fund of knowledge. By philosophical instruments, however, I do not here mean the globes, the orrery, and others, which are only the means that ingenious men have hit upon to explain their own conceptions of things to others; and which, therefore, like books, have no uses more extensive than the view of human ingenuity; but such as the air pump, condensing engine, pyrometer . . . which exhibit the operations of nature, that is of the God of nature himself.[2]

Priestley's distinction placed demonstration devices firmly in the context of artifice. He made philosophical instruments, in contrast, the unproblematic transmitters of God's messages. Others, such as the mathematical masters of Cambridge University, reckoned that singular showmanship was illustrative, but insufficient to convey demonstrative force. They required such force because their university was under pressure to reform the means it used to propagate the truths of mathematics and natural theology. The masters designed complex and elegant devices backed up with academic authority to instruct their students in Newton's truths. These differences, demonstration versus philosophy in Priestley's sense, or the Cambridge distinction between illustration and demonstration, indicate the varying roles that demonstration might play in the formation of different audiences for enlightened natural philosophy.[3]

Demonstration devices were often treated as means by which the esoteric truths of mathematical philosophy might be diffused through innumerate and polite social milieus. In 1718 Willem 'sGravesande, newly appointed professor of astronomy and mathematics at Leiden, told Newton: "I have had some success in giving a taste of your philosophy in this University; as I talk to people who have made very little progress in mathematics I have been obliged to have several machines constructed to convey the force of propositions whose demonstrations they had not understood." In 1720 the London lecturer J. T. Desaguliers translated 'sGravesande's book, profusely illustrated with the Leiden machines. He dedicated his edition to Newton, reiterating the points that instruments would diffuse the light of the *Principia* to the mathematically ill-informed and that machines "may not always prove, but sometimes only illustrate a proposition."[4] Illustrations, but potent ones, machines were

[2] Joseph Priestley, *History and Present State of Electricity,* 2 vols., 3rd ed. (London, 1775), Vol. I, p. xi. For the context see Willem Hackmann, "Scientific Instruments: Models of Brass and Aids to Discovery," in *The Uses of Experiment: Studies in the Natural Sciences,* ed. David Gooding, Trevor Pinch, and Simon Schaffer (Cambridge: Cambridge Univ. Press, 1989), pp. 31–65, on pp. 42–43.

[3] For a similar debate about "demonstration" between Thomas Beddoes and Lavoisier in 1792 see Jan Golinski, *Science as Public Culture: Chemistry and Enlightenment in Britain, 1760–1820* (Cambridge: Cambridge Univ. Press, 1992), pp. 153–155. For the pedagogic context of the later term *scientific instrument* see Deborah Jean Warner, "What Is a Scientific Instrument, When Did It Become One, and Why?" *British Journal for the History of Science,* 1990, *23:*83–93, on p. 86.

[4] Willem 'sGravesande to Isaac Newton, 13/24 June 1718, in A. R. Hall, "Further Newton Correspondence," *Notes and Records of the Royal Society,* 1982, *37:*7-34, p. 26. For the Leiden cabinet see C. A. Crommelin, *Descriptive Catalogue of the Physical Instruments of the Eighteenth Century* (Leiden: Rijksmuseum voor de Geschiednis der Naturwetenschappen, 1951), pp. 24–41. For Desaguliers's translation see I. B. Cohen, *Franklin and Newton* (Cambridge, Mass.: Harvard University Press, 1956), p. 236.

used to insinuate the very high philosophical status of public showmanship. They were also used to establish the authority of consulting engineers by representing their mastery over nature's powers. Desaguliers, the leading engineer of early eighteenth-century London, explained that "a great many Persons get a considerable Knowledge of Natural Philosophy by Way of Amusement; and some are so well pleas'd with what they learn that Way, as to be induc'd to study Mathematicks, by which they at last become eminent Philosophers." Desaguliers reckoned that philosophical eminence acquired through the study of demonstration devices would also promote commercial welfare and gentlemanly patronage for reliable engineering schemes.[5] The status problems of mathematics, philosophy, showmanship, and engineering were fundamentally concerned with the relations of nature and art. Demonstration devices were designed to teach truths about nature, and the gestures which accompanied them were supposed to be invisible. In 1713 the journalist Richard Steele puffed the newfangled orrery as "a Machine which illustrates, I may say demonstrates, a System of Astronomy to the meanest Capacity. . . . It is like the receiving a new Sense, to Admit into one's Imagination all that this Invention presents to it with so much Quickness and Ease."[6] Steele attributed these powers to the machine, not its demonstrator. This was the fundamental lesson which enlightened savants set out to teach, by making themselves nature's special representatives and then attributing their power to the machines.

I. ATWOOD'S MACHINE

Machinery did not simply transmit rational mechanics. It also helped to make it. Stipulations of the content of Newton's natural philosophy were as much a result of as a reason for this public culture.[7] Apparently univocal principles, such as those of Newtonian mechanics, are best seen as tools with local purposes. Their meaning is determined by the context of their use. Thus in his furious response to Leibnizian views about living forces in 1728, Samuel Clarke loyally attributed them to a foreign desire "to raise a Dust of Opposition against Sir Isaac Newton's Philosophy, the Glory of which is the Application of abstract Mathematicks to the real Phaenomena of Nature." Half a century later the young Cambridge mathematician Isaac Milner stated that this notorious controversy was based not on rival experiments but on a misunderstanding of Newton's original purposes: "The laws of motion, in certain cases, are incontestable, and no author of eminence contradicts them: it is from a mistaken application of these laws that a difference of opinion has arisen."[8] *Mistake*

[5] J. T. Desaguliers, *A Course of Experimental Philosophy,* 2 vols. (London, 1734, 1744), Vol. I, sig. c^r; for Desaguliers as engineer see Larry Stewart, *The Rise of Public Science: Rhetoric, Technology, and Natural Philosophy in Newtonian Britain, 1660–1750* (Cambridge: Cambridge Univ. Press, 1992), pp. 119–126.

[6] Richard Steele in *The Englishman,* no. 11 (27–29 Oct. 1713), cited in H. C. King and J. R. Millburn, *Geared to the Stars* (Bristol: Adam Hilger, 1978), p. 154.

[7] Simon Schaffer, "Natural Philosophy and Public Spectacle in the Eighteenth Century," *History of Science,* 1983, *23:*1–43.

[8] Samuel Clarke, "A Letter . . . occasion'd by the present Controversy among Mathematicians concerning the Proportion of Velocity and Force in Bodies in Motion," *Philosophical Transactions of the Royal Society,* 1727/8, *35:*381–388, on p. 382; and Isaac Milner, "Reflections on the Communication of Motion by Impact and Gravity," *Phil. Trans.,* 1778, *68:*344–379, p. 349. See Carolyn Iltis, "The Leibnizian-Newtonian Debates: Natural Philosophy and Social Psychology," *Brit. J. Hist. Sci.,* 1973, *6:*343–377, on pp. 372–376.

and *misunderstanding* were terms designed to make a partisan reading of Newton into the obvious one. Demonstration devices were used as part of the process of fixing and regulating the meanings natural philosophers gave to the doctrines which they taught. In his influential analysis of scientific discovery, N. R. Hanson argued of the laws of Newtonian mechanics that "few have appreciated the variety of uses to which law sentences can be put at any one time, indeed even in one experimental report." He was especially concerned with the demonstrative status assigned to Newton's laws. "The laws of physics are used sometimes so that disconfirmatory evidence is a conceptual possibility, and sometimes so that it is not." There could be no simple relation between scientific authority and experimental demonstration. Interpretations of the second law might range from a definition of force as rate of change of motion, in which case no possible machine could do more than illustrate this doctrine, to the claim that this relation was a contingent empirical result, in which case the performance of such a machine might confirm or refute the principle.[9]

Hanson chose an illuminating example: the changing uses of Atwood's machine, a device which might be seen either as administering a stiff empirical test of Newton's second law, or as clarifying the sense of this law as an analytic truth. The Cambridge mathematician George Atwood designed the machine in the late 1770s to demonstrate the motion of bodies under constant forces. The machine, described in his *Treatise on the Rectilinear Motion and Rotation of Bodies* (1784), consisted of two balanced cylinders linked by a thin silk cord suspended over a pulley (see Figure 1). Additional weights, in the form either of small perforated discs or of thin bars, could be attached to either cylinder to provide a net force. More than a dozen different trials were described, each designed to display the time and distance through which bodies would move under some constant and measured accelerating or decelerating force. Elastic and inelastic impact, and the effects of fluid resistance, could also be examined. "In books of mechanics," Atwood explained, "no account is found of methods by which the principles of motion may be subjected to decisive and satisfactory trials." His demonstration device was designed to offer precise numerical answers to this need.[10]

Atwood transformed a rather conventional pulley, worthy of the textbooks of classical mechanics, into a carefully scaled device capable of being subjected to mathematical analysis. He considered and condemned the traditional means natural philosophers had used to try mechanical principles: the fall of bodies through immense heights, where air resistance was a factor, or their motion on inclined planes, where friction counted. The pulley itself was made up of four friction wheels "to prevent the loss of motion, which would be occasioned by the friction of the axle if it revolved on an immoveable surface." He placed a long vertical divided rule behind the left-hand weight and a thirty-minute weight-driven pendulum clock on a column supporting the pulley stand. "The clock will be sufficiently exact if it keeps time with a common well regulated clock for this half hour." An ingenious hollow ring allowed the additional bar weights to be removed instantaneously from the accelerating cylinder as it dropped. Then the falling cylinder would move at a uniform speed

[9] N. R. Hanson, *Patterns of Discovery* (Cambridge: Cambridge Univ. Press, 1958, 2nd ed. 1965), pp. 97–98. For the second law see T. L. Hankins, "The Reception of Newton's Second Law of Motion in the Eighteenth Century," *Archives Internationales de l'Histoire des Sciences,* 1967, *20:*43–65.

[10] Hanson, *Patterns of Discovery* (cit. n. 9), pp. 100–102; and George Atwood, *A Treatise on the Rectilinear Motion and Rotation of Bodies* (Cambridge, 1784), p. ix.

Figure 1. *George Atwood's machine as depicted in his* Treatise on the Rectilinear Motion and Rotation of Bodies *(1784). The pulley mechanism, ring to interrupt the bar weights, and clock are illustrated in the enlargement at right. Courtesy of Whipple Library, Cambridge.*

that was easy to measure. Atwood's machine was represented as a high-precision device. He used a plumb line to level the stand. He calibrated the mass of the pulley wheel against the time of the variable weight's descent. The recipe demanded that the wheels' weights be precisely 2¾ ounces. This precision also required some manual dexterity. "Great attention should be paid to the adjustment of the weights used in these experiments as a very small error such as is scarcely perceivable in each, will tend greatly to affect the exactness of the experiments in which many weights are used." Crucially, his textbook gave exact numbers for the height of the machine, the size of the weights, and the mass of the friction wheels. The implication was that readers would commission replicas of the author's machine. Detailed instructions were provided for timing the weights' motion. These instructions indicated the practical skill the instrument required. Observers must not simply wait until they heard the clock's beat before releasing the weight, since then the fall would start too late. Instead, they must habituate themselves to the clock's rate: "The proper method is to attend to the beats of the pendulum until an exact idea of their succession is obtained; then the extremity of the rod being withdrawn from the bottom [of the weight] directly downward at the instant of any beat, the descent will commence at the same instant." By arranging the device so that "the time of motion shall be a whole number of seconds; the estimation of time . . . admits of considerable exactness." This exactness had to be mobilized for a specific reading of Newton's own purposes.[11]

Successive sections of this article describe some groups who worked with this device: mathematicians in Atwood's Cambridge, where the machine fitted the purposes of a conservative establishment whose status depended on the legitimacy of their role as interpreters of Newtonian mechanics; the community of civil engineers, led by John Smeaton, whose own instrument models displayed a radically different account of motion and friction; and instrument salesmen, who through agents such as the entrepreneur Jean Magellan made the machine for audiences beyond the management of and in many ways hostile to the orthodoxies of Georgian academic philosophy. Each use of the machine seemed authoritative. Recent sociology of scientific knowledge distinguishes between closed settings, in which all participants in a scientific task agree on the proper outcome of some instrumental trial, and open settings, in which they do not. In closed settings replication may proceed rather easily. Users have a reliable criterion to check when the instrument is working properly. Demonstration devices become authoritative in closed settings, when it seems hard to challenge a specific didactic implication of the show. The stories that follow reveal the tactics that help produce this remarkable situation. It is concluded that much of the apparent authority of these devices relies on the gestures performed by the demonstrator and the cultural resources available within the setting of the demonstration. These gestures and resources draw attention away from the workings of the machine and towards the philosophy which the machine is designed to demonstrate.[12]

[11] Atwood, *Treatise* (cit. n. 10), pp. 299, 310–311, 335–337. For precision instrumentation see J. L. Heilbron, "Introductory Essay," in *The Quantifying Spirit in the Eighteenth Century,* ed. Tore Frängsmyr, J. L. Heilbron, and Robin E. Rider (Berkeley/Los Angeles: Univ. California Press, 1990), 1–23, on pp. 4–10.

[12] For closed settings see H. M. Collins, *Changing Order: Replication and Induction in Scientific Practice* (London/Beverly Hills: Sage, 1985), pp. 73–74.

II. ATWOOD'S MACHINE AND NEWTON'S LAWS

The first audience for Atwood's machine was very carefully defined. Those who attended his lectures in the gatehouse of Trinity College were repeatedly instructed on the social function of the Georgian university. Isaac Milner defended Cambridge's stress on mathematics and natural philosophy because "we endeavoured, not only to fix in the minds of young students the most important truths, but . . . particularly to be on their guard against the delusions of fanciful hypotheses in every species of philosophy. . . . A judicious prosecution of the science of mathematics and natural philosophy is among the very best preparatives to the study of theology in general and of Christianity in particular." In a society of orders where rank hinged on the command of fixed property and patronage, these men possessed authority solely by right of their own advancement within the university and collegiate system.[13]

Atwood's closest colleagues, such as Milner, were brought up in humble jobs like bricklaying or weaving. As John Gascoigne has shown in his study of Georgian Cambridge, their spectacular academic success in the Cambridge mathematical examinations won these men high office. They were high wranglers, gaining top marks in these trials, and they became moderators, the managers of the examination system. Atwood and Milner became Fellows of the Royal Society in 1776. Five years later Atwood tried to persuade Joseph Banks, the society's president, that Cambridge mathematics was useful "in promoting the application . . . to practise." His paper on ship stability won him the society's Copley Medal in 1796. The Board of Longitude, another branch of Banks's regime, was an important target. The Christ's College mathematician Thomas Parkinson earned his student income by calculating the board's refraction tables. In 1779 Atwood tried unsuccessfully to use his new demonstration device to lobby Banks for the job of secretary to the board. In 1784 he joined in a revolt at the society against Banks's rule, and in the same year he at last won backing from his former student, the new Prime Minister William Pitt. Atwood was hired to a major post in the customs office as part of Pitt's campaign for administrative rationalization. The premier's power brokers also helped Parkinson and Milner to good church jobs. These successes were judged to be demonstrations of their public worth.[14]

[13] John Gascoigne, "Mathematics and Meritocracy: The Emergence of the Cambridge Mathematical Tripos," *Social Studies of Science,* 1984, *14:* 547–584, on p. 561; and Isaac Milner, *Strictures on some of the Publications of the Reverend Herbert Marsh* (London: Deighton Bell, 1813), p. 237. I thank Kevin Knox for help with this source.

[14] George Atwood to Joseph Banks, 7 Nov. 1779, British Library MSS Add. 33977, fol. 110; for his insistence on practical mathematics see Atwood to Banks, 24 Aug. 1781, British Library MSS Add. 33977 fols. 136–137. His Copley Medal paper is "The Construction and Analysis of Geometrical Propositions determining the Positions assumed by Homogeneal Bodies which float freely," *Phil. Trans.,* 1796, *86:*46–130. For the crisis of 1782–1784 see Russell McCormmach, "Henry Cavendish on the Proper Method of Rectifying Abuses," in *Beyond History of Science: Essays in Honor of Robert E. Schofield,* ed. Elizabeth Garber (London: Associated University Presses, 1990), pp. 35–51, on p. 37. Atwood's allies included Samuel Horsley and Nevil Maskelyne. Atwood's employment by Pitt is described in obituary of George Atwood, *Gentleman's Magazine,* 1807, *77*(2): 690. Pitt's fiscal reform is analyzed in John Brewer, *The Sinews of Power* (London: Unwin Hyman, 1989), pp. 85–87, 101–102. Pitt's patronage is described in John Gascoigne, *Cambridge in the Age of the Enlightenment* (Cambridge: Cambridge Univ. Press, 1989), pp. 218–220; Mary Milner, *Life of Isaac Milner* (London: John Parker, 1842), p. 71; and obituary of Thomas Parkinson, *Gentleman's Magazine*, 1831, *101:*85–89, on p. 86.

Those mathematicians who won official patronage helped weld the university curriculum to the authority of the established church and state. In the 1770s, at the period of the American war and the campaign for an end to discrimination against Dissenters, this establishment encountered fierce criticism from political activists within Cambridge who sought to revise the Anglican creed, change the forms of university examination, and challenge the link between crown, church, and university. The university countered with an overhaul of the public examination and a defense of the worth of public demonstration of orthodox natural philosophy and theology in order to reinforce tutorials and public lectures. Academic demonstration became a practice exercised in the exegesis of the texts of both rational mechanics and Scripture. A large group of new handbooks appeared for candidates for the tripos, notably Parkinson's *System of Mechanics* (1785) and the four volumes of *Principles of Mathematics and Natural Philosophy* (1795–1799) by the Plumian professor Samuel Vince and the Johnian tutor James Wood. Atwood's 1784 *Treatise,* and his much-expanded *Analysis of a Course of Lectures on Natural Philosophy* of the same year, were deliberate contributions to this enterprise.[15] As the successful wrangler John Herschel later recalled, in the early nineteenth century the mathematics examiners were still "fenced . . . in the tough bull-hide of Vince and Wood." The curriculum now helped demonstrate the truth of the propositions of Newton's mechanics and the legitimacy of the regime of the Georgian university. When he became senior wrangler "incomparabilis" in 1774, Milner "ordered from a jeweller a rather splendid seal, bearing a finely executed head of Sir Isaac Newton." Newton's work was reinterpreted and reinforced to make his legacy fit the university's newly established needs.[16]

Cambridge mathematical and natural philosophical instruction reinforced the management of fallible belief. Training involved gentlemanly conversation and closeted tutorials culminating, for the favored few, in stern face-to-face trials. Future mathematical duelists took tea with their rivals to prepare for the public challenge. Breakfasts with Parkinson included "a pretty strict examination in Euclid and algebra." He gave manuscript sheets of his *System of Mechanics* to his few serious students, such as the bright young Henry Gunning, third wrangler in 1788, who recalled that "I worked at them very hard and made myself master of that part of it which treated on elastic balls." James Wood, moderator of the 1788 tripos, used his own rooms to entertain Gunning and his colleagues while they were posed more advanced topics: "We were most hospitably regaled [with] an admirable dessert placed on the sideboard and some excellent wine."[17] In this milieu demonstrations of private mathematical skill were more common than those of public experimental philosophy. Milner was suspicious of the capacity of demonstration devices to produce sufficient conviction. Simple tutorials on "the collision of spherical bodies" were "better adapted to throw light on a disputable question than where the suppositions

[15] Christopher Wordsworth, *Scholae Academicae: Some Account of the Studies at the English Universities in the Eighteenth Century* (Cambridge: Cambridge Univ. Press, 1877), pp. 50–58; and Gascoigne, *Cambridge* (cit. n. 14), pp. 274–275. For the Dissenters' campaigns see D. A. Winstanley, *Unreformed Cambridge* (Cambridge: Cambridge Univ. Press, 1935), pp. 304–330.

[16] Herschel as quoted in W. W. Rouse Ball, *History of the Study of Mathematics at Cambridge* (Cambridge: Cambridge Univ. Press, 1889), pp. 119–120; and Milner, *Life of Milner* (cit. n. 14), p. 9.

[17] Henry Gunning, *Reminiscences of the University, Town, and County of Cambridge,* 2 vols. (London: George Bell, 1854), Vol. I, pp. 20, 88–89.

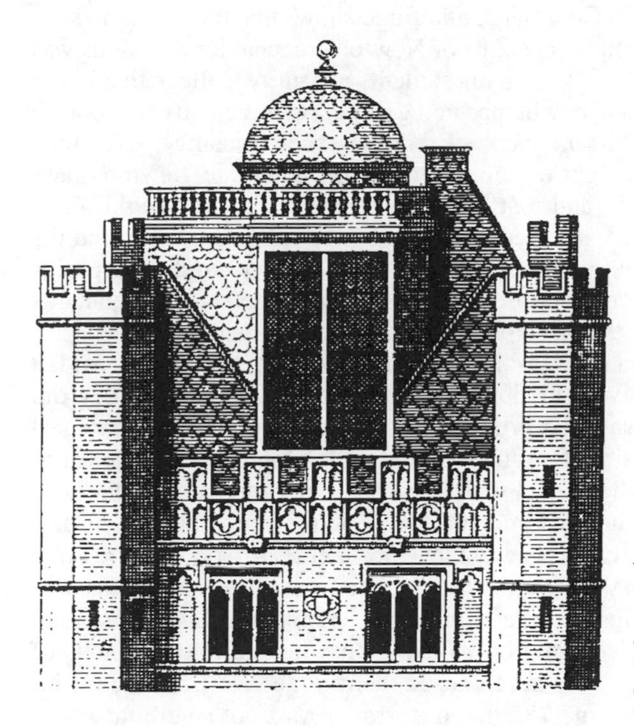

Figure 2. The observatory over the Great Gate of Trinity College, Cambridge, where Atwood lectured in the 1770s and 1780s. Courtesy of Whipple Library, Cambridge.

are more perplexed with mechanical contrivances." He wondered whether experiments would serve the didactic point of absolute certainty. "Even when experiments are produced which seem to prove the point, one is apt to suspect the universality of the conclusion." [18] But from 1776, Atwood, then a Trinity fellow, began lecturing on mechanics and natural philosophy in the college observatory (see Figure 2). The observatory was the result of a power grab in 1706 by the then Master of Trinity and staunch Newtonian Richard Bentley for his client Roger Cotes, who began using the rooms over the college gate as an observatory and lecture room. This group helped revive the Cambridge University Press as a source of philosophical matter and produced the second edition of the *Principia.* Without Cotes's machines but in his spirit, Atwood's lectures traded on an important aspect of Trinity's claim as a center of diffusion of Newtonian orthodoxy.[19]

Hanson claimed that eighteenth-century natural philosophers "universally" regarded Newton's second law as an empirical result, and so treated such devices as Atwood's as a public test of the law's truth in the context of the controversy with the apostles of living force. Hanson also noted that, in contrast, modern physics texts deploy the device as a means for measuring gravitational acceleration on the assumption of the truth of the law.[20] However, a closer scrutiny of the public

[18] Milner, "Reflections" (cit. n. 8), pp. 363, 369.

[19] For Bentley's program see John Gascoigne, "Politics, Patronage and Newtonianism: The Cambridge Example," *Historical Journal,* 1984, *27:*1–24, on pp. 18–20. For Atwood's courses see Winstanley, *Unreformed Cambridge* (cit. n. 15), pp. 151, 367n160.

[20] Hanson, *Patterns of Discovery* (cit. n. 9), p. 101; and Atwood, *Treatise* (cit. n. 10), p. 33. See also Thomas B. Greenslade, "Atwood's Machine," *Physics Teacher,* 1985, *23:*24–28. For a canonical use of Atwood's machine see A. Privat Deschanel, *Elementary Treatise on Natural Philosophy,* ed.

stipulations of mathematicians in Georgian Cambridge shows that there was no simple presumption of the purely empirical status of Newton's mechanics. Its status was a culturally produced artifact. Atwood told his students at Trinity College that it was "rational to suppose that any change, whether in the direction or velocity of a body's motion, should be proportional to the causes producing such a change." Newton's second law became a matter of right reason. Atwood conceded that Newton's laws were "incapable of direct proof" and that even the best experiments would differ from "the mathematical relation expressed in the laws." Atwood's machine, and the numbers it generated, "in some instances expressed to a greater number of places than may seem necessary," were supposed to make this agreement conspicuous without subverting the basis of the principle in reason and self-evident truth.[21]

This combination of mathematical reason and demonstrative experiment provided the Cambridge masters with a very flexible, tactically crucial, resource. They did not envisage the falsehood of Newton's principles. This presumption is to be understood as a social fact of university life, rather than simply a logical consequence of some rigorous syllogism. So when faced with apparent experimental challenges to their interpretation of the measure of force and the principle of the change of momentum, they used their laws to explain away the experiments. Atwood lectured that since "the evidence upon which the certainty of this theory rests is scarcely less than mathematical, it will be more eligible to refer mechanical experiments of every kind to the theory, as a means of discovering how far the unavoidable imperfections of construction and observation . . . may have caused them to deviate from the truth." Parkinson instructed his students in 1785 that experiment was not a sure means of demonstrating theory. Apparent experimental challenges "are too imperfectly understood to justify any general conclusions against . . . theoretic demonstrations, resulting from data which, perhaps, do not take place in the experiment."[22] Milner agreed that the principles of mechanics must be used to determine whether any apparent trouble should be taken seriously. When in 1779 he answered Jean d'Alembert's attempt to correct the Newtonian account of the precession of the equinoxes, for example, Milner stated that the French materialist's bravado "shews, that when we venture to differ from Sir Isaac Newton in these matters, it is with the utmost difficulty that we can arrive at certainty." In terrestrial mechanics, too, Milner urged that this authority must always be used to control any experimental challenge: "It is obvious that the laws of motion may, as described by Sir Isaac Newton, be founded on experiment, and yet, if they are extended to cases where they cannot be applied, the conclusions must still be erroneous."[23]

On the other hand, when they deployed demonstration devices, these mathemati-

---

J. D. Everett (London: Blackie, 1876), pp. 44–46: "Atwood's machine, however modified, gives only indirect evidence regarding the motion of bodies falling freely."

[21] Atwood, *Treatise* (cit. n. 10), pp. ix, 3–4, 358–359.

[22] *Ibid.*, p. 383; and Thomas Parkinson, *A System of Mechanics* (Cambridge, 1785), p. 70. Atwood was referring to John Smeaton's trials on percussion; Parkinson was discussing the impact of water on millwheels, which seemed proportional to the square of the water's speed.

[23] Isaac Milner, "On the Precession of the Equinoxes produced by the Sun's Attraction," *Phil. Trans.*, 1779, *69:*505–526, on p. 526; and Milner, "Reflections" (cit. n. 8), p. 349. For d'Alembert's pugnacious account of precession see Thomas L. Hankins, *Jean d'Alembert: Science and the Enlightenment,* 2nd ed. (New York: Gordon & Breach, 1990), p. 51; and Curtis A. Wilson, "Perturbations and Solar Tables from Lacaille to Delambre," *Archive for History of Exact Sciences,* 1980, *22:* 53–304, on pp. 104–107.

cians emphasized that precise experimentation was the only proper standard of physical truth. Milner announced that while the laws of motion were indubitably rational, "a law of nature is not merely a deduction of reason: it must be proved, either at once and directly, by some simple and decisive experiments." He reported with horror that some "young philosophers" in Cambridge "affirm that the third law of motion is nothing more than a definition." He attacked the Edinburgh mathematics professor Colin Maclaurin because in his 1748 textbook the Scotsman simply defined "elasticity" as the property of a body to preserve the size and reverse the direction of its motion in impact. Then, as Milner complained, "the principle that the relative velocity . . . is not altered by the stroke, is neither to be demonstrated nor confirmed by experience; it is a direct consequence of the definition of elasticity."[24] Parkinson also attacked Scottish arguments that seemed to make the principles of mechanics dependent on abstract definitions. He criticized John Keill's a priori definition of momentum as the product of mass and speed, claimed that momentum "is perhaps more properly and with more conviction demonstrated by experiments," and insisted that the truth of Newton's laws "is established by a number of concurring and uncontroverted observations." He had the master's blessing. In the *Principia,* according to Parkinson, Newton showed that the crucial fact that all matter gravitates in proportion to its bulk "is only demonstrable from experiments." James Wood helped out this argument by using Atwood's machine to prove that equal weights accelerated at the same rate.[25]

These arguments demanded a great deal from demonstration devices. They were supposed to prove the unchallengeable tenets of the academic faith. Atwood claimed that his machine was a tool that could actually produce such certainty, not simply a device for illustrating principles to which students might already assent on abstract grounds.

> If the experiments are designed only to assist the imagination by substituting sensible objects, instead of abstract and ideal quantities, an apparent agreement between the theory and experiment may be sufficient to answer this purpose, although it may be produced from an erroneous construction; but it must be allowed, that experiments of this kind are extremely defective, and entirely insufficient to impress the mind with that satisfactory conviction, which always attends the observation of experiments accurately made.[26]

The tutors taught that the university's principles had been challenged, so none could allege dogmatism against the established regime. But this regime had withstood the challenge so well that any further doubt was not only useless but irrational. Because conviction was supposed to quell criticism, the mathematicians had to make their experimental shows look like logical demonstrations. Parkinson opened his textbook with a defense of the prolixity and repetitiveness of his lectures, which gave many

[24] Milner, "Reflections" (cit. n. 8), pp. 252–254, on Colin Maclaurin, *An Account of Sir Isaac Newton's Philosophical Discoveries* (London, 1748), Book II.

[25] Parkinson, *System of Mechanics* (cit. n. 22), pp. 64–65, 72, 74. On p. 94 he refers to Principia 3.6 on the proportionality of gravitational weight to bulk. John Keill's argument is in his paper "In qua Leges Attractionis aliaque Physices Principia traduntur," *Phil. Trans.*, 1708–9, *26:*97–110. James Wood uses Atwood's machine in his *The Principles of Mechanics Designed for Students in the University* (Cambridge, 1796), p. 17n.

[26] Atwood, *Treatise* (cit. n. 10), p. 293.

"demonstrations" of the same proposition. "Clear and adequate ideas upon a new subject are not communicated by a transient impression. . . . Science implies something more than a mere ability to go through a demonstration."[27]

Singular classroom demonstrations, even if dressed up with geometrical reasonings, could not teach the politically crucial lesson of an established and unchallengeable disciplinary order. In 1796, at the height of state persecution of political and religious dissent, Wood argued that demonstrations would work if and only if they were backed up by established religion. Against Humean skepticism he declared that the divinely established order of the world made demonstration devices meaningful, since otherwise "experiment could only furnish us with detached and isolated facts, wholly inapplicable on other occasions, and that harmony, which we cannot but observe and admire in the material world, would be lost."[28] Parkinson agreed that Cambridge mechanics must teach the lawlike workings of God's nature. Demonstration devices should "exhibit proofs of unbounded power, consummate wisdom and paternal benevolence in the great Creator." Then the principles they demonstrated "would only be affected or subverted by a change in the constitution of nature." The powers of machines would be revealed as "many subordinate instruments in the government of nature, matter being impressed by its great Creator with several attributes which appear . . . ministerial to the continuance of existence and preservation."[29]

III. ATWOOD'S MACHINE AGAINST LIVING MATTER

These Cambridge tutors used Newtonian mechanics as a powerful weapon against dissent and in defense of "existence and preservation," both because it was a means of producing conviction and silencing opposition, and because it was a means of combating the materialism of dissenters and radicals. Central to this materialism was the claim that matter was innately active and capable of self-motion. Anglican divines judged that these views were conducive to irreligion since materialists identified the soul with an active form of matter and thus subverted sacramental religion. Priestley's *Disquisitions relating to Matter and Spirit* (1777) was publicly debated in London magazines throughout the late 1770s. He reckoned that since matter was endowed with attraction and repulsion, it lacked inertia. He announced: "I now consider the doctrine of the *soul* to have been imported into Christianity and to be the foundation of the capital corruptions of our religion." In autumn 1775 Priestley's erstwhile supporter William Kenrick joked in his *London Review* that Samuel Clarke, often cited as an authority on orthodox Newtonian matter theory, "was confessedly so merely a reasoning machine, that he would almost tempt one to think matter might think, and that he himself was a living proof of it."[30] Cambridge mathe-

[27] Parkinson, *System* (cit. n. 22), preface.

[28] Wood, *Principles* (cit. n. 25), pp. 9, 25; for his theology see E. Bushby, *The Very Reverend Dr. Wood* (Cambridge: C. E. Brown, 1839).

[29] Parkinson, *System* (cit. n. 22), pp. 61–62 (on laws), 3 (on the Creator's benevolence), 9 (on the constitution of nature), 1–2 (on subordinate instruments).

[30] Joseph Priestley to Caleb Rotheram, April 1778, in *The Theological and Miscellaneous Works of Joseph Priestley,* ed. John Rutt, 25 vols. (London: G. Smallfield, 1817–1835), Vol. Ia, pp. 314–315; and [William Kenrick] in *London Review,* 1775, 2:564, cited in John W. Yolton, *Thinking Matter: Materialism in Eighteenth-Century Britain* (Oxford: Basil Blackwell, 1983), p. 117. Compare Joseph

maticians had to hit back at this metropolitan ribaldry. Their mechanics was designed to make matter inanimate and to deny that motion was naturally conserved. Wood made Newton's laws depend on matter's passivity and used Atwood's machine to demonstrate this truth. The machine was useful because it created a situation in which bodies were solely moved by external and demonstrable forces. Parkinson also emphasized that "matter is totally inactive and incapable of communicating motion to itself." Whenever "new motions are observed without any sensible material impulse, resulting apparently from an innate tendency to motion," it was always the case that "these are the necessary result of established natural powers." These powers were all impressed in matter "by its great Creator."[31]

The Cambridge interpreters tried to make their philosophy of passive matter and an active God look like authoritative common sense. This was a hard task after the scandalous disputes within mechanical philosophy about the means by which the force of moving bodies should be estimated. Milner treated these *vis viva* disputes as political embarrassments because "as it generally happens in other disputes we do not hear of any conviction being produced on either side." This lack of conviction was a sign that "the violence of prejudice and party-spirit has so much clouded the reasonings of the best writers." Cambridge was supposed to be a tranquil oasis amidst the storms of dispute, allegedly a realm devoid of party feeling where conviction would be assured. Using the twin resources of analytic and experimental demonstration, Parkinson made sure that his students understood that Newton's mechanics was certain while Leibnizian views must be subjected to empirical test. "The doctrine advanced by Newton is universally true, according to his meaning of the term force; but whether the opinion of Leibnitz be true or not is best known from experiments, the result of which is generally repugnant to it."[32] So it became important to represent the history of the dispute as the effortless empirical defeat of the delusion that made matter intrinsically active, and then to associate materialism with this delusion.

The *vis viva* disputes have had as bad a press from historians as have Georgian Cambridge's mathematical achievements. But late-eighteenth-century Cambridge mathematicians worked hard to salvage a practical program from the detritus of the earlier debates. Atwood echoed the received view that these disputes had been "verbal, i.e. a dispute about a term or definition. . . . The permanency of motion, during its communication, should not be applied in the demonstration of mechanical propositions." Conservation principles reeked of materialism. But he also advertised his machine as a means of making the issue of matter's passivity and its lawlike behavior fit for experiment.[33] His machine's design was explicitly contrasted with the alleged futilities of trials of the 1720s on the resisted impact of falling bodies on potter's clay and candle wax. Then 'sGravesande had reckoned that since the size of an indentation depended on the distance fallen, these experiments manifested the

---

Priestley, *Disquisitions relating to Matter and Spirit* (London, 1777), p. xxxviii, for the denial of matter's inertia.

[31] Wood, *Principles* (cit. n. 25), pp. 17n, 19–20; and Parkinson, *System* (cit. n. 22), pp. 1–2, 62.

[32] Milner, "Reflections" (cit. n. 8), p. 347; and Parkinson, *System,* p. 68.

[33] Atwood, *Treatise* (cit. n. 10), p. 370. Against the futility of the disputes see David Papineau, "The *Vis Viva* Controversy: Do Meanings Matter?" *Studies in History and Philosophy of Science,* 1977, *8:* 111–142.

dependence of a body's effect on the square of its speed. He even alleged that his experiments meant "that what was before only a dispute about words now becomes a dispute about real things."[34] This was overoptimistic. Atwood could look back at a record of indecisive debate. In 1722 the London physician Henry Pemberton claimed that such impact trials were "much more fit to inform us of the Law, by which these yielding Substances resist the motion of Bodies striking upon them, than to shew the forces, with which Bodies strike." Desaguliers had tried to preserve the peace of the republic of letters: the Dutch "are too curious in making and too faithful in relating their Experiments, not to have us credit the Facts." He conceded the engineering uses of trials on impact which "give us a Principle to direct the Practice of some mechanical operations which was not very well known before." But he used arguments taken from Pemberton's essay and from an anonymous postscript by Newton to argue that forces must be estimated by their time of action, that larger impressions took longer to make, and so that the size of an impression did not measure a body's action. In 1740 he told Maclaurin that "in the Congress of soft Bodies, there is no Force at all lost in denting the Bodies." By 1744 Desaguliers publicly announced that no such trials could resolve the issue.[35] As a result the English began to identify living force with innate activity and perpetual motion. This was the interpretation of the dispute that suited Atwood, Parkinson, and Milner. Their revival of the story was helped by the posthumous publication of 'sGravesande's writings by his disciple Jean Allamand in 1774, including the anecdote of 'sGravesande's eureka experience when his trials on the indentations made by falling bodies turned him against the English measure of force. It was also aided by the journalistic activities of Kenrick, who preceded his endorsement of Priestley's materialism with a pamphlet (1770) that republished 'sGravesande's papers supporting perpetual motion and then publicized this work in the *Gentleman's Magazine* in 1772.[36]

The use of Atwood's machine to attack the notion that living force inhered in bodies demanded skillful reinterpretation of this tortuous history. Atwood noticed that the defenders of living force simply expelled perfectly inelastic bodies from nature and assumed that all forces acted continuously. Because he was fighting against the notion of self-moving matter, Atwood urged that no experimental mechanics should deal with such abstractions as inelastics: "Motion may be communicated or destroyed . . . by instantaneous impact . . . only in perfectly hard and inflexible bodies, which exist not in nature; and even in the abstract consideration of these, . . . difficulties arise hardly explicable by any method of reasoning." So Atwood

[34] Willem 'sGravesande, "Remarques sur la force des corps" (1729), quoted in T. L. Hankins, "Eighteenth-Century Attempts to Resolve the *Vis viva* controversy", *Isis,* 1965, *56:*281–297, pp. 287–288.

[35] J. T. Desaguliers to Colin Maclaurin, 10 April 1740, in Maclaurin, *Collected Letters,* ed. Stella Mills (Nantwich: Shiva Publishing, 1982), p. 334; Desaguliers, *Course* (cit. n. 5), Vol. I, p. 398–400, Vol. II, p. 63. Desaguliers's source was Henry Pemberton, "A Letter . . . concerning an Experiment, whereby it has been Attempted to shew the Falsity of the Common Opinion in relation to the Force of Bodies in Motion," *Phil. Trans.,* 1722, *32:* 57–66 (citation from p. 58). Newton was the "excellent and learned friend" who prepared Pemberton's postscript, pp. 67–68; see I. Bernard Cohen, *Introduction to Newton's* "Principia" (Cambridge: Cambridge Univ. Press, 1971), p. 266.

[36] William Kenrick, *An Account of the Automaton constructed by Orffyreus* (London, 1770), pp. 6–8, 19–23; [Kenrick], "Perpetual Motion Said to Be Discovered," *Gentleman's Magazine,* 1772, *42:* 172; and Jean Allamand, "Histoire de la vie et des Ouvrages de Mr 'sGravesande," in Willem 'sGravesande, *Oeuvres philosophiques et mathématiques,* 2 vols. (Amsterdam, 1774), Vol. I, p. xv.

argued that the proper sense of Newtonian mechanics would make all forces continuous agents. No instant changes of motion were allowed. "When finite velocity is communicated to any natural body, the time wherein it is communicated must be finite." Then his machine would be the perfect Newtonian system, and he could avoid the infelicities of impact trials. These older trials had never been able to reach any conclusion against Newtonian force theory because such a theory hinged on the time through which forces acted—here Atwood used Pemberton's conventional argument that the time to make an impression must be taken into account. Atwood explained that these experiments might seem "a more eligible (however imperfect) way of examining the principles of motion. This method is doubtless allowable, in order to estimate the force of moving bodies experimentally, if it be sufficiently accurate." But it was not. Atwood observed that "the resistances opposed to spherical bodies which impinge on a block of wood, bank of earth &c. depend not only on the tenacity or density of the parts, whereof the substances penetrated are composed, but upon the diameters of the impinging spheres." To get any general result from 'sGravesande's trials, it would have been necessary to get a general measure of the absolute resistance of bodies like clay, earth, and wood.[37] Atwood's machine was a better instrument, therefore, both because of its high claims to precision, and because it showed forces acting through time.

The machine allowed a very elegant demonstration of retarded motion. The experimenter could arrange the two cylinders so that they balanced, then add a small circular weight of unit mass to the right-hand cylinder and two bars each of similar weight to the other. He must set the circular frame so that it would intercept these bars and remove them from the falling left-hand weight as it moved past. At this point the weight would stop accelerating and start slowing down. Atwood gave his audience detailed instructions for these trials. They had to check the lowest point reached by the decelerating left-hand body before the greater weight of the right-hand one made it move back up. His handbook gave exact numbers. If one cylinder weighed 6⅛ ounces and the other ¼ ounce more, with two bar weights of ¼ ounce each riding on the lighter cylinder, once the two bar weights were removed, the lighter cylinder would drop to the 52-inch mark before pausing. The trials could be developed exactly to ape the old-fashioned impact experiments. If the weights in a pair of trials were set up so that the speed of the retarded body varied as the square of the retarding force, the drop in the two cases would be the same. This was an impressive trick: Atwood could give his audience exact numbers, especially for the retarding force. No messy and incalculable potter's clay was required. "These experiments are easily made, and the adjustments of the weights necessary may serve as an example to render the use of the instrument familiar." What had been a potentially disastrous episode in the history of mechanics became a textbook exercise in which the puzzle could safely be left to the ingenuity of the humblest Cambridge student.[38]

In order to make this phenomenon count as the same as previous, traditional, means of displaying the motion lost in resisted motion, Atwood and his colleagues insisted that all retardation was like continuously acting gravitational force. The

[37] Atwood, *Treatise* (cit. n. 10), pp. 8, 31, 40. For the time of the forces' action see pp. 34–35. Desaguliers recommends using soft clay balls as approximately inelastic bodies in *Course* (cit. n. 5), Vol. II, p. 9.

[38] Atwood, *Treatise* (cit. n. 10), pp. 312–313, 331–333.

complex phenomena of impact on soft bodies were rendered intriguing oddities rather than doctrinally basic. Atwood's colleagues agreed with this shift. Parkinson argued that in the Dutch indentation trials "the cavities formed are not the whole effect" because the falling body must also move the parts of clay or wax which it hit. There were no perfectly inelastic bodies. Wood showed that inked ivory balls left larger marks the greater their force of impact. Parkinson, Wood, and Atwood all agreed that 'sGravesande was correct: the depths of these pits varied as the square of the falling body's speed. They simply denied that this measure was relevant to the force of moving bodies. Furthermore, they changed the status of problems of elastic and inelastic impact. Atwood proposed the problem of working out the size of an elastic force that would preserve motion in impact for any given function of speed. Elasticity became a derivative issue for Cambridge mechanics, easily demonstrated with the gestures of Atwood's machine without supposing any special conservation law. In this way Atwood made the machine representative of Newtonian motion and himself one of the representatives of Newton's legacy.[39]

IV. ATWOOD AND SMEATON: THE SOCIAL FUNCTION OF FRICTION

Models of machinery certainly found a home within the philosophical cabinets of the English universities. But academic values did not sit easily with the entrepreneurial culture of metropolitan and industrial Britain. Historians of otherwise differing views have alleged that the Georgian university exhibited little interest in mechanic arts because the colleges depended upon the traditional regime of landed wealth: "the chapel and the hall, the coffee house and the common room," as Edward Gibbon notoriously described them, provided no home for machine philosophy. What has been called "a new kind of entrepreneurial and philosophical gentleman" seems to have emerged outside their walls.[40] Preeminent among these entrepreneurs were the innovative civil engineers, a group first publicly defined as a society in London in 1771. They could challenge academic machine philosophy by querying the security of inferences from Cambridge textbooks to working engines. Principles demonstrated in university lectures might fail in the mill, and the results gleaned from successful machines could be used to refute the platitudes of orthodox mechanics. These challenges were especially intense when leading engineers, devoid of academic training, designed demonstration devices to illustrate their own account of motion and to warrant the large-scale works they built. Engineers did not use their demonstration devices as part of a traditional and authoritative academic regime, but as a means to win credit from genteel clients. They sought patronage for their

[39] Parkinson, *System* (cit. n. 22), pp. 71, 185; Wood, *Principles* (cit. n. 25), pp. 114, 140; and Atwood, *Treatise* (cit. n. 10), pp. 377–378.

[40] For the failings of the universities see Nicholas Hans, *New Trends in Education in the Eighteenth Century* (London: Routledge & Kegan Paul, 1951), p. 52; A. R. Hall, "What Did the Industrial Revolution in Britain Owe to Science?" in *Historical Perspectives: Studies in English Thought and Society in Honour of J. H. Plumb,* ed. Neil McKendrick (London: Europa, 1974), 129–151, on pp. 143–144; Christopher Hill, *Reformation to Industrial Revolution* (Harmondsworth: Penguin Books, 1969), p. 251; and Roy Porter, *English Society in the Eighteenth Century* (Harmondsworth: Penguin Books, 1982), p. 178, who cites Gibbon. See Margaret C. Jacob, *The Cultural Meaning of the Scientific Revolution* (Philadelphia: Temple Univ. Press, 1988), p. 168, on "a new kind of . . . gentleman," and p. 151, on "the deficiencies of the universities."

schemes by demonstrating their understanding of the basic principles of mechanics and their practical success. The interest of the academic demonstration device, therefore, partly resides in its capacity to challenge or to reinforce the boundary that divided the university from the workshop. Atwood and his colleagues interpreted this boundary in a rather specific way. They developed the concept of "friction" into an all-embracing resource that helped explain the differences between different cultural settings for mechanics. Friction explained why the results of demonstrations needed, and deserved, further interpretation to make them fit mechanical principles. Atwood provided a lengthy analysis of the various ways in which friction would work in the different social settings of the academy and the workshop. He noted that there was a radical difference between trials in statics and in dynamics. Statics dictated that bodies would balance each other if their distances from a fulcrum were inversely proportional to their weights. But if there were any considerable friction at the fulcrum, the principle would seem to be confirmed even by imprecise trials. In dynamical trials the situation was reversed. Because of friction, Newton's laws were hard to make true: "The interference of friction and other resistances, which contribute to render the experiments on the equilibrium of forces apparently more perfect than they really are, causes the motion of bodies, which are the objects of experiments, to differ from the theory, with which it would precisely coincide were those obstacles removed." This problem governed the way Atwood designed his machine. Its pulley wheels, the limited distance through which its weights moved, and the lightness of the cord that held them were all designed to produce the unique situation in which Newton's mechanics worked, where "the bodies impelled should be conceived to exist in free space and void of gravity."[41]

Friction also mattered for the relationship between classroom and practical mechanics. Atwood expressed the hope that engineers might become interested in university mechanics. But he insisted that nothing that happened in engineering works could license criticism of this academic enterprise. "The obstacles which are occasioned by friction . . . which scarcely admit of precise estimation, must greatly obstruct, if not wholly prevent, a satisfactory and accurate comparison of the theory with practice, in any ordinary way of considering the subject." Nor should engineers be concerned if academics erred, since the improvement of practical devices derived not from university research but from "long experience of repeated trials, errors, deliberations, corrections, continued throughout the lives of individuals and by successive generations of them." This established a useful barrier between the world of Atwood's machine and that of Georgian engineering. Isaac Milner made the barrier a pragmatic one. It was just that theorists had been concerned with the laws of impact, while engineers used continuously acting forces, "in which motions are rather to be preserved by the gradual effects of weights and pressures. An accurate knowledge therefore of these is more essential to the interests of society."[42]

Atwood could not claim that his demonstration device was relevant to impact motion, since it was explicitly devoted to the gradual effects of weight. So he

[41] Atwood, *Treatise* (cit. n. 10), pp. 292–293, for "the interference of friction"; and George Atwood, *Analysis of a Course of Lectures on the Principles of Natural Philosophy* (London, 1784), p. 25, for "free space." For the new civil engineers see R. A. Buchanan, *The Engineers: A History of the Engineering Profession in Britain, 1750–1914* (London: Jessica Kingsley, 1989), pp. 38–45.

[42] Milner, "Reflections" (cit. n. 8), pp. 363–364.

deployed his useful distinction between the effects of friction in the cases of dynamics and in statics. Dynamical engineers worked with waterwheels and windmills. Atwood pointed out that such devices moved at constant speed under constant impulsion, so there friction must have a major role, rendering them irrelevant to the principles of friction-free mechanics. No waterwheel could inform philosophers of the status of Newtonian laws. Engineers used statics, in contrast, when they worked on bridges and arches. Atwood gained some experience of these structures at the end of the century when he served with engineers and mathematicians such as John Playfair, James Watt, and Nevil Maskelyne on the committee investigating Thomas Telford's scheme for a single-span London bridge made of iron. Atwood began corresponding with Telford. In his report he defended his use of friction-free academic experimental mechanics in analyzing Telford's plan, precisely because, as in the case of the balance, were any friction to act in the bridge structure, this would always aid its stability. "It seems certain that in whatever degree friction and the other impediments to motion may act on the models it is by rendering the whole structure more secure from disunion."[43] In both waterwheel and bridge design, therefore, Atwood used his model of the effects of friction, and their management in his machines, to define the role that he reckoned his machine philosophy should play. It could judge the efforts of bridge builders, but it could not be challenged by millwrights. Milner concurred with this useful conclusion: "It is acknowledged that the experiments which have been made to determine the effects of wind and water-mills do not agree with the computations of mathematicians, but this is no objection to the principles here maintained." It was crucial for the social place of Cambridge mechanics that its demonstration devices should make the established academic regime unobjectionable. Declarations of the principles of this mechanics were situated, tactical moves in defense of the regime.[44]

An important occasion for this defense emerged in spring 1776, when the eminent engineer and instrument maker John Smeaton published work on impact and millwork of direct concern to the Cambridge mathematicians. There is at least a chronological reason to suggest that Atwood's development of his own machine between 1776 and 1779 was in immediate response to Smeaton's techniques. Smeaton's work on millwheels shifted Atwood's favorite distinction between impact and continuous forces from rational mechanics to practical economics. Georgian England was short of high waterfalls and useful power. Waterwheels were commonly either undershot, using the impact of water against paddles, or overshot, using water's gravity to drive buckets from above. To determine whether impact or gravity were the better means of driving them, Desaguliers built a demonstration device that used the reaction of ejected water to turn a vertical tin cylinder. In summer 1743 he showed it to the Royal Society. The model was designed to show the superiority of reaction over impact for waterwheels. At the end of the year he claimed that overshot wheels

[43] George Atwood, *A Dissertation on the Construction and Properties of Arches* (London, 1801), "Supplement", pp. vii-viii. For Atwood and Telford see A. W. Skempton, "Telford and the Design for a New London Bridge," in *Thomas Telford: Engineer,* ed. Alastair Penfold (London: Thomas Telford, 1980), pp. 62–83, on pp. 71, 74.

[44] Milner, "Reflections" (cit. n. 8), p. 371n. For Milner and engineers see T. J. Hilken, *Engineering at Cambridge University, 1783–1965* (Cambridge: Cambridge Univ. Press, 1967), pp. 37–38. For millwrights and the "theoretical confusion" of machine philosophy see Terry S. Reynolds, *Stronger than a Hundred Men: A History of the Vertical Water Wheel* (Baltimore: Johns Hopkins Univ. Press, 1983), pp. 242–247.

were almost four times more efficient than undershot.[45] In the 1750s, inaugurating a lucrative private practice, Smeaton designed a series of workshop machines to model the behavior of such mills. These were not demonstration devices for classroom use. They were intended to make "the outlines" of millwork calculable, and they were confessedly provisional. Smeaton told the Royal Society, and his genteel customers, that "the best structure of machines cannot be fully ascertained, but by making trials with them, when made of their proper size." This was why he delayed publishing his analysis of the efficiency of waterwheels until 1759 and his contemporary work on the relation between power and speed until 1776.[46]

Atwood was trying to design frictionless machines, while Smeaton set out to measure friction's effects. In Cambridge reality was defined in Newtonian terms as void and resistanceless; in Smeaton's engineering works it was modeled through friction and mechanical effect. Smeaton's most important technique was the method of estimating the friction of the wheel (see Figure 3). He held that previous engineers and philosophers had mistaken the role that friction played. This was why his work interested the Cambridge mathematicians. Smeaton reckoned that because of friction he could not assume that the wheel rate in his model was the same as the rate of water hitting the wheel. So he began his runs by hanging a counterweight from the wheel to drive it with the scalepan empty, and increased this counterweight until the wheel revolved at the same rate whether the water was running or not. The size of the counterweight would then give him the size of friction.[47] His results were presented as conclusive: overshot wheels were twice as efficient as undershot wheels, and were driven by weight alone. Impact played almost no role in the most efficient machines. Undershot wheels were inefficient because in collisions inelastic bodies, such as water, "communicate only a part of their original power; the other part being spent in changing their figure in consequence of the stroke."[48]

He needed new resources to make this argument stick, so in 1759 he built a new experimental model to measure the relation between his estimates of power and the machine's speed (see Figure 4). The tabletop machine included a scalepan suspended over a pulley by a cord attached to a vertically rotating spindle. Weights in the pan drove the spindle, whose motion could be altered both by changing the diameter around which the cord was wound and by moving two lead cylinders attached to the spindle by a horizontal arm. In this device, unlike his millwheel model, Smeaton aimed to minimize friction, not to measure it. Considerable care was needed: he had to make sure the weight fell slowly, lest gravitational acceleration play a role, and he had to judge when the cylinders' "velocity is apparently diminished," for from then on friction would be a factor.[49] Once again, when he

[45] Desaguliers, *Course* (cit. n. 5), Vol. II, pp. 450–453, 459–461; see also Reynolds, *Stronger than a Hundred Men* (cit. n. 44), pp. 214–216. See Stewart, *Rise of Public Science* (cit. n. 5), pp. 326–333, for background.

[46] John Smeaton, "An Experimental Enquiry concerning the Natural Powers of Water and Wind to turn Mills," *Phil. Trans.*, 1759, *51*:100–174, p. 101; and Smeaton, "An Experimental Examination of the Quantity and Proportion of Mechanic Power necessary to be employed in giving different Degrees of Velocity to Heavy Bodies from a State of Rest," *Phil. Trans.*, 1776, *66*:450–475. See Norman Smith, "Scientific Work," in *John Smeaton,* ed. A. W. Skempton (London: Telford, 1981), 35–57; and Jacob, *Cultural Meaning* (cit. n. 40), pp. 160–163, for Smeaton's ambitions.

[47] Smeaton, "Natural Powers of Water and Wind" (cit. n. 46), pp. 107–111.

[48] *Ibid.*, p. 130.

[49] Smeaton, "Quantity and Proportion" (cit. n. 46), p. 460.

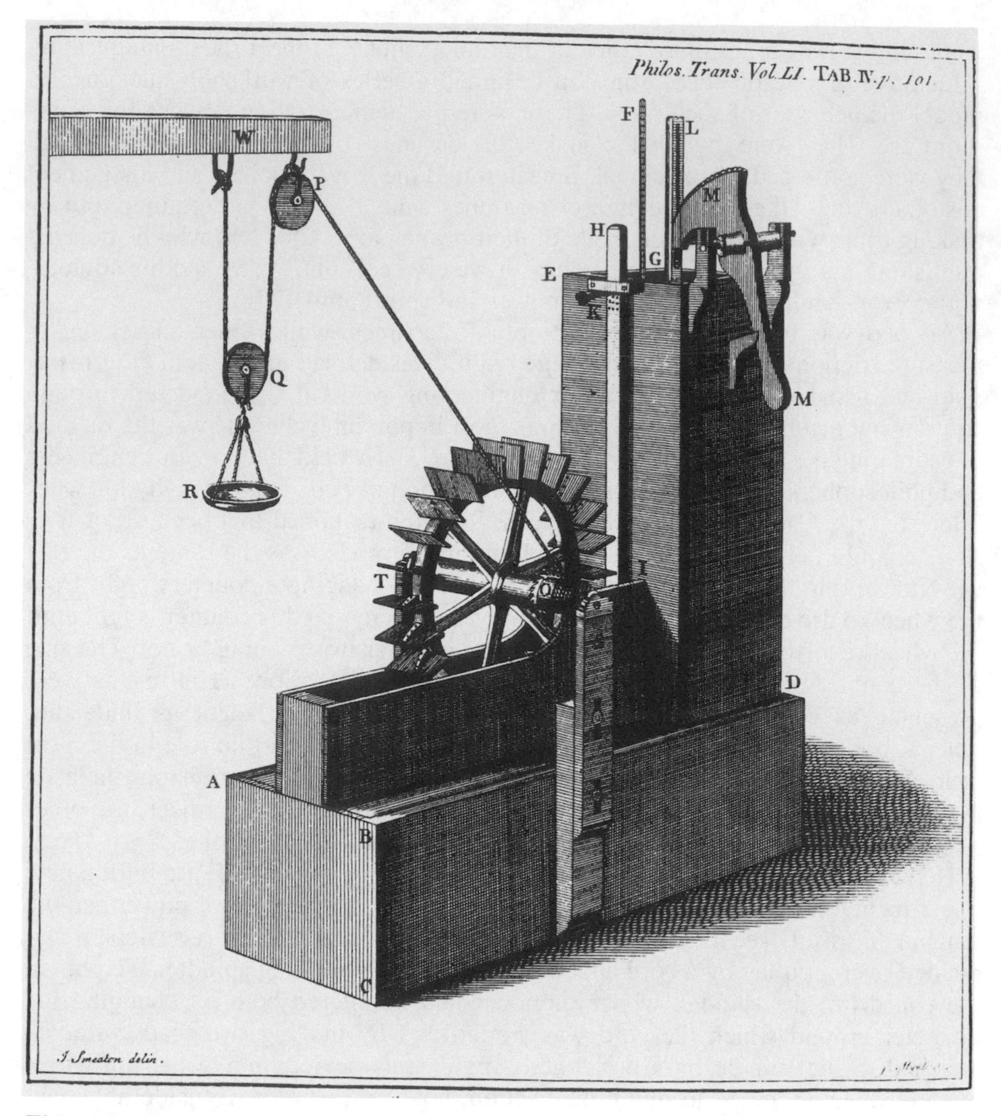

Figure 3. *John Smeaton's device for estimating the efficiency of waterwheels. From* Philosophical Transactions, *1759,* 51:*101. Courtesy of Whipple Library, Cambridge.*

published the results in April 1776 at the Royal Society, his numbers were presented as decisive. He reckoned that the mechanic power was proportional to the square of the cylinders' speed; that the product of the weight and its time of fall was proportional to their speed; and that "this is the universal law of nature respecting the capacities of bodies in motion to produce mechanical effects."[50]

Baldly expressed as mathematical proportionalities, Smeaton's results could easily have been absorbed into academic rational mechanics. However, the gentlemanly engineer presented them as major challenges to received machine philosophy. In stark contrast to Milner and Atwood, he insisted that the errors of the philosophers

[50] *Ibid.*, p. 469.

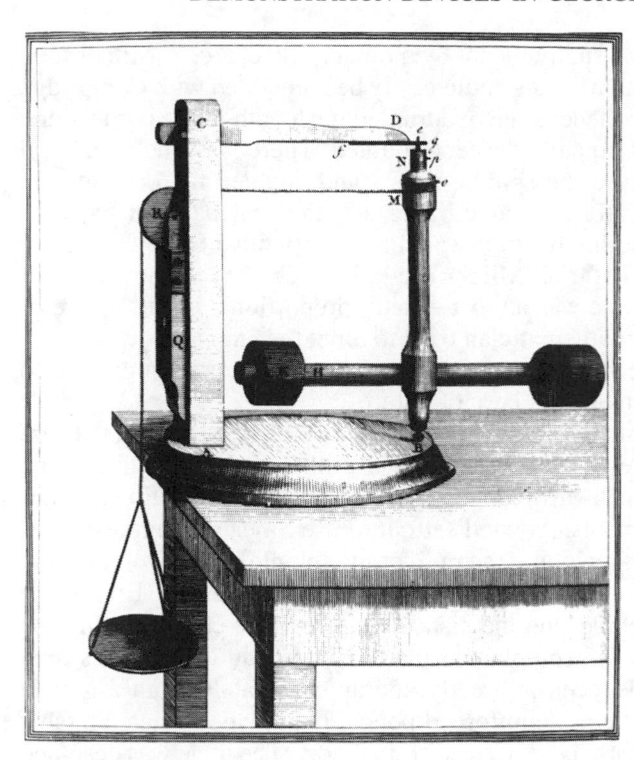

Figure 4. *Smeaton's device for measuring the relation between applied weight and speed of rotation. From* Philosophical Transactions, *1776,* 66:*460. Courtesy of Whipple Library, Cambridge.*

could be judged by engineers. "Some of these errors are not only very considerable in themselves, but also of great consequence to the public, as they tend greatly to mislead the practical artist in works that occur daily, and which often require great sums of money in their execution."[51] Smeaton's experimental demonstrations licensed novel accounts of Newtonian mechanics between the 1770s and 1790s. His work was also recast into the Cambridge vernacular. Milner and Atwood agreed that his authority gave his trials immense value. Atwood reckoned that "from the known abilities and ingenuity of the author" Smeaton's experiments "might be assumed as standards," were it not for the crucial fact that Newtonian mechanics must be used to judge all such trials and could not be tested by them.[52] So Atwood and his colleagues deployed their models of academic mechanics and the role of friction to manage the interpretation of Smeaton's work. Thus Parkinson pointed out that Smeaton should have taken friction into account before declaring in favor of the Leibnizian measure on the basis of his millwheel model. Parkinson also judged that the claim that mechanic power varied as the square of the speed was simply a tautological restatement of Smeaton's definition of power as the product of weight and distance. "But this

[51] *Ibid.,* p. 452. For Smeaton's later crucial demonstration of the motion lost in elastic collision, see John Smeaton, "New Fundamental Experiments upon the Collision of Bodies," *Phil. Trans.,* 1782, *72:*337–354; and commentaries in William Hyde Wollaston, "Bakerian Lecture on the Force of Percussion," *Phil. Trans.,* 1806, *96:*13–22, p. 18 (delivered 1805), and Peter Ewart, "On the Measure of Moving Force," *Memoirs of the Literary and Philosophical Society of Manchester,* 1813, 2nd ser. *2:*105–228, on p. 161 (delivered 1808).

[52] Milner, "Reflections" (cit. n. 8), p. 373; and Atwood, *Treatise* (cit. n. 10), p. 383 (quotation). For Smeaton's role in American engineering see Edwin T. Layton, "Newton Confronts the American Millwrights," in *Beyond History of Science*, ed. Garber (cit. n. 14), 179–193.

power is different from what is usually meant by moment, force, &c."[53] Milner took the same line. Almost all Smeaton's trials could easily be reconciled with Cambridge theory. Smeaton reported that whenever he quadrupled the length of the cord wound round the spindle, the time of equable descent doubled. There was one puzzle: in the second trial, Smeaton halved the spindle radius and doubled the cord length. Though the space of fall therefore remained constant, the scalepan fell twicc as slowly as in the first trial. The implication was that the effective force must have been four times larger in the first trial. Milner used his mechanics to show that for a given weight the effective force was almost exactly proportional to the square of the spindle radius. The young mathematician tried to blunt the engineer's challenge: "The nice agreement of Mr. Smeaton's experiments with the theory cannot fail to add fresh evidence to these established laws of nature."[54]

Atwood turned Smeaton's data to the same end. He unfavorably compared the care with which friction had been estimated in the earlier millwheel trials with the fact that Smeaton had ignored such "unavoidable imperfections of construction and observation" in the 1776 experiments. Atwood set out to use Smeaton's figures to get the dimensions of his machine, including the radii of the spindle and the total weight of the system. Then he could derive the time of the scalepan's descent from these weights, the radius of the machine, and the distance between the center of gyration and the center of gravity. Atwood complained that Smeaton had ignored this final value, which could only be deduced indirectly and approximately from a trial in which Smeaton used the spindle as an unforced pendulum. Atwood printed a table comparing Smeaton's figures with those derived from theory. The table was designed to show that the machine moved more slowly than it should because of the significant friction of the spindle. Finally, he argued, as had Parkinson, that Smeaton's claims about mechanic power were tautological. They were only meaningful, allegedly, if understood as referring to the "intensity" of the power, the ratio of the spindle radius to the distance of the center of gyration from the machine's axis. So Smeaton was wrong to argue "some deficiency in the theory." The true problem was "the obstacles which have prevented the application of it to the complicated motion of engines."[55] Atwood reckoned he had mastered this problem—armed with a rival demonstration device and a successful measure of friction, he was able to subordinate engineering works to the authority of the established order and its performances.

V. ATWOOD'S MACHINE IN THE MARKETPLACE

Atwood's machine, in some respects, occupies a similar cultural niche to that of the London lectures and instrument shops where men like Smeaton first established themselves. It was "executed with great mechanical skill" by the eminent London mathematical instrument designer George Adams the younger, maker to the king and master of the country's largest instrument shop. He owned copies of all of Atwood's publications, including his lecture courses, and was obviously a close ally of the Cambridge don. Adams was also a high flyer in religion, a fierce enemy of

[53] Parkinson, *System* (cit. n. 22), p. 70–71.
[54] Milner, "Reflections" (cit. n. 8), pp. 375–376.
[55] Atwood, *Treatise* (cit. n. 10), pp. 382–399, esp. p. 382.

Priestley's radical cosmology: "Natural philosophy affords no support to the wretched system of materialism but concurs with religion in endeavouring to enlighten the mind, to comfort the heart, to establish the welfare of society and promote the love of order."[56] Distribution of the book and of the machine itself relied upon the excellent network of sales and information centering in the London shops, at this period the world leaders in the instrument trade. For his *Treatise,* Atwood commissioned a very fine picture of his device from the celebrated London architectural draftsman and Royal Academy gold medalist Thomas Malton; the engraver was James Basire, the leading artist in the city. The device stood out precisely because of the quality of its representation, especially in contrast with the host of more conventional geometrical diagrams that accompanied it. Equally graphic was the speed with which the machine became eponymous. By 1796 Wood's mechanics textbook already assumed its Cambridge readers would know of "Mr ATWOOD's Machine" and would understand how to use it "to prove the truth of [Newton's second] law of motion."[57] The London makers William and Samuel Jones purchased Adams's instruments and manuscripts at his death in 1795. They arranged the sale of the important philosophical cabinet of the wealthy virtuoso and Tory magnate James Stuart, the Earl of Bute, in 1793, including Bute's copies of Atwood's handbooks and of his machine. In 1799 Harvard College ordered an example of the machine from them. Within a decade Atwood's demonstration device had become standard and widely distributed in the workshops, salons, and lecture rooms of natural philosophy. By the early nineteenth century such makers as John Newman in London and Nicolas Pixii in Paris sold them worldwide.[58]

A good example of the multiplication of the milieus in which such devices were made to work is provided by the entrepreneurship of the celebrated Portuguese agent Jean Magellan, who spent the 1770s and 1780s commissioning London instruments for enlightened Europe. One of Magellan's chief clients was Alessandro Volta, the new natural philosophy professor at Pavia, where he acted for the state purposes of the Hapsburg governor Carlo Firmian, whose autocratic education policies were used to back the expansion and high status of Volta's new cabinet against clerical opposition.[59] In selling Atwood's machine to Volta, Magellan had chosen well. Volta

[56] Atwood, *Treatise* (cit. n. 10), p. 337. Adams's antimaterialism is evident in George Adams, *Lectures on Natural and Experimental Philosophy,* 5 vols. (London, 1794), Vol. I, pp. ix-x. Adams owned many pietist works; see John R. Millburn, *The Library of George Adams* (Aylesbury: For the author, 1988). See also Maurice Daumas, *Scientific Instruments of the Seventeenth and Eighteenth Centuries and Their Makers* (London: Portman, 1989), pp. 238–239.

[57] Wood, *Principles* (cit. n. 25), p. 27n. London makers' market leadership is discussed in J. A. Bennett, "The Scientific Context," in Bennett *et al., Science and Profit in Eighteenth Century London* (Cambridge: Whipple Museum, 1985), pp. 5–9. In 1801 Malton and Basire produced a famous engraving of Telford's projected bridge when Atwood served on the committee to assess the project. See Skempton, "Telford and London Bridge" (cit. n. 43), pp. 71–72.

[58] David P. Wheatland, *The Apparatus of Science at Harvard, 1765–1800* (Cambridge, Mass.: Harvard Univ., 1968), pp. 97–98; and G. l'E. Turner, "The Auction Sales of the Earl of Bute's Instruments, 1793," *Annals of Science,* 1967, *23:*213–241, on p. 239, lot 236. Jones sent Harvard a copy of Atwood's *Treatise* to accompany the machine. The Deutsches Museum possesses a version of the machine designed by the Eichstadt instrument maker Johan Wisenpaintner in 1795: Daumas, *Scientific Instruments* (cit. n. 56), p. 256. The Whipple Museum, Cambridge, owns a machine made in Paris around 1820. Pixii sold machines to the United States from before 1819; Newman sold one to the Quebec seminary in 1836 (information generously supplied by Deborah Warner, National Museum of American History, and by Maryse Tellier, Musée de Séminaire de Québec).

[59] For Magellan's interests in London see Joaquim de Carvalho, "Correspondencia científica dirigida a Joao Jacinto de Magalhaes," *Revista da Faculdade de Ciencias da Universidade de Coimbra,*

recognized that because of the public's "passion for novelty and wonders," the embodiment of a theory in a graphic instrument was crucial for that theory's public reception. He gave the example of his own development of the electrophorus in 1775: "As soon as my apparatus came on the scene, its effects, which were all the greater and more surprising as they were easily obtained, must have struck all observers and blinded them." Volta helped support Magellan's program by cultivating such surprise once again. In his turn, and brushing aside Atwood's tardiness in publicizing the machine, Magellan ingeniously transformed a Cambridge demonstration device, designed to make a frictionless, precision determination of the motion of bodies under continuous forces, into a dramatic piece of showmanship. Where Atwood stressed that his machine was designed to avoid the troubles of free fall and the inclined plane, and thence to generate precise numbers about accelerated motion, Magellan said much more about its use in public lectures. He told Volta that "you know that observations on the fall of bodies and the acceleration of their speeds demand operations which are very delicate, highly difficult and rather laborious, and, what is more, absolutely impracticable in a regular Course of Experimental Physics."[60] Magellan tailored his presentation to the tastes of continental rational mechanics by adding a new trial to demonstrate that the force acting on a body falling through a fixed distance was proportional to the square of its final speed. He simply omitted Atwood's comments on the errors of the doctrine of living force.[61] The bulk of his pamphlet was taken up with instructions for these operations. Three new techniques were introduced: a trick for starting the falling weight using a small copper handle to hold the weight before releasing it without shaking the pulley; the replacement of Atwood's wooden cylinders with stacks of discs linked with a three-inch brass ring; and an added set of balance screws for stabilizing the machine's base, explicitly based on techniques worked out by Adams and by Gowin Knight for stabilizing compasses in measures of magnetic dip. Magellan claimed that each of these techniques made the performance much more effective than in Atwood's original design.[62]

The most important changes were in the use of the clock and the setting of the stage that stopped the weight's fall. Magellan replaced Atwood's original "common

---

1951, *40:*93–283, on pp. 98–101; and Stephen F. Mason, "Jean Hyacinthe de Magellan and the Chemical Revolution of the Eighteenth Century," *Notes Rec. Royal Soc., 45:*155–164, on pp. 158–161. My thanks to Palmira Costa and Stephen Mason for help with these sources. For Volta and Firmian's projects see Dino Carpanetto and Giuseppe Ricuperati, *Italy in the Age of Reason* (London: Longman, 1987), pp. 163–166; Volta to Firmian, 12 Aug. 1779, Magellan to Volta, 28 Sept. 1779, and Volta to Firmian, 7 Jan. 1780, in Alessandro Volta, *Epistolario,* Edizione Nazionale, 5 vols. (Bologna: Nicola Zanichelli, 1949–1955), Vol. I, pp. 378–379, 388–390, 394; and Volta, "Nota di macchine singolari," *ibid.,* Vol. II, p. 467.

[60] For Volta on the electrophorus see Marcello Pera, *The Ambiguous Frog: The Galvani-Volta Controversy on Animal Electricity* (Princeton: Princeton Univ. Press, 1992), p. 46. For Magellan on lecture courses see Jean Magellan, *Description d'une Nouvelle Machine de Dynamique inventée par Mr G Atwood* (London, 1780), p. 259. This text was continuously paginated with Magellan's separate *Collection* (see below, n. 63). His diagram of Atwood's machine was printed on the last sheet, plate 6, of the *Collection* and bound as the frontispiece of the *Description.* The despatch to Italy is discussed in Magellan to Volta, 21 Nov., and 29 Dec. 1780, and Firmian to Volta, 8 May 1781, in *Epistolario* (cit. n. 59), Vol. II, pp. 14–15, 17, 37.

[61] Magellan, *Description* (cit. n. 60), pp. 262, 277–278. Magellan's mathematical physics, especially his version of the conservation of motion, came from Etienne Bézout, *Cours de mathématiques,* 5 vols. (Paris, 1764–1769), Vol. V, pp. 440–441.

[62] Magellan, *Description,* pp. 265–266, 287–288.

clock" with a specially commissioned timepiece which he had designed himself to resist temperature changes and to beat seconds as audibly and reliably as possible. He had already sent such a device to the Spanish court and recommended it to Volta as the perfect accompaniment for Atwood's device. This clock allowed Magellan to make the trials on accelerated motion as dramatic as possible. First the demonstrator must follow Atwood's instructions and become habituated to the clock's beat. Then he must set up a trial so that the falling cylinder would be released at one beat and strike the lower stage at a later beat. Distances of fall had to be managed so that the sound of this impact and that of the clock's beat would come together. This was one indication that Magellan did not envisage the machine as a harsh trial of the laws of motion. Rather, it was supposed to dramatize them. Thus the textbook outcome of experiments was presented not as a testable prediction, but as a recipe for putting on a good show: "Do not lose courage if you do not succeed at first in some experiment. For by repeating it several times you will end up performing it well." For example, with given weights in the two cylinders and an extra bar of ¼ ounce on the falling weight, the ring should be set at 16 inches and the lower stage at the bottom of the scale. Then at exactly two seconds the bar would be removed by the ring and at the third second the remaining cylinder would hit the stage. "This is all that is needed to be assured of the exactitude of the experiment by the evidence of the senses. A little exercise and habit, together with patience, and a natural disposition of the observer for these objects, will not fail to make all these operations very easy for him in a very short time."[63]

Magellan's comments showed that in order to make the machine's "exactitude" visible to an audience it was important that its demonstrator become habituated to the machine. The work of "exercise and habit" deserves further historical examination. Very different uses of Atwood's machine in lecture rooms, laboratories, and public shows demanded rather different patterns of conduct from its users. Variations in what Magellan called demonstrators' "natural disposition" accompanied variations in the doctrine demonstrated. Atwood and Smeaton were both capable of extracting "universal laws of nature" from the performance of their devices and models. This helps explain the capacity of robust instruments to work outside the setting of their invention and to be deployed for new purposes. By changing the set of gestures with which they managed their devices, and making these gestures an apparently obvious accompaniment of the device in question, users could transform an instrument's message while preserving its authority. In disciplining their audiences, they also disciplined both the machine and themselves. The material culture of natural philosophy, its instruments and models, was a vital part of its doctrinal authority. This helps explain the important, and neglected, role of museums of apparatus in the propagation of machine philosophy, as the work of Cambridge demonstrators such as Samuel Vince, William Farish, and Robert Willis shows. From the 1790s "working Models of almost every kind of Machine" were on show "in such

[63] *Ibid.,* pp. 266–267, 286. For the new clock and its despatch to the Spanish colonies see Jean Magellan, "Notice des Instrumens d'Astronomie, de Géodesie, de Physique" (1780), in *Collection de différens Traités sur des Instrumens d'Astronomie, Physique, &c.* (London, 1775–1780), pp. 193–254, on pp. 204–207. For Volta's enthusiasm about Atwood's machine see Volta to Firmian, 1 May 1781, in Volta, *Epistolario* (cit. n. 59), Vol. II, pp. 34–36: "The laws of fall of weights are presented so clearly and distinctly to the eyes that anyone who knew nothing of the theory would soon be driven to a marvellous understanding of it."

a way as to make them in general do the actual work of the real Machine on a small scale." In these showrooms, real work was represented through machine philosophy's devices.[64] As recent commentators on museum displays have urged, the objects of material culture are incapable of transmitting a single, sempiternal meaning that survives throughout all their possible uses. But neither are they so weak that their meaning can be completely determined by a local, verbal description. Practical interaction with material devices temporarily associates them with a culturally specific sense. This sense can be changed by transforming cultural practice; it can be secured by building new patterns of disciplined behavior.[65]

[64] For Vince see Rouse Ball, *History of the Study of Mathematics* (cit. n. 16), p. 104. For Farish (whose lecture is quoted here) see Hilken, *Engineering at Cambridge* (cit. n. 44), pp. 38–40. For Willis see Robert Willis and John Willis Clark, *Architectural History of the University of Cambridge,* 3 vols. (Cambridge: Cambridge Univ. Press, 1886), Vol. III, pp. 162–163.

[65] Susan Pearce, ed., *Objects of Knowledge* (London: Athlone Press, 1990); and George W. Stocking, ed., *Objects and Others: Essays on Museums and Material Culture* (Madison: Univ. Wisconsin Press, 1985).

INSTRUMENTS IN THE LIFE SCIENCES

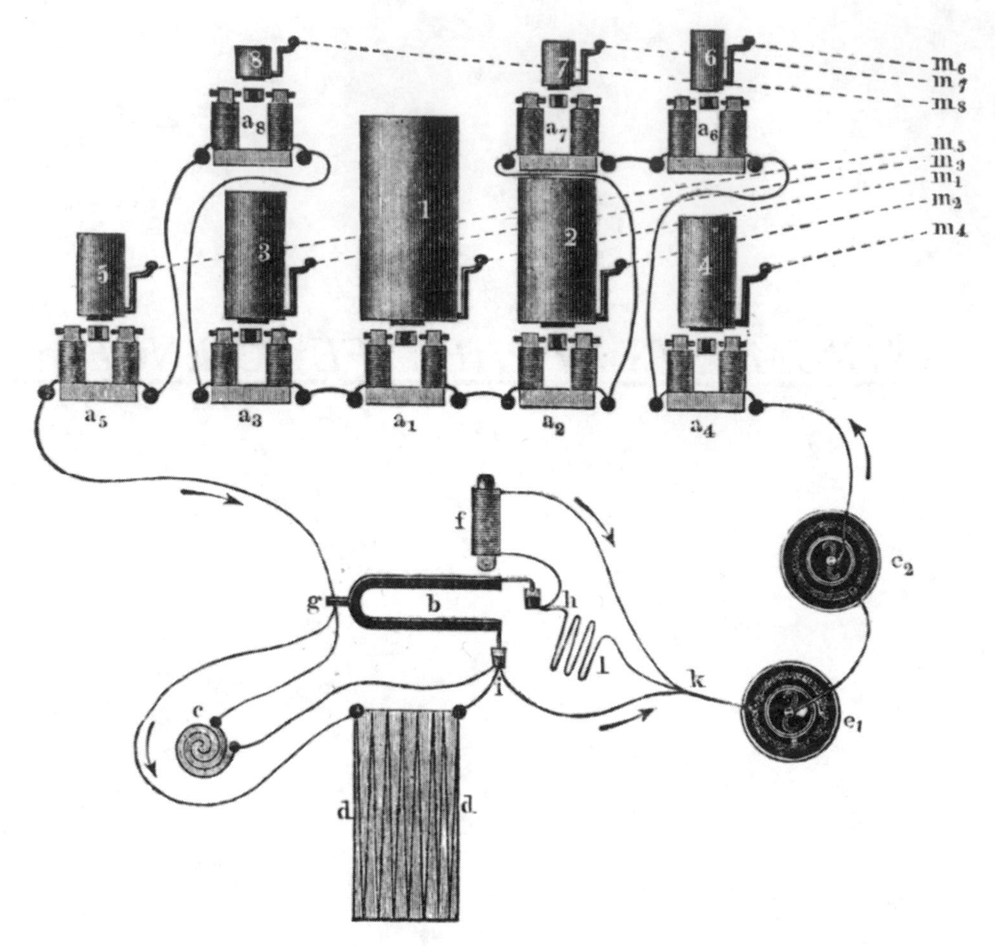

Figure 1. *The connected series of tuning-fork resonators that Helmholtz used to combine partial tones, creating tones indistinguishable from those produced by musical instruments. See the discussion on page 200. From Hermann Helmholtz,* Die Lehre von den Tonempfindungen als physiologische Grundlage für die Theorie der Musik, *5th ed., (Brunswick: Vieweg & Sohn, 1896), p. 633.*

Helmholtz and the Materialities of Communication

*By Timothy Lenoir**

ONE OF HERMANN HELMHOLTZ'S PRIMARY CONTRIBUTIONS to physiological optics was the experimental elaboration of the three-receptor hypothesis for explaining color vision, a theory first proposed by Thomas Young. Early in his career, however, Helmholtz had publicly rejected the Young hypothesis. My concern in this paper is with the role of experiment and instrumentation in Helmholtz's reversal of position on this issue and more specifically with the positive contribution made by a variety of new media technologies, particularly electric, photographic, and telegraphic inscription devices. These media devices served Helmholtz as analogues and as models of the sensory processes he was investigating. They did not merely assist him in understanding the operation of the eye and ear through measurement; more to the point, Helmholtz conceived of the nervous system as a telegraph—and not just for purposes of popular presentation. He viewed its appendages—sensory organs—as media apparatus: the eye was a photometer; the ear a tuning-fork interrupter with attached resonators. The output of these devices was encoded in the form of an *n*-dimensional manifold, a complex measure to which a sign, such as *Rot, Blau-grün,* or *ü* was attached. These materialities of communication were important not only because they enabled theoretical problems of vision and hearing to be translated, externalized, and rendered concrete and manipulable in media technologies but furthermore because in this exteriorized form analogies could be drawn between devices; linkages could be made between different processes and between various aspects of the same process.[1] Crucial to Helmholtz's theorizing were analogies between sound and color perception. Indeed, I will argue that a crucial step in his development of the trichromatic receptor theory of color sensation came through analogies between the technologies of sound and of color production. The juxtaposition of media enabled by the materialities—the exteriorized forms—of communication was a driving force in the construction of theory.

Essentially my claim is as follows: Helmholtz's model of representation was that of an abstract system of relations among sense data. Like Bernhard Riemann, working independently at almost exactly the same time, Helmholtz treated the mental

* Program in History and Philosophy of Science, Stanford University, Stanford, California 94305-2024.

I am grateful to M. Norton Wise, Bernhard Siegert, Hans-Jörg Rheinberger, and Peter Galison for helpful suggestions on earlier drafts of this paper.

[1] The notion of "materiality of communication" I refer to throughout this study is an adaptation of proposals made by Hans Ulrich Gumbrecht and Ludwig Pfeiffer in the introduction to their edited volume *Materialität der Kommunikation* (Frankfurt am Main: Suhrkamp, 1988).

representation of sensations as *n*-dimensional manifolds. Different modalities of sense were characterized as manifolds obeying different metric relations.[2] The sense data were organized into symbolic codes by a system of parameters due to the physical properties of each sense and adjusted by experience. These ideas about representation emerged out of three fields of investigation—color, sound, and electrotelegraphy—familiar to Helmholtz during 1850–1863, the period spanning his work on the speed of nerve transmission, physiological color mixing, and physiological acoustics and culminating with the publication of Part II of the *Physiological Optics*. For my purposes the most innovative character of Helmholtz's work derived from his adapting a number of interrelated technical devices employed in telegraphy to the measurement of small intervals of time and the graphic recording of temporal events in sensory physiology. From as early as 1850 he drew analogies between the electrical telegraph and the process of perception. The telegraph began to serve as a generalized model for representing the processes of sensation and perception. In light of this telegraph analogy Helmholtz, so I hypothesize, imagined the virtual image cast on the retina as dissolved into a set of electrical impulses, data to be represented by symbols as an "image" in the brain through a perceptual analogue of Morse code.[3]

Between 1850 and 1855 Helmholtz was working intensively with the myograph and a variety of electrical devices adapted from the telegraph industry to measure the speed of nerve transmission and other features connected with nerve action and muscle contraction. Telegraphy was not simply a useful model for representing and thinking about vision and hearing. Experiments involving those devices were also crucial in advancing his own program of sensory physiology. This role of telegraphic devices and a variety of imaging devices became particularly important between 1855 and 1860, when, reacting to a critique of his theory of spectral color mixing by Hermann Grassmann, Helmholtz retracted his earlier (1852) rejection of the Young trichromacy theory of physiological color mixtures. Helmholtz suggests that he arrived at this view via a comparative analysis with hearing,[4] and I pursue this suggestion in depth in the third section of this article. Helmholtz pursued a similar research strategy of representing tone production and reception in terms of a variety of components of electrical telegraphic circuitry combined with several techniques for graphic display of wave motion, particularly sound waves. These devices were crucial in his investigation of combination tones, the analogue to forming color mixtures from primary colors. Helmholtz postulated retinal structures—three receptors sensitive primarily to wavelengths in the red, green, and violet ranges respectively—analogous to the arches of Corti in the ear. The analogy provided the path to accepting the Young trichromatic theory, previously rejected. Once again, new media

[2] On Helmholtz's sign theory of perception and the relation of his views to the work of Herbart see Timothy Lenoir, "The Eye as Mathematician: Practice, Instrumentation, and Helmholtz's Construction of an Empiricist Theory of Vision," in *Hermann von Helmholtz and the Foundations of Nineteenth-Century Science,* ed. David Cahan (Berkeley: Univ. California Press, 1993), pp. 109–153; and Lenoir, "Helmholtz, Müller und die Erziehung der Sinne," in *Johannes Müller und die Philosophie,* ed. Michael Hagner and Bettina Wahrig-Schmidt (Berlin: Akademie Verlag, 1992), pp. 207–222.

[3] See Hermann Helmholtz, "Über die Methoden, kleinste Zeittheile zu messen, und ihre Anwendung für physiologische Zwecke," in *Wissenschaftliche Abhandlungen,* 2 vols. (Leipzig: Barth, 1883), Vol. II, pp. 862–880, esp. p. 873.

[4] See Hermann Helmholtz, "Die neueren Fortschritte in der Theorie des Sehens," in *Hermann von Helmholtz: Selected Writings,* ed. Russell Kahl (Middleton, Conn.: Wesleyan Univ. Press, 1971), p. 181.

technologies associated with communication and representation were crucial in this transition; for Helmholtz drew upon processes of photonegative production in providing a physiological explanation of positive and negative afterimages, crucial to refining the three-receptor hypothesis.

I. THE BERLIN PHYSICAL SOCIETY AND THE NEW MEDIA TECHNOLOGIES

Helmholtz's researches, including his treatment of color, were situated within a context of interest in telegraphy and technologies of representation. Several of Helmholtz's closest associates during the late 1840s and early 1850s were deeply involved in electrotelegraphy and photography. These individuals had founded and sustained the Berlin Physical Society; its journal, *Die Fortschritte der Physik,* a review of the literature on measurement technologies and physics applied to various fields, was intended as a program statement of their so-called physicalist school. The proceedings of the Berlin Physical Society provide an overview of the members' interests in measurement and technologies of representation. The first volume of the *Fortschritte der Physik* (1845), for instance, devoted a seventy-two-page review by Gustav Karsten to literature on photography and daguerreotypy, and a twenty-two-page review by Werner Siemens to telegraphy.

Methods for measuring small intervals of time and for graphically representing processes taking place in times too brief to experiment with directly were high on the list of the society members' interests—Helmholtz referred to these graphic methods as "Mikroskopie der Zeit" (time microscopy) in his paper of 1850 summarizing these different developments.[5] In the meeting of 25 July 1845 Siemens presented a paper on measurement of the velocity of mortar shells using marks on graph paper made by sparks triggered by the projectile moving through the cannon bore; in the same meeting Karsten discussed employing daguerreotypes to measure solar spectra. Emil du Bois-Reymond discussed methods for measuring the speed of nerve transmission and muscle contraction in the meeting of 7 March 1845; and in that same meeting Ernst Brücke gave the first of several papers he would deliver in 1845 on the subjects of retinal cones and on the inability of infrared light to penetrate the optical media of the eye to the retina. Brücke's experiments were assisted by Karsten, who prepared extrasensitive photographic paper to use as light detectors. Brücke concluded his discussion of the sensitivity of the retina to light in the range of wavelengths between red and violet as "the most sensitive of all known Actinoscopes."[6] Among the topics discussed in the meeting of 31 October 1845 was H. L. d'Arrest's treatment of various methods for determining the isochrony of pendula. On 20 February 1846 a certain Leonhardt, an instrument maker in Berlin, discussed his new electrical telegraph.[7]

Helmholtz delivered his first paper in the Physical Society on 23 July 1847, on the conservation of force. In the previous meeting, of 9 July, Brücke had discussed afterimages and physiological-contrast colors. He followed this paper with another on 15 October on methods for making the motion of a vibrating string visible. In 1848 Brücke discussed further aspects of color theory, namely, the origins of

[5] Helmholtz, "Über die Methode, kleinste Zeittheile zu messen" (cit. n. 3), p. 870.
[6] See Ernst Brücke's review of the recent progress in physiological optics in *Fortschritte der Physik,* 1846, 2:227.
[7] *Ibid.*, pp. xv-xviii.

"brown" and the order of colors in Newton's rings. Du Bois-Reymond presented work on his "multiplicator," a galvanometer capable of registering bioelectric currents; and he and Johannes Halske presented their related work on the magnetoelectromotor. In the first meeting of 1850 Siemens presented his new work on telegraph apparatus. At the next meeting Helmholtz presented work on the speed of nerve transmission, to which three sessions of 1850 were devoted. One session, 18 July, was devoted to explaining and demonstrating the operation of Helmholtz's myograph in graphically recording nerve transmission and muscle contraction. In addition to telegraphs and recording devices for nerve and muscle action, several sessions in 1850 were devoted to Wilhelm Beetz's work in acoustics, in particular tones produced by rotating tuning forks and work on combination tones. The three years 1854–1857 were probably the most active years in the Berlin Physical Society for presentations on telegraphy. Siemens presented nine lectures on several different topics, including the design and operation of the electromagnetic telegraph, his system for sending multiple messages or messages in opposite directions over the same cable, and the problems of laying underwater cables. Halske presented papers on improvements he and Siemens had made to the Morse telegraph. Halske also discussed his new polarization kaleidoscope and work on moving stereoscopic images. Helmholtz contributed refinements on his earlier work on the speed of nerve transmission. Telegraphy, imaging devices, electromagnetic devices for time measurement, and graphic display of temporal processes connected with light, sound, or neurophysiological phenomena were the most enduring interests of the active members of the Berlin Physical Society.

II. HELMHOLTZ'S REACTION TO GRASSMANN

Helmholtz's decision to reconsider his earlier rejection of Young's three-color hypothesis came not as a result of the technologies explored at the Physical Society, but in response to the extremely abstract criticism of Hermann Grassmann. Helmholtz first encountered Grassmann's work in 1852. As part of the procedure connected with his appointment at Königsberg, Helmholtz chose as the subject for his *Habilitationschrift* a critique of David Brewster's theory of color, which was based on the view that the spectral colors are mixtures of three elementary colors, red, yellow, and blue. Grassmann found Helmholtz's paper wanting in certain respects, but his primary interest lay in using it to illustrate once again the general applicability of the methods of his *Ausdehnungslehre*. For Helmholtz, the interest lay in Grassmann's geometrical theory of color-space as a sensory manifold, but he insisted, as Grassmann had not, that the color-space be demonstrated by experiment as well as by logic. The exchange through papers in *Poggendorff's Annalen der Physik und Chemie* between 1852 and 1855 led Helmholtz to revise his own early ideas.

Helmholtz's papers on color mixtures featured the adaptation and refinement of existing instruments and experimental practice that characterized all his early papers and aroused the admiration of his contemporaries.[8] In his first paper of 1852 the arrangement for mixing pure spectral colors consisted of two slits forming a V in a

[8] See James Clerk Maxwell's praise of Helmholtz's approach in "Experiments on Colour, as Perceived by the Eye, with Remarks on Colour-Blindness," *Transactions of the Royal Society of Edinburgh,* 1855, *21*:275–298.

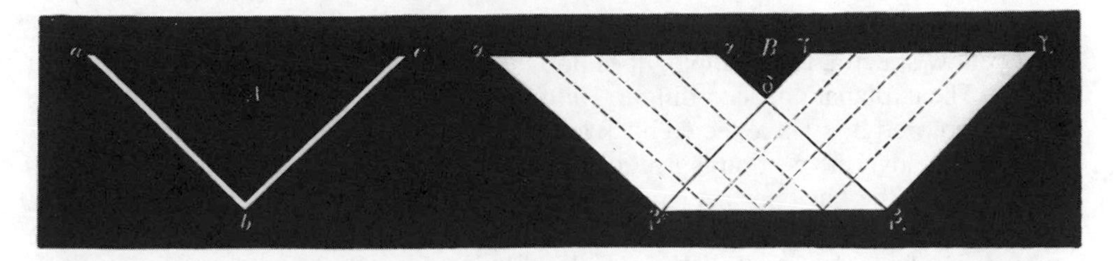

Figure 2. *Helmholtz's V slit for mixing colors. From Hermann Helmholtz,* Handbuch der physiologischen Optik *(Hamburg/Leipzig: Leopold Voss, 1896), p. 352.*

black blind (see Figure 2).[9] The slits were inclined by 45 degrees to the horizontal and were at right angles to one another. Different spectral colors passed through each slit were combined at the point of intersection. To generate the possible combinations of these two colors, a flint-glass prism was placed vertically in front of the objective lens of a telescope, which was focused on the intersection of the slits, both prism and telescope at a distance of twelve feet from the V-slit blind. Helmholtz noted that a similar arrangement with a single vertical slit instead of a V slit produces a rectangular spectrum in which the different color bands and Frauenhofer lines run vertically parallel to one another. The spectrum of an inclined slit is a parallelogram with two parallel horizontal sides and two sides parallel to the slit. In this inclined situation, the color bands and Frauenhofer lines run parallel in the direction of the slit. In the case of Helmholtz's V slit the two spectra overlapped, with the two sets of color bands and Frauenhofer lines running in the directions of the slits. When viewed through the telescope, the area of overlap of the two spectra was a triangle, and within the triangle all the combination colors resulting from the mixture were visible. Beyond the edges of the triangle, in the remaining portion of each parallelogram, the spectral color admitted through each slit was visible. In this first approach to the problem Helmholtz did not attempt to provide a quantitative determination of the wavelengths of his color mixtures. He fixed crosshair lines in his telescope which he oriented at 45 degrees so that they ran parallel to the Frauenhofer lines. This enabled him to provide a qualitative estimate of the proximity of a color mixture to the dark lines in the spectra of the color bands entering the specific mixture. To compare the relative intensities of the two colors entering a particular mixture, Helmholtz noted that if the prism was rotated about the axis of the telescope, the surface area of the illuminated parallelogram changed, being greatest when the slit and prism were parallel. In that position the illuminated area was a rectangle. As the prism was rotated relative to the slit, therefore, the same quantity of light would illuminate a larger or smaller surface area, and appear correspondingly less or more intense. In the original position of the V slit the intensities of the two spectral colors were equal. By rotating the prism one could achieve all combinations of relative intensities of the two spectra.

Using this experimental design Helmholtz arrived at several remarkable conclusions. The first was that color mixtures formed from pigments or powders differ markedly from color mixtures formed from pure spectral colors. In contrast to the

[9] Hermann Helmholtz, "Über die Theorie der zusammengesetzten Farben," *Annalen der Physik und der Chemie,* 1852, *87*:45–66, repr. in *Wissenschaftliche Abhandlungen* (cit. n. 3), Vol. II, pp. 1–23, on pp. 15–16.

experience of painters for a thousand years, Helmholtz wrote, the mixture of blue and yellow spectra, for example, does not yield green but rather a greenish shade of white. The explanation, according to Helmholtz, is that a portion of the light falling upon a colored body is reflected back as white light, while of the portion that penetrates the body, one portion is irregularly absorbed while the remaining light is reflected from the back surface and taken by the observer as the color of the body. Similarly, in the case of powders most of the reflected light comes not from the surface but from deep within the body. If a blue and a yellow powder are mixed, the surface particles will reflect blue and yellow light, which will combine to form a greenish white (as in the case of the mixture of the pure spectral colors). This will be a small portion of the total reflected light, however. Only light that is not absorbed by the blue and yellow particles will be reflected back from within the body. Blue bodies allow green, blue, and violet light to be transmitted; yellow, on the other hand, only red, yellow, and green. Hence, the mixture of blue and yellow particles will allow only green light to be transmitted, and only green light will be reflected to the eye. Green light will predominate in the total mixture of reflected light and not greenish white as the mixture of the pure spectra. What happens when pigments are mixed, Helmholtz concluded, is that different rays of colored light are *lost,* rather than that colors combine. This is why when two pigments of apparent equal intensity are mixed, the mixture is darker. Only mixtures of pigments standing close to one another in the spectral series will yield colors close in intensity and hue to the mixture of the same spectral colors. The combination of spectral colors and the mixing of pigments rest on two different physical processes.[10]

One aspect of the paper that attracted Grassmann's attention was the range of results achieved from mixing pairs of spectral colors. "The most striking of these results," Helmholtz wrote, "departs widely from the heretofore accepted facts; namely, that among the colors of the spectrum only two combine to produce white, being therefore complementary colors. These are yellow and indigo blue, two colors which were previously almost always thought of as producing green." It was because previous investigators had based their theories on mixtures of pigments rather than on the mixture of pure spectral colors that this incredible error had been propagated and reinforced. In his investigation of combinations of three colors (done by replacing the V slit with a lazy-Z slit) Helmholtz observed that it was possible to produce white from red, violet, and green, which in turn could be represented as three pairs of complementary colors: namely, red and a mixture of dim blue-green; green and a mixture of purple-red; violet and dim yellow. He noted that these results agreed with those Newton had reported in his *Opticks.*[11]

Helmholtz also concluded that the least number of colors out of which the entire spectrum could be generated was five. He arrived at this result by trying to construct a color circle. As the best method of construction he favored Newton's procedure of producing each simple color by combining it from the neighboring colors on either side, but he restricted Newton's approach even further by adding that the distance between the two combining colors should not be too great. Otherwise, he said, the resulting intermediate shades would not match those of the spectrum. Proceeding in this manner, Helmholtz concluded that the minimal list of colors required to imitate

[10] *Ibid.,* p. 17.
[11] *Ibid.,* pp. 12–13 (quotation), 15.

the spectrum was red, yellow, green, blue, and violet. "We must therefore drop the theory of the three primary colors as primary qualities of sensation proposed by Thomas Young."[12]

Grassmann took issue with Helmholtz's claim that there is only one pair of complementary colors. Contrary to this assertion, Grassmann set out to show that Newton's view was indeed correct: *Every* color has a complement with which it combines to produce white light. Grassmann's mathematical demonstration of this claim was remarkable for being built directly upon certain structural features of the *perception* of color. He thus examined the purely phenomenal, mental side of color relations; here was exactly the analogue of the problem of generating the spatial components of visual experience from the phenomena. I cannot follow the tortuous details of Grassmann's argument here. It suffices to observe that the "proof" must have seemed strange to an experimentally oriented empiricist such as Helmholtz, for it proceeded as a *reductio ad absurdum.* Instead of experimentally demonstrating that every color has its complement, Grassmann proceeded in an abstract mathematical fashion by attempting to show that if this were not the case, our concept of continuity and our experience of the closure in the continuous transition of colors would be violated.[13]

Grassmann concluded the paper with a discussion of the rule for combining colors. His purpose in this discussion—indeed, the primary objective of his entire critique of Helmholtz's work on color—was to show that the rule for combining colors was a straightforward application of an operation he called *geometrical addition* (for all practical purposes equivalent to our vector addition), which was one of the central operations of the new mathematical calculus developed in his *Ausdehnungslehre* (sec Figure 3). Grassmann went on to show that the method of geometrical addition was fully equivalent to the method for mixing colors employed by Newton in his *Opticks,* where the procedure is analogized to the problem in statics of finding the center of gravity of two arbitrary weights. In Newton's method hue is represented as a radius directed to the outside of a circular band of pure spectral colors, while the intensity of the color entering a mixture producing white light is represented by the weight hanging from the radius.

Grassmann's discussion of color mixtures led Helmholtz to rethink his own approach to the subject. Helmholtz believed that Grassmann scored several crucial objections. Yet he still felt that Grassmann's paper contained a number of loose ends. For one thing, he would have to reexamine his own treatment of complementary colors. Inadequate instrumentation seemed to be the primary weak point. Indeed, in his own paper Helmholtz had noted that a more refined instrumental arrangement for projecting the color mixtures onto a larger surface area and an improved method for measuring the distance of the color mixture from the nearcst Frauenhofer line might lead to different results concerning the composition of the whitish hues.[14] But Grassmann's argument was totally inadequate. Among the community of measurement physicists Helmholtz respected, it was insufficient to establish a claim based

[12] *Ibid.,* p. 21.

[13] Hermann Grassmann, "Zur Theorie der Farbenmischung," *Ann. Physik Chemie,* 1853, *89*:69–84, repr. in *Gesammelte mathematische und physikalisch Werke,* ed. F. Engel, 3 vols. (Leipzig: Teubner: 1894–1911), Vol. II, Pt. 2, pp. 161–173, esp. pp. 161–162. See Paul D. Sherman, *Colour Vision in the Nineteenth Century: The Young-Helmholtz-Maxwell Theory* (Bristol: Adam Hilger, 1981), for the details of Grassmann's argument.

[14] Helmholtz, "Über die Theorie der zusammengesetzten Farben" (cit. n. 9), pp. 13–14.

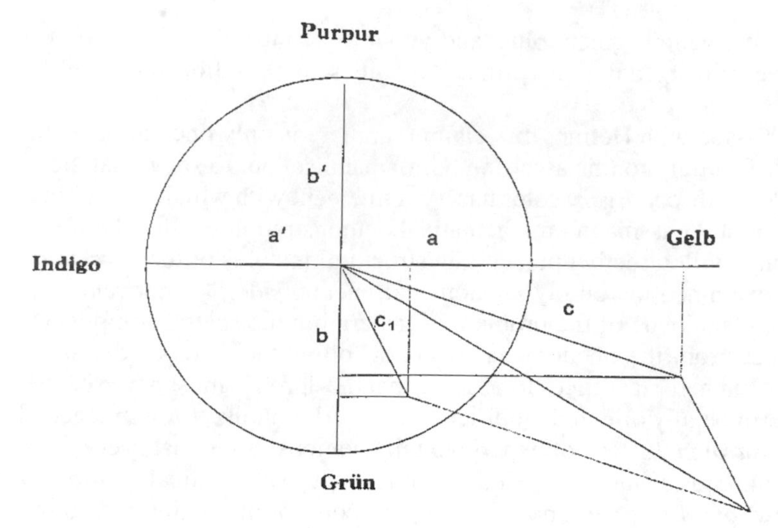

Figure 3. Grassmann's method for representing color mixtures. From Hermann Grassmann, "Zur Theorie der Farbenmischung," Gesammelte mathematische und physikalische Werke, *ed. F. Engel, 3 vols. (Leipzig: Teubner, 1894–1911), Vol. II, Pt. 2, p. 173.*

on an abstract mathematical argument without also demonstrating the result empirically. The production of white light from complementary spectral colors had to be demonstrated by an experiment. Furthermore, Grassmann's *reductio* argument depended upon assumptions about the discriminating power of the eye as a measuring device. These were extremely interesting physiological assumptions. But having made them, Grassmann left the plane of physiological argumentation altogether. A particularly glaring problem from Helmholtz's point of view was that Grassmann's mathematical technology began to drive his physiological argument: his mathematical methods led him to assume that *four* colors would be required to perform the parallelogram construction. Yet even in Grassmann's ingenious construction one of the colors, purple, was itself a mixture of red and violet. Perhaps *three* colors would indeed suffice. By wedding himself to a circle as the method for representing color spaces instead of a triangle, Grassmann had introduced some unnecessary, perhaps even unwarranted, assumptions. Furthermore, casting the argument in terms of a triangle might correspond more adequately to the physiology of the color perception. This too needed to be checked empirically. In any case, the entire issue of Young's theory was reopened and the theory of color mixtures given a decisively physiological dimension as a result of the encounter with Grassmann.

III. HELMHOLTZ'S RECONSIDERATION OF COLOR MIXTURES

Helmholtz took up these issues in a paper entitled "Über die Zusammensetzung von Spectralfarben," published in *Poggendorff's Annalen der Physik und Chemie* in 1855, and in much expanded form in the *Physiologische Optik* in 1860. While experimental in character, the 1852 paper criticizing Brewster was qualitative in its approach. In his renewed attack on the problem Helmholtz sought to produce a fully quantitative theory of color mixtures by refining each component of his apparatus for viewing and mixing spectral colors. With his new instrument, in place of the qualitative mixtures of hues in the earlier experimental arrangement, Helmholtz could now measure the wavelengths of the complementary colors entering a mixture of white light.[15] He could also experimentally control and quantitatively measure the intensities of the light entering a mixture. With this improved capability Helmholtz determined anew the series of complementary spectral colors. Consistent with Grassmann's prediction he now was able to produce white from violet and greenish yellow, indigo blue and yellow, cyan blue and orange, greenish blue and red. He was, however, not able to produce white from mixing green with any other simple color, but only from mixing it with purple, "that is with at least two other colors, red and violet."[16]

Helmholtz represented these results in a graph relating complementary colors to one another as a function of their wavelengths (see Figure 4). From the graph it was immediately obvious why his V-slit arrangement could not have been expected to reveal colors complementary to red and violet: the transitions between the blue and green color bands proceed extremely rapidly, being represented as a nearly vertical line in the graph: thus these colors form extremely narrow bands difficult to detect in Helmholtz's experimental arrangement. Indigo blue and yellow, on the other hand, had the advantage of being relatively wide bands of color.

These results also had certain consequences for the geometrical representation of the color table. Helmholtz now praised Newton's use of the center-of-gravity method to represent the colors in a plane as one of the most ingenious of all his creative ideas.[17] But Newton himself had proposed the rule as a mere aid for summarizing the phenomena in a qualitative manner and had not defended its correctness as a quantitative explanation. Grassmann's contribution had been primarily to call attention to the mathematical assumptions underlying the center-of-gravity method, and his treatment had convinced Helmholtz that this was indeed the appropriate quantitative method to use. But he was not convinced that the color table should be represented as a circle. Grassmann's analysis simply duplicated Newton's assumptions. But in fact, the center-of-gravity method for mixing colors was compatible with many geometrical representations.[18] To determine which of those representations fit the causal picture most closely, it was necessary to interpret the parameters in Grassmann's model in terms of empirical measurements. Subjected to this requirement, the choice of a circle would no longer adequately fit the refined data Helmholtz had derived. In fact, the result was a completely different shape for the color space. Instead of a circle or a circular lumen, a representation consistent with the experimental measurements turned out to be a hyperbola-like curve with violet, green, and red at the vertices (see Figure 5).

Grassmann had stimulated Helmholtz to revise fundamentally his approach to the theory of subjective colors. Grassmann's abstract, structural mathematical approach to these problems was indeed impressive. But while acknowledging that Helmholtz

[15] For details on this innovative apparatus see Sherman, *Colour Vision* (cit. n. 13), pp. 81–90, 111–115.

[16] Hermann Helmholtz, "Über die Zusammensetzung von Spectralfarben," in *Wissenschaftliche Abhandlungen* (cit. n. 3), Vol. II, pp. 45–70, on p. 51, and *Handbuch der physiologischen Optik,* Vol. II (Leipzig: Voss, 1860), pp. 272–309, on p. 277 (hereafter cited using *WA* or *PO*).

[17] *Ibid., WA*, p. 64; *PO*, p. 288.

[18] See Helmholtz's discussion and mathematical treatment *ibid., PO*, p. 287.

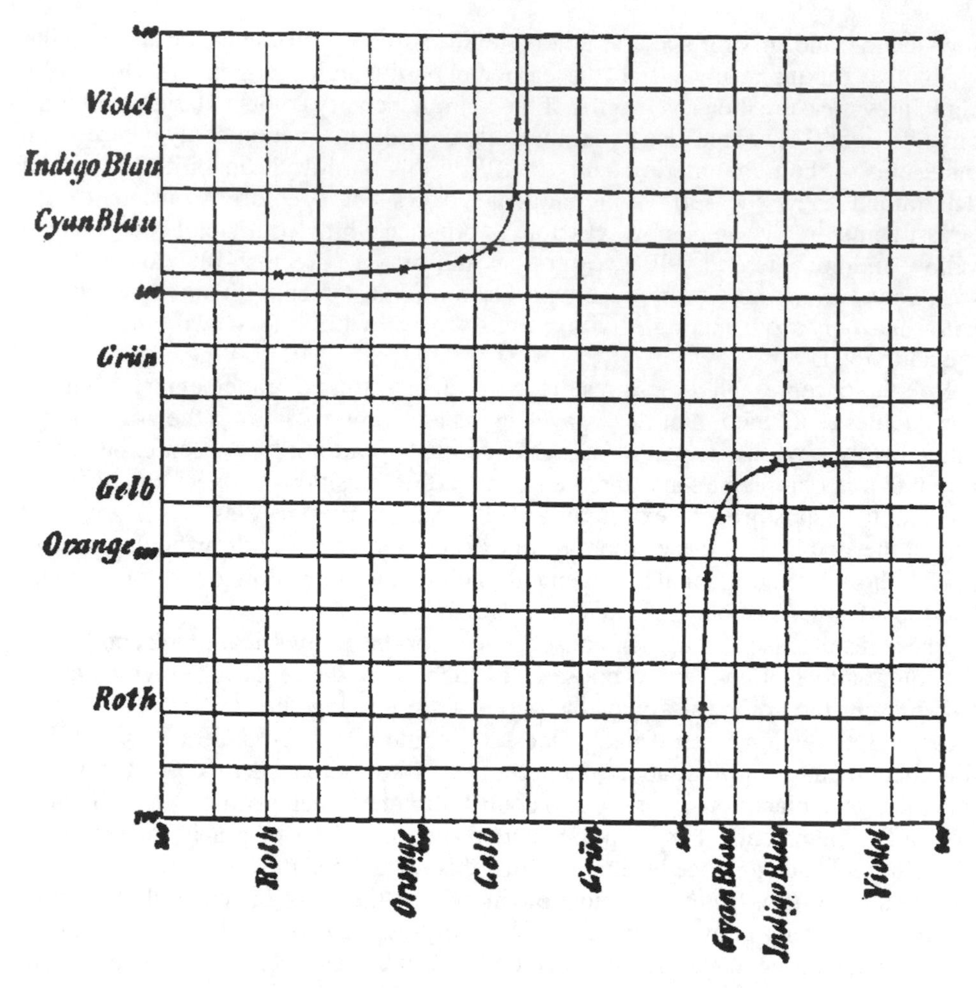

Figure 4. A graph relating complementary colors in terms of wavelengths. From Helmholtz, Handbuch der physiologischen Optik, *p. 317.*

had profited from the interaction, we should also not overlook the object lesson Helmholtz was prepared to give Grassmann in relating abstract mathematical structures to the requirements of physics and physiology. A point Helmholtz would state explicitly in the next stage of his researches in sensory physiology was that to be meaningful in reference to a physical problem, an abstract structure had to be embedded in measurements, and its internal logic adapted to the requirements of that problem. That process of adaptation was achieved through a dialogue with the instruments, and the dialogue would result in a model of the physical system, in this case the physiological apparatus of the eye responsible for the production of color sensations. In his reference to Newton's approach, Helmholtz made it clear that the proper mathematical or graphical representation of a system was not merely convenient or useful but rather corresponded to the causal properties of the physical system. Helmholtz's color space was offered as such a representation of the physiological apparatus employed by the eye in measuring color.

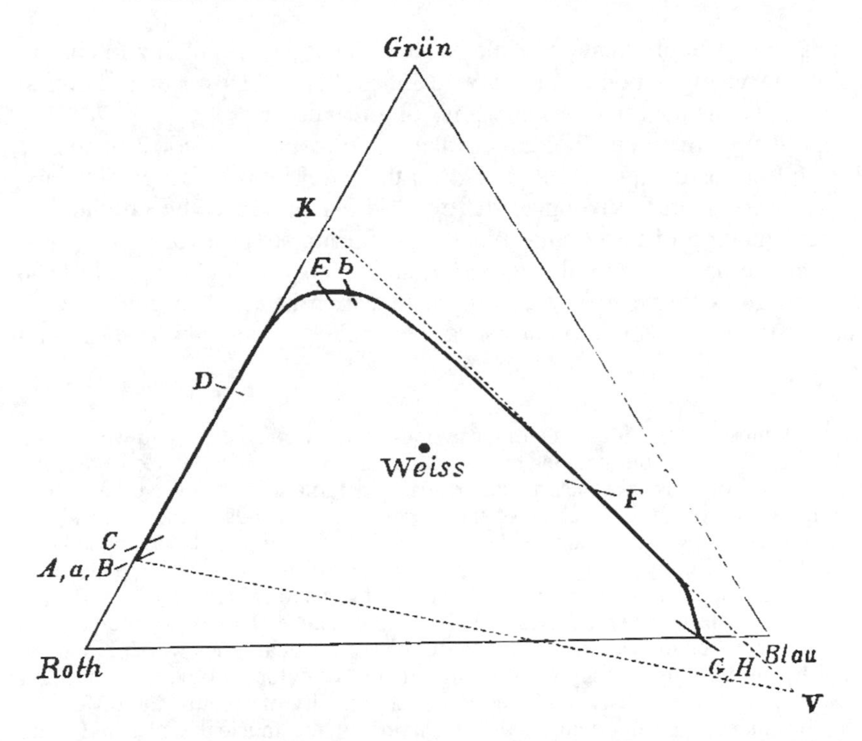

Figure 5. Helmholtz's color space for objective (spectral) colors. From Helmholtz, Handbuch der physiologischen Optik, *p. 340.*

IV. MUSIC TO THE EYE: PHYSIOLOGICAL ACOUSTICS, VISUALIZATION DEVICES, AND THE RECEPTOR HYPOTHESIS

It is tempting to assume that Helmholtz reversed his stand on Young's theory of color vision in the course of writing up the new experiments central to the 1855 paper. However, the original paper says nothing about the three-receptor theory. Indeed, Helmholtz stated in a footnote for the version included in his *Wissenschaftliche Abhandlungen* that his first recorded support of the Young hypothesis appeared in the third part of the *Physiologische Optik,* published in 1860.[19] Nothing about the encounter with Grassmann would have forced him to reverse his position on the physiological bases of colors. That dispute had confirmed Newton's center-of-gravity method for determining color mixtures but led Helmholtz to conclude that the color chart should be represented geometrically as a shape approaching a conic section rather than as a circle, as both Newton and Grassmann had assumed. Nothing concrete had been determined about the physiological causes of color mixtures. Helmholtz had shown that *any* three colors would suffice to generate the color chart, but those methods did not specify which three colors must be associated with the color receptors in the retina. Helmholtz had established, furthermore, that the center-of-gravity method for representing color combinations was indeed useful, but he wanted to go further and establish that its underlying principles embodied a more general calculating apparatus for the representation of sensations; that is, he wanted

[19] Helmholtz, "Über die Zusammensetzung von Spectralfarben," *WA* (cit. n. 16), p. 70.

to establish it as a psychophysical principle as well. What led Helmholtz to change his mind about the Young hypothesis between 1855 and 1860? How did he arrive at red, green, and violet as the three primary physiological colors?

Neither unpublished materials, nor experimental notebooks, nor correspondence answers these questions unequivocally. I suggest that work in physiological acoustics and a concerted effort to analogize the eye and ear provided the grounds for Helmholtz's reevaluation of the Young hypothesis. Strong support for this suggestion comes from a series of popular lectures Helmholtz delivered in Cologne in 1868 on the recent progress in the theory of vision, in which he tried to convince his audience that the three-receptor hypothesis was plausible by analogy with the sensations of tone:

> I have myself subsequently found a similar hypothesis very convenient and well suited to explain in a most simple manner certain peculiarities which have been observed in the perception of musical tones, peculiarities as enigmatic as those we have been considering in the eye. In the cochlea of the internal ear, the ends of the nerve fibers, which lie spread out regularly side by side, are provided with minute elastic appendages (the rods of Corti) arranged like the keys and hammers of a piano. My hypothesis is that each of these separate nerve fibers is constructed so as to be sensitive to a definite tone, to which its elastic fiber vibrates in perfect consonance. This is not the place to describe the special characteristics of our sensations of musical tones which led me to frame this hypothesis. Its analogy with Young's theory of colors is obvious, and it explains the origin of overtones, the perception of the quality of sounds, the difference between consonance and dissonance, the formation of the musical scale, and other acoustic phenomena by as simple a principle as that of Young.[20]

The passage presents the order of discovery as if the Young three-receptor hypothesis suggested exploring a similar mechanism for hearing, but my claim is that the actual order was just the reverse. The internal evidence of Helmholtz's papers between 1855 and 1860 argues instead that his work in physiological acoustics supported the receptor hypothesis before he took up the line of investigation that established the receptor theory in physiological optics. In addition, the papers published in this period indicate that the juxtaposition of these two sensory modalities, the back-and-forth comparison of models in one domain with those in the other, guided Helmholtz toward the reversal of his earlier position on Young's trichromatic receptor hypothesis. Moreover, it was his work in physiological acoustics that drew most directly on the new telegraph technologies, and it was through the sound-light analogy that these technologies influenced Helmholtz's theory of color vision.

The analogy between physiological acoustics and color vision presumed that just as in the ear a set of fundamental or primary tones is objectively based in the rods of Corti, so in the eye a set of primary colors is based in specific nerve endings in the rods and cones. Neither assumption could be established in humans, although some evidence from comparative anatomy supported the analogy on the side of physiological acoustics. The analogy between eye and ear was a salient feature of Helmholtz's work on physiological acoustics. Comparisons between eye and ear are prominent in his first extended acoustical study in 1856, and they abound in the

[20] Hermann Helmholtz, "Die neueren Fortschritte in der Theorie des Sehens," in *Selected Writings,* ed. Kahl (cit. n. 4), p. 181.

first edition of his *Tonempfindung,* or *Sensations of Tone,* published in 1863.[21] Such analogies were by no means new, and Helmholtz was certainly familiar with similar comparisons made by other authors. He had followed work on physiological acoustics for several years, writing the reports on advances in the field for the *Fortschritte der Physik* in 1848 and 1849. Central among the works he reported on in those years, and influential for all his later work in acoustics, were the papers of August Seebeck on the siren and on resonance phenomena. Although the comparison was not a main concern of his researches, Seebeck assumed that under certain circumstances transverse (light) and longitudinal (sound) waves behave similarly, and he had proposed an optical analogue to acoustical resonance, suggesting that resonance of spectral colors with vibrating molecules in groups of nerves in the retina was the mechanism for the sensation of brightness.[22]

The specifics of earlier uses of the analogy between vision and hearing and their speculated mechanical and physiological bases do not concern me here. It is striking, however, that Helmholtz took up physiological acoustics in earnest in 1855, precisely at the time he was searching for anatomical and physiological bases for the trichromatic theory. The analogy seems to rely on the discovery in 1851 by Alfonso Corti of the cochlear membrane that bears his name. Helmholtz set forth the central elements of his approach to physiological acoustics in two papers on combination tones published in 1856. His goal was to establish that all musical tones are compounded from a set of fundamental, simple tones. In the report of this research to the Berlin Academy of Sciences Helmholtz explicitly deployed an analogy between spectral colors and primary tones: "In analogy to the primary colors of the spectrum we intend to call such tones *simple tones* in contrast to the compound tones of musical instruments, which are actually accords with a dominant fundamental tone."[23]

The objective Helmholtz set for his acoustical investigations turned on issues much like those that had prevented him from further developing the Young three-receptor hypothesis: He wanted to determine that the mathematical form of the physical description of hearing had a material, physical basis in the physiology of the ear. Helmholtz wanted to show not just that Fourier analysis is a useful mathematical tool for representing the phenomena, but rather that the ear itself is a Fourier analyzer; that, like spectral colors, primary tones have an independent objective existence; and furthermore that combination tones as well have an objective existence, that they are not simply a psychological phenomenon.

> The theorem of Fourier here adduced shows first that it is mathematically possible to consider a musical tone as a sum of simple tones, in the meaning we have attached to the words, and mathematicians have indeed always found it convenient to base their acoustic investigations on this mode of analysing vibrations. But it by no means follows that we are obliged to consider the matter in this way. We have rather to inquire, do these

[21] Hermann Helmholtz, "Über Combinationstöne," *Ann. Physik Chemie,* 1856, *99*:497–540, on p. 526, repr. in *Wissenschaftliche Abhandlungen* (cit. n. 3), Vol. I, pp. 263–302, on p. 290. For *Tonempfindung* see quotation below (n. 24).

[22] August Seebeck, "Bemerkungen über Resonanz und über Helligkeit der Farben im Spectrum," *Ann. Physik Chemie,* 1844, *62*:571–576. This was a response to Macedonio Melloni, "Beobachtung über die Färbung der Netzhaut und der Krystall-Linse," *ibid.,* 1842, *56*:574–587, which proposed treating the sensation of color as a resonance phenomenon analogous to acoustical resonance.

[23] Hermann Helmholtz, "Über Combinationstöne," *Monatsbericht der königlichen Akademie der Wissenschaften zu Berlin,* 1856, pp. 279–285, repr. in *Wissenschaftliche Abhandlungen,* Vol. I, pp. 256–262, especially p. 257. See also Helmholtz, "Über Combinationstöne" (cit. n. 21).

> partial constituents of a musical tone, such as the mathematical theory distinguishes and the ear perceives, really exist in the mass of air external to the ear? . . . [T]herefore, we shall inquire whether the analysis of compound into simple vibrations has an actually sensible meaning in the external world, independently of the action of the ear, and we shall really be in a condition to show that certain mechanical effects depend upon whether a certain partial tone is or is not contained in a composite mass of musical tones. The existence of partial tones will thus acquire a meaning in nature, and our knowledge of their mechanical effects will in turn shed a new light on their relations to the human ear.[24]

It was possible to show that theory corresponded to experiment when one examined a vibrating string, such as a piano wire. Helmholtz noted that in most other cases, however, the "mathematical analysis of the motions of sound is not nearly far enough advanced to determine with certainty what upper partials will be present and what intensity they will possess."[25]

This inadequacy of theory to analyze a given wave form into its components was particularly evident in the determination of tone quality or timbre, the character that distinguishes a violin from a flute or clarinet. The same tone produced at the same intensity will sound characteristically different depending on which instrument produced it. Helmholtz made the plausible assumption that tone quality was determined by the form of the sound wave,[26] and devoted much of his subsequent investigations to elucidating the relation between the form of the sound wave and tone quality. Noting that the problem had been solved for only a few isolated cases,[27] he compensated for the defect in theory by resorting to resonators and a variety of devices for visualizing the form of a sound wave and mechanically analyzing it into its constituent primary tones. He introduced this empirical method of attacking the problem in the first edition of his *Sensations of Tone* in 1863:

> No complete mechanical theory can yet be given for the motion of strings excited by the violin-bow, because the mode in which the bow affects the motion of the string is unknown. But by applying a peculiar method of observation proposed in its essential features by the French physicist [Jules] Lissajous, I have found it possible to observe the vibrational form of individual points in a violin string and from this observed form, which is comparatively very simple, to calculate the whole motion of the string and the intensity of the upper partial tones.[28]

Helmholtz's investigations in these years were guided by the ever-present analogy of eye and ear. The encounter with Grassmann had netted the important point that the sensation of color is determined by three variables: brightness, hue, and saturation. Tone was analogous in being similarly determined by three variables: loudness, pitch, and tone quality (timbre). Although light and sound were different types of wave phenomena, brightness and loudness were both associated with amplitude, and

[24] Hermann Helmholtz, *Die Lehre von den Tonempfindungen als physiologische Grundlage für die Theorie der Musik* (Brunswick: Vieweg, 1863), trans. Alexander J. Ellis, as *On the Sensations of Tone as a Physiological Basis for the Theory of Music* (London: Longmans, Green, 1885), 2nd ed., pp. 35–36.

[25] *Ibid.,* p. 55.

[26] *Ibid.,* pp. 19, 21.

[27] Hermann Helmholtz, "Über die physiologischen Ursachen der musikalischen Harmonie," in *Selected Writings,* ed. Kahl (cit. n. 4), p. 85.

[28] Helmholtz, *Sensations of Tone* (cit. n. 24), p. 80.

hue and pitch were dependent on frequency.[29] To pursue the analogy further, tone quality would not depend on wave form. If, as we have seen, saturation was due to the mixture of colors entering the composition, then tone quality might plausibly be expected to result from the combination of primary tones. All tones are combinations of primary tones, and the timbre of musical instruments is characterized by a primary tone and numerous upper partial tones modulating it to produce a specific quality. The difference between these acoustic phenomena and the sensation of colored light is that the eye is incapable of distinguishing the components of any compound color in the sensation. The ear, by contrast, can be trained relatively easily to distinguish the component elements of a compound tone: "The eye has no sense of harmony in the same meaning as the ear. There is no music to the eye."[30] Helmholtz's teacher, Johannes Müller, had argued that each sensory modality owes its particular quality—color in the case of the stimulated retina, sound in the case of the stimulated auditory nerve—to a "specific sense energy." Helmholtz sought to locate this difference in sensibility in the organization and mechanical functioning of the two organs; namely, in the case of the ear in the thousands of different fibers embedded within the membrane of Corti. He conceived of these fibers as each resonating with a different primary tone. Musical tones, then, would be combinations of these primary sensations.

To demonstrate that sound waves are composed in the way he hypothesized, Helmholtz resorted to instruments that "artificially" produced and visually displayed primary tones. He then compared the wave forms of, for example, notes bowed on a violin with the wave form of the primary tone. In similar fashion he compared the tones produced by horns or clarinets and the vowels produced by the human voice. The device he employed for producing the primary tones was a modification of the self-regulating current interrupter patented by his friend Werner Siemens as the basis for his improved telegraph. Described in the various patent applications as a kind of oscillating fork, the interrupter was meant to insure the synchronous operation of different parts of the telegraphic apparatus and the constancy of the translation of the message relayed between several telegraph instruments.[31] In place of the spring and oscillating lever by which Siemens's instrument interrupted a continuous direct current, Helmholtz introduced a tuning fork (see Figure 6). Electromagnets near the ends of the fork alternately attracted the ends of the fork, made contact, and transmitted pulses of current at the frequency of the fork. These current pulses were transmitted to a second apparatus in which a tuning-fork was placed between the poles of an electromagnet activated by the incoming pulses (see Figure 7).

[29] *Ibid.,* p. 19.

[30] Hermann Helmholtz, "Über die physiologischen Ursachen der musikalischen Harmonie," *Selected Writings,* ed. Kahl (cit. n. 4), p. 107.

[31] The Siemens patent application on the interrupter in Prussia was dated 1 May 1847. See Werner Siemens, "Application for a Patent for a New Kind of Electric Telegraph and Combined Arrangement for Printing Messages," in *The Scientific and Technical Papers of Werner von Siemens,* 2 vols. (London: Murray, 1892–1895), Vol. II, pp. 13–26, esp. pp. 15–20. A much-elaborated version of the same discussion is presented in the English patent application of 3 April 1850; see *ibid.,* pp. 30–74, esp. pp. 49–53, "Translating Apparatus," for the analogy to Helmholtz's interrupter. A similar presentation was made to the Paris Academy of Sciences on 15 April 1850 on Siemens's behalf by Emil du Bois-Reymond, who translated and read the memoir in an effort to secure a royal privilege on the Siemens telegraph in France. See Siemens, "Memoir on the Electric Telegraph," *ibid.,* Vol. I, pp. 29–64, esp. pp. 44–45, where the interrupter is characterized as an oscillating fork. The firm of Siemens & Halske manufactured Helmholtz's tuning-fork interrupter.

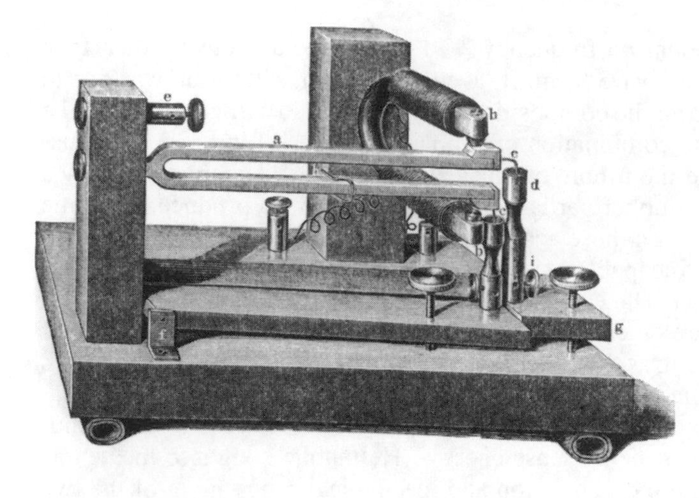

Figure 6. The tuning fork interrupter. From Helmholtz, Tonempfindungen, *p. 196.*

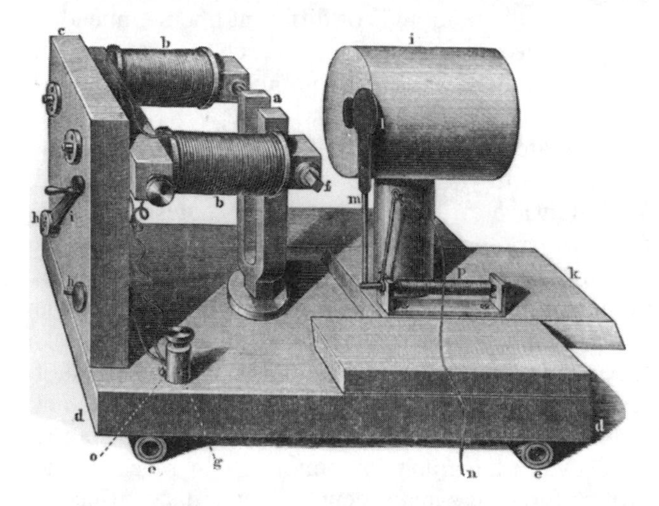

Figure 7. The tuning-fork resonator. From Helmholtz, Tonempfindungen, *p. 633.*

Extremely low prime tones were required, which entailed using forks whose tones were barely audible. These tones were amplified by placing a resonator tuned to the proper frequency near the fork. By connecting these resonator devices, Helmholtz was able to combine numerous partial tones into tones indistinguishable from those produced by musical instruments (see Figure 1, p. 184 above).

Helmholtz described several methods for making auditory vibrations visible. The first he termed the "graphic method," to "render the law of such motions more comprehensible to the eye than is possible by lengthy verbal descriptions."[32] He illustrated the graphic method with the phonautograph, which consisted of a tuning fork with a stylus on one prong of the fork. The vibrating fork produced a curve on paper blackened with lampblack and attached to a rotating drum, the same arrangement as in Carl Ludwig's kymograph or in Helmholtz's own myograph. The most dramatic of these visualization devices was the so-called vibrational microscope, an instrument embodying methods described first by Jules Lissajous for observing com-

[32] Helmholtz, *Sensations of Tone* (cit. n. 24), p. 20.

Figure 8. *The vibration microscope. From Helmzholtz,* Tonempfindungen, *p. 138.*

pounded vibrational motions. The microscope was constructed so that the objective lens was mounted in one of the prongs of a tuning fork. The eyepiece of the microscope was mounted on a plate so that the tube of the microscope was attached to the backing of the bracket holding the tuning fork. The prongs of the fork were set in vibration by two electromagnets just as in the interrupter and resonator described above. When the tuning fork was set in motion, the object lens would vibrate vertically in a line. When the microscope was focused on a stationary grain of white starch and the forks set in motion, a white vertical line would be seen. If the grain of starch was placed on a vertical string so that the grain was vibrating horizontally while the lens was moving vertically, the image viewed in the field of the microscope would be a line compounded of both motions inclined at 45 degrees (see Figure 8).

The principal use Helmholtz made of these instruments was to demonstrate that phase differences in primary tones making up a compound tone had no effect on the perceived quality of the tone. To make this argument he used the vibration microscope to study the form of waves compounded from primary tones out of phase with one another. Phase changes were produced in two ways: (1) by putting the tuning fork of the resonator out of phase with the interrupter; and (2) by putting the resonator out of phase with the resonator fork. Both the tuning fork of the resonator and the fork of the vibration microscope were set in motion by the same interrupter. Thus the pitch of the two forks was the same, both being determined by the number of interruptions of current per second. To change the phase of the fork in the resonator, Helmholtz placed small clumps of wax on the fork. He placed the fork of the vibration microscope in a horizontal position. The tuning fork of the resonator was placed vertically, with grains of white starch on one prong. Thus the object glass of the microscope vibrated vertically, while the grain of starch on which the microscope was focused vibrated horizontally. As the phase of the resonator fork altered,

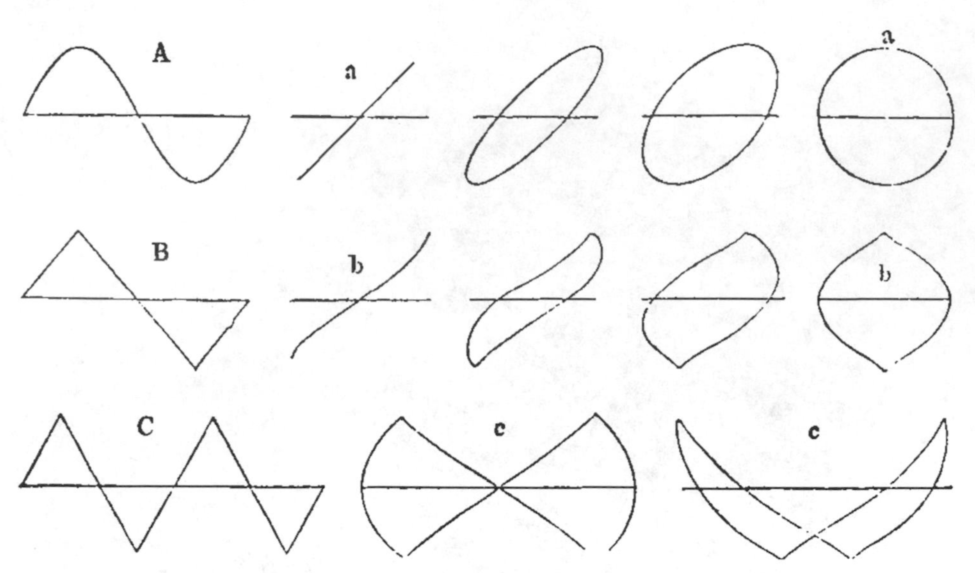

Figure 9. *Lissajous figures created by the vibration microscope. From Helmzholtz,* Tonempfindungen, *p. 140.*

the line visible in the field of the microscope shifted from a straight line inclined by 45 degrees (when the two forks were in unison) through various oblique ellipses until the phase difference reached one quarter of the period, and then passed through a series of oblique ellipses to a straight line inclined 45 degrees in the other direction from the vertical when the phase difference reached half a period—the first series of waveforms depicted in Figure 9. If the resonator fork chosen was the upper octave of the interrupter fork, phase alterations generated the second series of wave forms shown.[33]

Helmholtz's point in these experiments was that the form of the wave, here visibly evident in the vibration microscope, did not determine the quality of the tone. That quality was determined by the force of the impression on the ear, that is, on the amplitudes and primary tones entering the composition of the tone. As long as the relative intensities of the partial tones compounded into a musical tone remained the same, the tone would sound the same to the ear no matter how the alteration of phases of the partial tones affected the form of the wave. This last point was made evident through phase alterations controlled by shading the aperture of the resonator, the second method of altering phase mentioned above. In the experiments with the vibration microscope described above, for instance, as long as the resonator chambers were fully open, no difference of tone quality could be heard even though the phase differences produced by the wax clumps on the tuning fork would visibly alter the wave form. In a second series of experiments, without the microscope, Helmholtz brought the resonator out of tune with its fork by partially closing the lid on the aperture. He had shown in an 1859 paper that narrowing the aperture of the

[33] *Ibid.*, p. 126.

chamber altered the phase of the wave.[34] For the ear the effect of closing the lid was to diminish the loudness of the tone, it being loudest with the aperture fully open. The same effect of reducing the loudness of the partial tones could be achieved by leaving the apertures fully open and adjusting the movable resonator platform, *k* in Figure 6, distancing the resonator from the tuning fork. Thus a compound tone could be produced either by combining partial tones and weakening the resonance by closing the aperture, or by weakening the partial tones through moving the chamber further away from the fork. The former approach altered the phase of combining tones, whereas the latter did not:

> In this manner every possible difference of phase in the tones of two chambers can be produced. The same process can of course be applied to any required number of forks. I have thus experimented upon numerous combinations of tone with varied differences of phase, and I have never experienced the slightest difference in the quality of tone. So far as the quality of tone was concerned, I found that it was entirely indifferent whether I weakened the separate partial tones by shading the mouths of their resonance chambers, or by moving the chamber itself to a sufficient distance from the fork. Hence the answer to the proposed question is: *the quality of the musical portion of a compound tone depends solely on the number and relative strength of its partial simple tones, and in no respect on their differences of phase.*[35]

These experiments and the crucial role of the vibrating microscope in particular were the basis of a direct comparison between the eye and ear. As the imaging device revealed, the eye is capable of detecting differences, even relatively minute differences, between wave forms. The ear is not. "The ear, on the other hand, does *not* distinguish *every* different form of vibration, but only such as when resolved into pendular vibrations, give different constituents."[36] This conclusion emphasized several of the differences as well as the fundamental similarities in the mechanism of color vision and hearing. The reason the ear is able to distinguish the partial tones in a compound tone is that among the 4,500 or so different nervous fibers in the arches of Corti specific nerve fibers resonate in sympathetic vibration with the spectrum of primary tones composing musical tones. Simple tones of determinate pitch will be felt only by nerve fibers connected to the elastic bodies within the cochlear membrane whose proper pitch corresponds to the various individual simple tones. The fact that even amateurs paying minimal attention are able to distinguish the partial tones in a compound tone, and that trained musicians can distinguish differences of pitch amounting to half a vibration per second in a doubly accented octave, would thus be explained by the size of interval between the pitches of two fibers. Similarly the fact that changes in pitch can take place continuously rather than in jumps finds its explanation in the sympathetic vibration of arches with proper tones most nearly identical, while the elastic bodies in the membrane with more distantly separated proper tones were incapable of vibrating in resonance.

The physiological organization Helmholtz envisioned as the basis of tone

[34] See Hermann Helmholtz, "Theorie der Luftschwingungen in Röhren mit offenen Enden," *Journal für die reine und angewandte Mathematik,* 1859, *57*:1–72, repr. in *Wissenschaftliche Abhandlungen* (cit. n. 3), Vol. I, pp. 303–382.

[35] Helmholtz, *Sensations of Tone* (cit. n. 24), p. 126 (italics in original).

[36] *Ibid.,* p. 128 (italics in original).

sensation was modeled directly by the tuning fork and resonator apparatus. Indeed, the full set of resonators and connected tuning forks was a material model of the ear in reverse. The resonator apparatus was used to produce compound tones artificially out of the simple tones generated by the tuning-fork interrupter. But this transmitting device could also be imagined to run in reverse as a recording device. In this sense it was a material representation of the functioning ear, its resonators being the material analogue of the fibers in the Corti membrane, its tuning fork and acoustical interrupter being the device for translating, encoding, and telegraphing the component primary tones of the incoming sound wave analyzed by the resonators.

Moreover, the representation rendered material by these communication technologies also provided a resource for understanding the differences between the eye and ear as well as a suggestion for further development in the theory of color vision: "The sensation of different pitch would consequently be a sensation in different nerve fibres. The sensation of a quality of tone would depend upon the power of a given compound tone to set in vibration not only those of Corti's arches which correspond to its prime tone, but also a series of other arches, and hence to excite sensation in several different groups of nerve fibres." Thus some groups of nerves would be stimulated through shared resonance, whereas other nerves would remain silent. The analogy to the eye and to Young's hypothesis was obvious:

> Just as the ear apprehends vibrations of different periodic time as tones of different pitch, so does the eye perceive luminiferous vibrations of different periodic time as different colors, the quickest giving violet and blue, the mean green and yellow, the slowest red. The laws of the mixture of colors led Thomas Young to the hypothesis that there were three kinds of nerve fibres in the eye, with different powers of sensation, for feeling red, for feeling green, and for feeling violet. In reality this assumption gives a very simple and perfectly consistent explanation of all the optical phenomena depending on color. And by this means the qualitative differences of the sensations of sight are reduced to differences in the nerves which receive the sensations. For the sensations of each individual fibre of the optic nerve there remains only the quantitative differences of greater or lesser irritation.
>
> The same result is obtained for hearing by the hypothesis to which the investigation of quality of tone has led us. The qualitative difference of pitch and quality of tone is reduced to a difference in the fibres of the nerve receptive to the sensation, and for each individual fibre of the nerve there remains only the quantitative difference in the amount of excitement.[37]

The analysis of hearing seemed to suggest that just as the myriad of musical tones capable of being distinguished is rooted in an organic Fourier analyzer that yields specific elemental sensations, so the eye in similar fashion could be conceived as generating color from a primary set of sensations rooted in specific nerve fibers. The delicate power of distinction the ear is capable of was explained in the large number of specific nerve fibers and related elastic resonating bodies in the Corti membrane. The inability of the eye to resolve colors into elemental sensations would be explained accordingly as a result of the small number of different types of sensitive nerve fiber, and by the assumption that all three nerve types respond in different degrees to light stimulation. The assumption that fibers predominantly sensitive to red, green, or violet light nonetheless respond weakly to light of other wavelengths

[37] *Ibid.*, p. 148.

would explain the continuity of transitions in the sensation of color as well as the inability of the attentive mind to analyze compound light into its elements. There is no music to the eye, because the eye has only three rather than the roughly 1,000 "resonator" types of the Corti membrane. Here was a remarkable feature of the Helmholtz-Young hypothesis that could be tested empirically. From the point of view of Grassmann's theory of color mixtures, any three colors that could combine to produce white would be satisfactory choices for producing the color space. From the perspective of the assumed physiological mechanisms of the Helmholtz-Young hypothesis, the most-saturated colors should be those generated by the primary receptors. Experiments aimed at diminishing or eliminating the activity of one or two of the color-sensitive nerve endings in the retina would thus provide support for the distinction between three different nerve types as well as dramatic evidence for the choice of red, green, and violet as the three primary physiological colors.[38]

VI. CONCLUSION

I have argued that the new media technologies that captivated the interest of Helmholtz and his young friends in the Berlin Physical Society, particularly the telegraph and a variety of new audio and visual inscription devices, were crucial in establishing the boundaries for his experiments in physiological optics and physiological acoustics. These new devices provided the means for delimiting the domain of scientific objects in a form in which they could be differentiated (and hence characterized), manipulated, and recombined. We miss the significance of these devices for producing traces, however, if we regard them simply as instruments for testing theoretical claims and resolving disputed issues, such as the dispute over Young's hypothesis, or the debate on whether musical tones have an objective existence independent of the ear. I have attempted to show that the new technologies were a resource for representing the scientific object, and that in their material form they were not just "representatives" of an object described by theory; rather they created the space within which the scientific objects, "eye and ear," existed in a material form. These technologies of representation exceeded the power of theory. By this I do not intend that theory carried Helmholtz only so far and that the technologies of representation then provided a mere supplement enabling the extension of research into areas where theory was insufficient to tread. Rather, I have been suggesting that we need to pay closer attention to the materiality of the inscription devices themselves and to the manner in which they are actually constitutive of the signifying scene in technoscience.

We are used to speaking about the relation of theory to its object as if the scientific instrument and experimental system were somehow a passive and transparent medium through which the presence of the object is to be achieved. The instrument, in our traditional manner of speaking, is simply an extension of theory, a mere supplement, useful for exteriorizing an ideal meaning contained within theory. When we treat the experimental system as a model of the theory, we tend to regard it simply as an expression, an unproblematic translation of the ideal relations and entities of the theory into the representative hardware-language of the experimental system. But this manner of proceeding neglects the empirical, material character of the

[38] See Helmholtz, "Über die Zusammensetzung von Spectralfarben," *PO* (cit. n. 16), pp. 292–293.

experimental system as a graphic trace, a *grapheme,* in Jacques Derrida's sense. This exterior materiality is not molded to the demands of theory; it resists and imposes its own constraints on the production of meaning. As Derrida notes, "the outside of indication/indexicality does not come to affect in a merely accidental manner the inside of expression. Their interlacing (*Verflechtung*) is originary."[39] In Derrida's view, there is never a "mere" supplement; the supplement, in this case, the materiality of the experimental system, is a necessary constituent of the representation.

The Derridean perception is useful in understanding Helmholtz's experimental strategies. In the quoted passage and elsewhere Derrida refers to the material character of the signifier, the set of marks on the page of written text. In order to extend this line of thinking to Helmholtz's scientific theorizing, I suggest that we consider the centrality to Helmholtz's work of the graphic method and of the products of technologies for visualization such as spectral collimators (the V- and Z-slit experiments) and the Lissajous figures in the vibrational microscope; particularly important is the manipulability of elements of physical models such as the system of tuning-fork resonators, which both physically and graphically (via the vibrational microscope) represent the sensation of tone. In Derrida's terms, the materiality of these signifiers was nonaccidental in structuring the content of Helmholtz's sensory physiology. As Helmholtz noted, mechanical theory was insufficient for depicting in detail how the ear functions when sensing tone quality. The mathematization of the theory of resonance in terms of Green's function and Fourier analysis did not indicate whether phase difference and wave form should affect tone quality. By using the tuning fork and resonator device coupled with the interrupter and vibrational microscope, however, one can depict visually the relationships central to the production of timbre. The device did not merely enable Helmholtz to provide a more detailed graphic representation of the elements involved in producing tones of different timbre; more important, the representation of the ear as tuning-fork resonator linked to a telegraphic device became the object upon which he experimented to correct an assumption about wave form held by most physicists, enabling him to arrive at a deeper understanding of the sensations of tone. The representation thus fundamentally affected the articulation of theory.

I have also suggested that the selection of telegraphic apparatus and its modification in various ways to achieve the ends of experiment had a significance beyond the fact that these devices were readily available and familiar objects of investigation for Helmholtz. Telegraphic devices were not only important as means for representa-

[39] Jacques Derrida, *La voix et le phénomène* (Paris: Presses Universitaires de France, 1967), pp. 89–90, as quoted in David E. Wellbery, "The Exteriority of Writing," *Stanford Literary Review,* 1992, *9*:11–23, on p. 16. Wellbery's essay provides an excellent overview of the relationship of the materiality of communication to poststructuralist philosophy. See also Wellbery's introduction to Friedrich Kittler, *Discourse Networks 1800/1900* (Stanford: Stanford Univ. Press, 1988). Also relevant to this theme are the chapters "Tympan," "Differánce," and "Signature Event Context," in Jacques Derrida, *Margins of Philosophy,* trans. and ed. Alan Bass (Chicago: Univ. Chicago Press, 1982); and Bachelard's notions that scientific instruments are materialized theories and theories are idealized machines: see Gaston Bachelard, *L'activité rationaliste de la physique contemporaine* (Paris: Presses Universitaires de France, 1951). For a full exposition of the relevance of Derrida's work to studies of experiment see Hans-Jörg Rheinberger, *Experiment, Differenz, Schrift: Zur Geschichte epistemischer Dinge* (Marburg: Basiliskinpresse, 1992). For the notion that the material character of signification introduces an uncontrollable "supplement" which becomes a structuring element in a semiotic system, see Brian Rotman, *Ad Infinitum: The Ghost in Turing's Machine. Taking God Out of Mathematics and Putting the Body Back In* (Stanford: Stanford Univ. Press, 1993).

tion and experiment; telegraphy embodied a system of signification that was central to Helmholtz's views about mental representations and their relationship to the world. I have suggested that apart from the challenge to his own early work and the adaptation of Grassmann's approach needed to make it relevant to physiological modeling, Grassmann's proposal of a formalism that operated in terms of three quantifiable measures was of interest because it was a system, based on the notion of *n*-dimensional manifolds, for constructing spatial representations that operated similarly to the way messages were encoded in the telegraph. Viewed in this light, the telegraphic system with which Helmholtz was familiar in his daily experience and upon which he and his friends frequently reflected was a materiality conditioning the choice and development of his own ideas about representation:

> Nerves have been often and not unsuitably compared to telegraph wires. Such a wire conducts one kind of electric current and no other; it may be stronger, it may be weaker, it may move in either direction; it has no other qualitative differences. Nevertheless, according to the different kinds of apparatus with which we provide its terminations, we can send telegraphic dispatches, ring bells, explode mines, decompose water, move magnets, magnetise iron, develop light, and so on. So with the nerves. The condition of excitement which can be produced in them, and is conducted by them, is, so far as it can be recognised in isolated fibres of a nerve, everywhere the same, but when it is brought to various parts of the brain, or the body, it produces motion, secretions of glands, increase and decrease of the quantity of blood, of redness and of warmth of individual organs, and also sensations of light, of hearing, and so forth. Supposing that every qualitatively different action is produced in an organ of a different kind, to which also separate fibres of nerve must proceed, then the actual process of irritation in individual nerves may always be precisely the same, just as the electrical current in the telegraph wires remains one and the same notwithstanding the various kinds of effects which it produces at its extremities. On the other hand, if we assume that the same fibre of a nerve is capable of conducting different kinds of sensation, we should have to assume that it admits of various kinds of processes of irritation, and this we have been hitherto unable to establish.
>
> In this respect then the view here proposed, like Young's hypothesis for the difference of colors, has still a wider signification for the physiology of the nerves in general.[40]

Helmholtz's work in physiological acoustics relied on the materiality of the representation of the ear as tuning-fork resonator. The juxtaposition and comparison of differences between media, between hearing and vision, was a positive resource for revisiting Young's hypothesis. Indeed, as the above passage implies, the analogy between the Young hypothesis for color vision and Helmholtz's model for hearing and the assimilability of both to the telegraph as apparently generalizable for sensory physiology provided convincing support for the analogical approach. For Helmholtz the construction of the color chart as a graphical representation embodied the very principles used by the eye in encoding signals perceived as color in the brain. In his studies on tone sensation Helmholtz had constructed a mechanical simulacrum for advancing his theory. In the final stages of his work on color vision the graphic trace itself became both the material embodiment of theory and the source of its improvement.

[40] Helmholtz, *Sensations of Tone* (cit. n. 24), p. 149.

Instruments, Nerve Action, and the All-or-None Principle

*By Robert G. Frank, Jr.**

TO PHYSIOLOGISTS WORKING HALF A CENTURY AGO it must have seemed that there was scarcely a problem of their science more hopeful of solution than that of the physicochemical nature of the nervous impulse." So began the thirty-three-year-old Cambridge physiologist Keith Lucas in June 1912, as he delivered the Croonian Lecture before the Royal Society of London. His subject, "The Process of Excitation in Nerve and Muscle," accorded well with the seventeenth-century founder's wish to promote the physiological study of movement. But if the society's biologists and medical men expected the feted lecturer to sketch a picture of satisfying recent progress and exciting future opportunities, they must have been disappointed. "The problem," Lucas stated bluntly, "is still unsolved," and he cited the laments of distinguished physiologists, such as Wilhelm Biedermann, Jacques Loeb, Max Verworn, and Jacob von Uexküll, that so many of the "shrewdest heads and the most talented investigators" had little to show for "decades of work."[1] A field that had started so brilliantly with the work of Hermann Helmholtz and Emil du Bois-Reymond in the 1840s, and which since that time had occasioned tens of thousands of experiments, thousands of articles, and scores of models, hypotheses, and "laws," appeared to many scientists around 1910 to have few firm and fundamental facts extractable from the "turbid and contentious" current of electrophysiology.[2]

Within two decades that pessimism had been dispelled. In January 1930 A. V. Hill, a protégé of Lucas and a Nobelist in 1922 for his work on heat production in muscle contraction, could write to Stockholm to propose that the older English physiologist Charles Scott Sherrington and another Lucas pupil, Edgar Douglas Adrian, be recognized "for their work on the nature of nervous reactions." Sherrington's vivisectional investigations on reflex arcs dated from before World War I, but new light, Hill wrote, "has recently been thrown upon the subject, new problems have been visualized, and the importance of Sherrington's work has been emphasized, by the researches of Adrian upon the nature of sensory and motor response." Adrian's work, Hill argued, had brought afferent nervous impulses, those coming inward from sensory organs, "within the range of quantitative and precise investiga-

* Medical History Division, Department of Anatomy, School of Medicine, University of California, Los Angeles, California 90024.

[1] Keith Lucas, "Croonian Lecture: The Process of Excitation in Nerve and Muscle," *Proceedings of the Royal Society, Series B,* 1912, *85:*495–524, quoting from p. 495.

[2] "Francis Gotch," *Lancet,* 2 Aug. 1913, p. 348.

tion." Similarly, "the efferent nervous impulses, the motor impulses to the effector organs, have been successfully observed, and rendered objective and quantitative, by the same technique." There could be no doubt, Hill concluded, that both kinds of nervous activity were built up from unit impulses that "are propagated in an 'all-or-none' way, or that the intensities of sensation and response depend simply upon the *number* of nerve impulses which travel to or from the nervous system, per unit of time." What Adrian had shown for the peripheral nervous system was almost certainly true of the brain and spinal cord as well, thus providing a "basis of further investigations into, and future theories of, the working of the central organ." Two years later Hill reiterated his case for Sherrington and Adrian, emphasizing that "Adrian's recent research has made clear and rendered susceptible to direct investigation the 'atomic' nature of the motor and afferent systems upon which that reflex activity is based." The work itself was "of great beauty and of mixed simplicity and subtleties," Hill believed, and represented "one of the greatest achievements in physiology of the last quarter century."[3]

Adrian and Sherrington did indeed receive the Nobel Prize in physiology or medicine in 1932, for exactly those accomplishments that Hill had cited. Adrian, for his part, had no doubt about what had made his discoveries possible: improved instrumentation. At the outset of his Eldridge Reeves Johnson Lectures in Philadelphia in 1930, later published as *The Mechanism of Nervous Action* (1932), Adrian stated flatly that "the history of electrophysiology has been decided by the history of electric recording instruments."[4] The vacuum-tube amplifier had made his own Nobel Prize–winning research possible, and it, coupled with the cathode-ray oscilloscope as a display device, was already by the early 1930s revolutionizing the study of the nervous system. A single nerve impulse in a single nerve axon, of invariable height, form, and duration, could be displayed on a screen and captured on film. Trains of such impulses could be seen to vary in frequency, responding to the intensity and time course of stimulation of sense organs to which the axon was connected—whether it be a pressure receptor in the skin, a position receptor in muscle, the retina in the eye, a chemoreceptor in the bloodstream, or an olfactory receptor in the nose. Similarly, in carrying messages out from the central nervous system to effector organs such as muscle, signals were coded by the frequency of trains of identical impulses; the greater the frequency, the more intense the effect.

The innovations of the 1920s provided the technological, conceptual, and experimental basis for an even more stunning set of discoveries made from the mid 1930s to the late 1950s. Amplifiers and cathode-ray oscilloscopes were further harnessed with ingenious experimental techniques, carried out on unusual biological materials such as squid axons, to show that the nerve impulse, or action potential, results from mathematically describable ion flows: sodium ions first moving inward down their concentration gradient to depolarize the nerve membrane, and then potassium ions flowing outward down their gradient to repolarize the membrane and make it ready for the next impulse—all taking place within about a millisecond. Similar instrumentation and analogous techniques showed that when the impulses arrive at the

[3] Hill to the Nobel Committee, 27 Jan. 1930, 18 Jan. 1932, A. V. Hill papers 4/15, Churchill College, Cambridge.

[4] E. D. Adrian, *The Mechanism of Nervous Action: Electrical Studies of the Neurone* (Philadelphia: Univ. Pennsylvania Press, 1932), p. 2.

next cell in the chain, they cause, via a chemical transmitter, ion flows that either excite the target cell so that it fires, or inhibit it. Neurons could be shown to modulate each other's activities, millisecond by millisecond, and thereby to create the massive complexity of nervous activity. Not only could the action of the nervous system be pictured, it could be explained on physicochemical grounds.[5] A hundred-year-old dream was coming to fulfillment.

On the surface, the first half of this revolution in neurophysiology, the quarter century 1905–1930, was a closely articulated and tidy sequence of progress summarizable almost in textbook form:

ca. 1910	Classical nerve physiology is frustrated by its inability to detect or display nerve impulses.
1904–1908	Lucas demonstrates that muscle fibers contract in an all-or-none way.
1911–1914	Adrian shows that the all-or-none principle applies to nerve fibers, expressing it in verbal, qualitative terms.
1919–1925	Alexander Forbes uses a single-stage vacuum-tube amplifier to record clusters of nerve impulses produced both naturally and artificially, thereby demonstrating the potential of the new instrumental approach.
1920–1928	Herbert Gasser and colleagues build a much more powerful three-stage amplifier, and use it to drive a cathode-ray tube displaying artificially stimulated nerve impulses.
1925–1932	Adrian and associates use a three-stage amplifier and a capillary electrometer to record naturally occurring single nerve impulses, thereby proving graphically the all-or-none concept and showing that nervous activity is coded by frequency modulation.

Yet such a schematic recapitulation, while true to the most general outlines of the historical phenomena, obscures as much as it illuminates. It presents technology as the unproblematic servant to, and at the same time adjudicator of, conceptual schemes. More particularly, it hides the often critical differences that competing instrumental systems have at any given time, and how these relative differences change over time. It gives no indication of the sources and processes by which instruments are brought into the biological laboratory; biologists are after all not engineers, whose full-time occupation it is to conceive, design, build, and test instruments. It conceals the gradations that always exist in the quality of the instrumental output, such as graphic records, and their importance to the credence accorded experimental results. It leaves little scope for experimental technique—the actual manipulation on the laboratory bench—in contributing to the success or failure of a line of investigation. It neglects entirely the often crucial and determining nature of the biological substrate—exactly which nerve or muscle, in which animal, with which features, is the object of investigation. And finally, it obscures the way in which instrument, concept, experimental technique, and biological material are embedded in an ongoing laboratory program that has both an internal logic and external social connections, which very much shape the direction of those programs.

[5] An overview of the accomplishments of mid-twentieth-century neurophysiology can best be gained from a trio of books by students respectively of Sherrington, Adrian, and Hill: John C. Eccles, *The Physiology of Nerve Cells* (Baltimore: Johns Hopkins Univ. Press, 1957); Alan Lloyd Hodgkin, *The Conduction of the Nervous Impulse* (Liverpool: Liverpool Univ. Press, 1964); and Bernard Katz, *The Release of Neural Transmitter Substances* (Liverpool: Liverpool Univ. Press, 1969).

The following analytical narrative retells that bald summary in such a way that instrumentation emerges in all its dimensions and points of contact. It is a story known only in part, and that largely in the memories and hand-me-down tales of older physiologists; hence it needs to be told in greater detail than might be the case for other twentieth-century research traditions, such as population genetics or molecular biology, that have had the benefit of scholarly attention.

I. THE FRUSTRATIONS AND INSTRUMENTAL RESOURCES OF TRADITIONAL NERVE PHYSIOLOGY

The problems of traditional neurophysiology were fundamentally dimensional. Late-nineteenth-century histology had made it clear that neurons transmitted their messages over great distances, as far as two to three meters, via a long cellular extension, the nerve fiber or axon. For vertebrates these axons were recognized to be extremely small and fragile, on the order of 0.02 millimeter (20 microns) in diameter, and they functioned only because hundreds or thousands of fibers were bound together, like an intercity telephone cable, into the "nerves" that a dissector saw with the naked eye. Within each axon traveled those hypothetical entities, the nerve impulses. Artificial and indirect evidence had suggested that they lasted about a millisecond. They passed down the fiber at a speed of about 20 to 30 meters per second, and seemed to consist of, or be accompanied by, electrical changes on the order of 0.1 volt (100 millivolts). Thus the biological entities studied were in three ways miniscule—in space, in time, and in electrical effect—rendering them almost invisible to experimentation.

What had tantalized investigators since Luigi Galvani in the 1790s was that the almost ineffable nerve impulse could nonetheless be detected. This was done most easily by observing the contraction of a voluntary muscle, such as the gastrocnemius of a frog, when the researcher stimulated the attached motor nerve, such as the sciatic. This biological detector, the "rheoscopic frog," had the advantage of exquisite sensitivity, but gave no clue how the signal was carried in the nerve. During 1820–1860 the generation of Carlo Matteucci and Emil du Bois-Reymond, using the newly invented moving-coil galvanometer, fashioned an instrumental approach that bypassed the muscle and detected instead the electrical signs of nervous action. But the instrument took 5 to 20 seconds to react to an electrical event that took only a small fraction of a second in the nerve, as the muscle contraction made clear. In the 1860s Julius Bernstein, then Helmholtz's assistant in the Physiological Institute at Heidelberg, showed how ingenuity could to some degree circumvent the limitations of the galvanometer, essentially by taking readings of the magnitude of successive "slices" of the standing wave created in the nerve by repetitive stimulation.[6] Yet although Bernstein could thus plot out the set of amplitudes yielded by his "rheotome" to show the presumed shape of the nerve impulse, points in this diagram represented an average of magnitudes for hundreds of nervous impulses, not the real value of a single one.

[6] On the rheotome see H. E. Hoff and L. A. Geddes, "The Rheotome and its Prehistory: A Study in the Historical Interrelation of Electrophysiology and Electromechanics," *Bulletin of the History of Medicine,* 1957, *31:*212–234, 327–347.

The great instrumental hope of the 1870s and 1880s was the capillary electrometer of Gabriel Lippmann, as improved by the Parisian physiologist Étienne Jules Marey and the Oxford physiologists John Burdon Sanderson, George J. Burch, and Francis Gotch. Mercury and dilute sulfuric acid were put into a capillary glass tube and a change of electrical potential was applied across the phase boundary. This caused the meniscus to move rapidly, and with suitable lighting and lenses the deflection could be captured on film; one could record unique, not averaged, electrophysiological events such as the contraction of skeletal muscle and of the heart. Although much quicker than the coil galvanometer, the capillary electrometer still had enough inertia that a muscle action current lasting 25 to 100 milliseconds was distorted in form. Fortunately, in the 1890s Burch and others developed a laborious but accurate algorithm for correcting the raw photographic records. Unfortunately, such corrections proved impossible for recording from nerve, which had electrical events of 1 to 2 seconds, and current flows two orders of magnitude smaller than muscle. The best records obtainable, by Gotch in 1898–1902, showed a featureless blip 1 millimeter high and 2 millimeters long.[7]

By 1905 yet another recording instrument captured the fancy of electrophysiologists: the string galvanometer, developed by Willem Einthoven, physiology professor at Leiden.[8] He had calculated that to make a galvanometer sensitive enough for physiological purposes, one had to increase the strength of the magnetic field enormously by replacing the usual permanent magnet with an electromagnet, and to diminish the mass of the coil by reducing it to a single thin, light, electrically conducting filament, or "string." When a physiological action current passed through the filament, the resulting deflection, as in the case of the capillary electrometer, could be projected onto rapidly moving film. Einthoven's original instrument at Leiden—which weighed several tons, filled an entire room, and needed constant running water to cool the electromagnet—was soon scaled down and made hardier. By 1908 Edelman & Sons in Munich and the Cambridge Scientific Instrument Company in England were producing instrument setups commercially. A flood of articles on electrocardiology and electromyography soon began to pour forth from the laboratories and clinics of Western Europe.

Yet despite the expenditure of endless hours, great ingenuity, and not a little money, not one of the instrumental resources of traditional nerve electrophysiology—rheoscopic frog, moving coil galvanometer, differential rheotome, capillary electrometer, or string galvanometer—was equal to the task. If they were sensitive, they were either too slow (galvanometer) or responded in an uninformative way (rheoscopic frog). If they were relatively quick (capillary electrometer or string galvanometer), they were not sensitive enough. What served well to record electrical

[7] On the capillary electrometer see Robert G. Frank, Jr., "The Telltale Heart: Physiological Instruments, Graphic Methods, and Clinical Hopes, 1854–1914," in *The Investigative Enterprise: Experimental Physiology in Nineteenth-Century Medicine,* ed. William Coleman and Frederic L. Holmes (Berkeley/Los Angeles: Univ. California Press, 1988), pp. 211–290, esp. 226–257. See also Francis Gotch, "The Electrical Response of Nerve to Two Stimuli," *Journal of Physiology,* 1899, *24:*410–426; Gotch, "The Effect of Local Injury upon the Excitatory Electrical Response of Nerve," *J. Physiol.,* 1902, *28:*32–56; and Gotch, "The Submaximal Electrical Response of Nerve to a Single Stimulus," *J. Physiol.,* 1902, *28:*395–416.

[8] On the string galvanometer see Frank, "The Telltale Heart," pp. 257–263, and the sources cited there.

impulses in muscle and heart was inadequate when brought to bear on the much quicker and much smaller nerve impulses. Moreover, using galvanometers or differential rheotomes one could construct representations of averaged events, but without knowing whether the average represented numerous events of the same magnitude or a range of magnitudes clustered around a mean. If the galvanometer deflected further when the frequency of stimulation was increased, was it because there were more impulses of the same size, or the same number of bigger impulses? Or a mixture of big and small?

More frustrating was the inability of such instruments even to approach the activity of a single axon in a nerve bundle that physiologists knew contained hundreds or thousands of active fibers, all of whose individual impulses added up to the single tiny blip that he could record. To create even this, they had to administer an artificial electric shock to the bundle that in no way corresponded to the natural activity of the nervous system. It was much like putting the population of a large village on their telephones, pinching them all simultaneously, and trying to puzzle out the English language from the chorus of profanity recordable on the telephone cable leading out of town. Little wonder that many physiologists saw much more potential in the reflexes program that Sherrington had been pursuing at Liverpool since the early 1890s: Stimulate a skin area on a decerebrate animal so as to create one or several natural incoming messages, and then, by means of the natural outgoing muscle action that resulted from nervous integration in the spinal cord, try to work out the complex relationships of excitation and inhibition. Such a procedure was open to the objection that the reactions of muscles introduced complexities of their own, but Sherrington and his colleagues believed that these difficulties could be overcome by multiple controls and clever experimental techniques. The ascendancy of this program is perhaps symbolically best represented in Sherrington's election in 1913 to the Waynflete Professorship of Physiology at Oxford—as successor to Francis Gotch.

II. LUCAS, ADRIAN, AND THE ALL-OR-NONE PRINCIPLE

Keith Lucas took a very different approach from Sherrington's, one that reflected the unusual background from which he entered nerve and muscle physiology. His father was a self-taught engineer who designed numerous improvements in submarine telegraph cables and cable laying and rose to be managing director of the Telegraph Construction and Maintenance Company, outside London. The son loved mechanical apparatus and electrical instruments, especially those he could build with his own hands. Although Lucas came up from Rugby School to Trinity College, Cambridge, with a classics scholarship, he at first took preliminary examinations in the medical course and then drifted into the natural sciences. The death of a close friend prompted him to leave Cambridge after his B.A. in 1901 and live in New Zealand. When he returned in 1903, he had decided on physiology—not medicine or engineering—as a career, and immediately set to work on muscle experiments that won him the Walsingham Medal and the Gedge Prize a year later, as well as a prize fellowship at Trinity and a demonstrator's position at the physiological laboratory. These he held for the next decade, progressively building a reputation as one of the brightest young physiologists in England. He became a director of the Cambridge

Scientific Instrument Company in 1906 and a lecturer in natural science at Trinity in 1908; he was elected Fellow of the Royal Society in 1913, at the age of thirty-four.[9]

His first brilliant piece of research, in 1904–1905, asked the simple question of how a skeletal muscle achieves a gradation in its contraction. Is this done by the gradation of contraction of all its constituent muscle fibers; or by the variation of the number of fibers called into play; or by a combination of both mechanisms? Earlier experiments by Germans over the period 1871–1891 had pointed to the latter two possibilities, but no one had ever proved it. Lucas reasoned that if variation in number was the operative principle, and the contraction of a small enough number of muscle fibers could be recorded, then a gradually increasing direct electrical stimulus should show the contraction increasing discontinuously in definite steps. As his experimental material he chose the cutaneus dorsi of the frog, a small muscle from which he could cut strips containing some twelve to thirty muscle fibers. The movable end of the strip was attached to a recording lever, which rotated a minute galvanometer mirror, from which a beam of light reflected back onto moving photographic film, thus magnifying the movement of the fibers some fourteen times. As direct stimulation was gradually increased, the contraction of the muscle strips did indeed show a staircase effect. The actual number of steps was irregular—sometimes three, four, five, or six—but always less than the number of fibers in the preparation. The individual muscle fiber, when stimulated directly by electricity, therefore contracted in an all-or-none fashion—completely or not at all.[10]

Because he had stimulated the muscle directly, Lucas's technique, ingenious and delicate though it was, left his experiments open to the objection that perhaps muscle fibers in the intact animal behaved differently. In 1908 a histologist colleague at the Cambridge Physiological Laboratory prompted a new approach. H. K. Anderson pointed out to Lucas that an unusually small number of motor fibers innervate the skin muscles of the frog. On examining the nerve that supplied his own experimental material, the cutaneus dorsi, Lucas found that it contained only nine to ten large medullated (motor) fibers. Since he had previously determined that the muscle itself has about 150 to 200 muscles fibers, then on average every nerve fiber must innervate about twenty muscle fibers. If each of the twenty fibers responded differently to impulses in its innervating axon, then of course the possible number of steps in contraction would be on the order of two hundred. But if each nerve fiber, when excited at all, caused all of its attached muscle fibers to contract, then the number of steps could not be greater than the number of motor nerves. Lucas dissected out the muscle *in situ* under Ringer solution, attached it to his myograph as before, and then delicately isolated the branch of the iliohypogastric nerve that led to the cutaneus dorsi. Upon stimulating this nerve with gradually increasing strength at intervals of 30 seconds, he found that the maximum contraction was reached in two or three steps, sometimes more—but never more steps than there were motor nerve fibers. The all-or-none property therefore applied not only to the individual muscle

[9] On Lucas see Horace Darwin and William M. Bayliss, "Keith Lucas, 1879–1916," *Proc. Roy. Soc., Ser. B,* 1917–1919, *90:*xxxi-xlii.

[10] Keith Lucas, "On the Gradation of Activity in a Skeletal Muscle-Fibre," *J. Physiol.,* 1905, *33:*124–137.

fiber, but also to the motor unit supplied by one axon. The strong inference was that it did so because that nerve itself either fired or did not.[11]

Lucas's work establishing the all-or-none principle for skeletal muscle was characterized by his ingenuity in finding an appropriate biological material, by his skill in delicate manipulations, but most of all by the clear, logical thinking with which he cut to the core of a problem. As Lucas turned his attention increasingly to nerve after 1909, he clearly hoped that it was this last characteristic that would open up new approaches. The key feature of the nerve impulse, he reasoned, was that it propagated. Physiologists knew that it did so at a fixed rate determinable for the kind of nerve. Moreover, two impulses could not be propagated arbitrarily close to each other. There was a refractory period after the passage of an impulse within which a second could not travel; the nerve had to recover before being able to propagate another. And lastly, physiologists knew that when a section of nerve was narcotized with alcohol vapor or carbon dioxide, propagation through that segment would eventually fail. What Lucas proposed was to reanalyze the nerve impulse on the basis of these fundamental properties—from the ground up.

But in so attempting to rebuild nerve physiology on a strictly analytical basis, Lucas did not wish to dispense with the recording tools that classical electrophysiology had delivered to him. Rather, with the conceptual talent and manual skill of an engineer, he set out to improve them. He designed and built an electric capillary-tube drawer that produced tubes of greatly superior characteristics. He developed a pendulum that opened and closed electrical switches as it fell, and which could be adjusted to deliver precise stimulating shocks of between 0.1 and 120 milliseconds in duration. He improved the projection system and the inertia characteristics of his capillary electrometer until it could reach half excursion in 8.7 milliseconds, a sevenfold improvement in speed over the instruments of twenty years before. Perhaps the most elegantly engineered piece of all was a machine whose gears embodied Burch's algorithm for correcting points on a capillary electrometer photographic record, in order to derive the true paper record of potential changes. The operator had only to arrange a cross-hair over a point on the photographic record tangent to the curve, and the machine did the rest.[12] As might be expected, all these improved instruments were also available for sale from Cambridge Scientific Instruments, for the design of whose physiological apparatus Lucas was responsible.[13] Although he had been consulted about the string galvanometers manufactured commercially by CSI for medical purposes, he felt that for laboratory research that instrument was "vastly inferior" to the more rigorous and useful capillary electrometer, which was more

[11] Keith Lucas, "The 'All or None' Contraction of the Amphibian Skeletal Muscle Fibre," *J. Physiol.,* 1909, *38:*113–133.

[12] For the tube drawer see Keith Lucas, "A Method of Drawing Tubes for the capillary electrometer," *J. Physiol.,* 1908, *37:*xxviii-xxx; for the "Lucas pendulum," Lucas, "On the Rate of Development of the Excitatory Process in Muscle and Nerve," *ibid.,* pp. 459–480, on pp. 460–464; for the capillary electrometer and projection system, Lucas, "On the Relation between the Electric Disturbance in Muscle and the Propagation of the Excited State," *J. Physiol.,* 1909, *39:*207–227, on pp. 210–218; and for the machine embodying Burch's algorithm, Lucas, "On a Mechanical Method of Correcting Photographic Records Obtained from the Capillary Electrometer," *J. Physiol.,* 1912, *44:*225–242, esp. p. 237.

[13] The Cambridge Scientific Instrument Company, *Some Physiological Apparatus,* July 1913, List No. 124, pp. 2–11.

rapid and whose records were susceptible to an "exact method of correction." Lucas designed this ensemble of improved instruments explicitly to produce more exact records of the nerve or muscle impulse as an electrical signal, and thus to complement and extend the concepts derived from studying the impulse as a purely phenomenological "propagated disturbance."[14]

Into this program Lucas inducted a brilliant young scholar, Edgar Douglas Adrian. Born into an upper-class, civil servant family in 1889, raised in London and schooled at Westminster, Adrian had come up to Trinity College in 1908. He read medical subjects, performed brilliantly in the natural sciences tripos, and gained a first class degree in physiology in 1911.[15] Lucas launched his protégé into research by handing him a prime problem: the conduction of the nervous impulse.

Since even the improved recording devices in Lucas's laboratory could still do little to record a picture of the nerve impulse, they agreed that their elusive quarry had to be analyzed using the rigorous logic and manipulative skills that had served Lucas so well in establishing the all-or-none principle for muscle action. Physiologists, largely German, had demonstrated as early as the 1890s that when a nerve was cooled or anesthetized locally, the conducted disturbance underwent a change in rate and time relations that seemed to indicate a diminution of the effect. It underwent, as it came to be called, a *decrement.* When the depth of narcosis reached a threshold point in time, or when the length of nerve not yet narcotized to threshold was increased, the impulse would fail to propagate through the obstacle. Yet it could be started again if the nerve were stimulated somewhere in the middle of the narcotized segment. Finally, when propagation through such a segment had just failed, no amount of stronger stimulation could push an impulse through the obstacle. Lucas and Adrian concluded that, taken together, these phenomena indicated that the nerve impulse started with a constant magnitude and underwent a gradual decrement as it traversed the narcotized strip. The nerve impulse, in other words, was also of the all-or-none nature hinted by Lucas's experiments of 1908–1909. It was a fixed quantity representing the full activity of which the axon was capable at that point. If the axon's capacity were diminished, as in narcosis, then so was the nerve impulse, gradually declining to nothing in such regions. But if the impulse did reach a segment of normal nerve, then it should return to normal magnitude. The nerve impulse was thus like a burning train of gunpowder rather than a concussive wave or a bullet; the energy of propagation came not from the previous history of any given impulse, but from the state of the nerve at that point. Although later neurophysiology was to visualize the all-or-none principle as the unchanging waveform of the action potential, in its original form the principle was a verbal, qualitative statement, rather than an objective, quantitative picture.

To test this principle, Adrian designed an elegant *experimentum crucis,* which he carried out in the spring and summer of 1912.[16] Suppose a nerve is narcotized over a segment *AB* just sufficiently that it cannot conduct an impulse to the other side. Then intercalate a normal segment of nerve into the middle of the narcotized region,

[14] Keith Lucas, "On the Recovery of Muscle and Nerve after the Passage of a Propagated Disturbance," *J. Physiol.,* 1910, *41:*368–408; quoting from p. 370.

[15] The best source on Adrian's life is Sir Alan Hodgkin, "Edgar Douglas Adrian, Baron Adrian of Cambridge, 1889–1977," *Biographical Memoirs of Fellows of the Royal Society,* 1979, *25:*1–73.

[16] E. D. Adrian, "On the Conduction of Subnormal Disturbances in Normal Nerve," *J. Physiol.,* 1912, *45:*389–412.

and two outcomes are possible. Either the impulse will be diminished by the first region, conducted in the normal segment at a subnormal intensity, and then extinguished by the second region of narcotized nerve. Or it will regain intensity in the normal segment and thus be able to pass through the second portion of narcotized nerve, just as if the experimenter had actually stimulated within the intercalated segment.

Adrian's apparatus, extremely simple, mirrored the conceptual design. Alcohol vapor was made to flow into two parallel sets of chambers containing frog nerves. The nerve of one preparation traversed first a narcotizing chamber 4.5 millimeters long, then passed outside the alcohol vapor, and then traversed another 4.5-millimeter chamber. The nerve of the second preparation passed through a single 9-millimeter chamber. To measure the depth of narcosis Adrian used the amount of time that the alcohol vapor flowed before failure. He found that the double-chambered preparation always took longer to fail, up to 75 percent longer, than the single-chambered preparation. Moreover, the time until failure of a nervous impulse traversing the two separate short chambers differed hardly at all from the time until failure in traversing a single short one. In other words, the nerve impulse did seem definitely to recover its amplitude, or intensity, and the all-or-none principle held.

Adrian's experiments, carried out at the age of twenty-two, albeit with the apparatus, encouragement, and intellectual support of Lucas, brought him the Walsingham Medal in the autumn of 1912 and a prize fellowship at Trinity in October of 1913. That same year he brought out an extended defense of his findings against some anomalous evidence, and in 1914 rounded out this phase of his research with additional experimental results contained in a more general treatment entitled "The All-or-None Principle in Nerve."[17] Six months after that paper was published, war broke out, and Adrian left off physiological work in order to do his clinical training in London and thereby be able to contribute to the national effort. Lucas's laboratory career came to an equally abrupt halt; he at first attempted to volunteer in the artillery, but was eventually seconded to the research staff of the Royal Aircraft Factory at Farnborough, where his engineering talents were put to work designing and testing such items as bombsights and aerial compasses.

III. YANKEE INGENUITY: FORBES, GASSER, AND VACUUM-TUBE AMPLIFICATION

It is ironic that, for all of Lucas's engineering skill and expertise, it was not his prize pupil Adrian, but rather a visiting American, Alexander Forbes, who took up the instrumental side of Lucas's program. Forbes came from a wealthy, distinguished, and civic-minded Boston shipping family; his father was the president of the Bell Telephone Company, and his mother was the daughter of Ralph Waldo Emerson.[18] He had become interested in the nervous system while still an undergraduate at Harvard College and had learned the "rudiments of electrophysiology" as a graduate student there, with the zoologist George H. Parker in 1904/5. Thereafter he had made physiology, especially neurophysiology, his primary elective activity while

[17] E. D. Adrian, "Wedensky Inhibition in Relation to the 'All-or-None' Principle in Nerve," *J. Physiol.*, 1913, *46:*384–412; and Adrian, "The All-or-None Principle in Nerve," *J. Physiol.*, 1914, *47:*460–474.

[18] On Forbes see Wallace O. Fenn, "Alexander Forbes: May 14, 1882–May 27, 1965," *Biographical Memoirs of the National Academy of Sciences,* 1969, *40:*113–141.

getting an M.D. at Harvard Medical School in 1906–1910. Having little interest in, or financial need to go into, practice, Forbes immediately joined the department of physiology as an instructor. Walter Bradford Cannon, the professor, assigned him responsibility for the medical student lectures in neurophysiology, and so he spent his first year doing myographic experiments on reflexes and educating himself intensively in the literature, reading especially the works of Sherrington, Lucas, and Hill. Both reading and experiments convinced him that he had more laboratory technique to learn; he therefore wrote Sherrington in February 1911, requesting to spend the academic year 1911/12 in Liverpool.[19]

The Liverpool sojourn provided Forbes with the expected opportunity to hone his experimental skills in cat surgery and reflex myography, as well as the unexpected pleasure of meeting Lucas, Hill, and Adrian in a weekend visit at Cambridge in late November. Forbes was impressed "at once with the homelike, congenial atmosphere of the laboratory" and "his wonderful apparatus," and so taken with Lucas's program that, before going home in April 1912, he squeezed in a three-week stay at the Cambridge Physiological Laboratory to join Lucas in a series of experiments.[20] Forbes was captivated by "the charm of Lucas's personality" and intellect, and by a mechanical ingenuity devoted not to shiny appearance in apparatus but "wholly to what really counted in yielding the result." Forbes felt that Lucas's experimental ability, joined to his critical faculties, had "thrown more light on the fundamental functional properties of the excitable tissues, nerve and muscle, than has been thrown by the combined efforts of all other investigators."[21] In his brief Cambridge sojourn Forbes had found his own new program: investigating reflex and other Sherrington-like phenomena using the concepts and instruments of Lucas. Forbes may well have seen fate operating in yet another way in his Cambridge visit; his decision to work there caused him and his wife to give up their booking on the maiden voyage of the *Titanic* and sail home instead on the *Lusitania.*[22]

Forbes immediately began to duplicate Lucas's instrumental setup in his own laboratory at Harvard Medical School; or, as Cannon wrote to Hill: "Dr. Forbes had a delightful experience in Cambridge, and returned eager to fit a special room for electrical methods of recording."[23] At Cannon's suggestion the physiology department had already bought one of Cambridge Scientific's string galvanometers, which had been delivered in March 1911, and with which Forbes had recorded a nerve action, probably with direct stimulation, in June 1911.[24] Forbes brought a capillary electrometer back with him from England, and he ordered a capillary drawing ma-

[19] Alexander Forbes to Charles Sherrington, 11 Feb. 1911, Alexander Forbes papers, boxes 119–120, Countway Library, Harvard Medical School. The archive is described in Robert G. Frank, Jr., and Judith H. Goetzel, "The Alexander Forbes Papers," *Journal of the History of Biology,* 1978, *11:*429–435.

[20] Forbes pocket diary for 1911, 25–27 Nov.; Forbes to Alys Lucas, 12 Dec. 1916, in the possession of Prof. David Keith-Lucas, Emberton, Olney, Buckinghamshire, eldest son of Keith Lucas (quotations); and Forbes, pocket diary for 1912, 8–27 April. A complete set of pocket diaries, from 1910 on, are in the possession of Forbes's daughter, Mrs. Andrew Locke, Milton, Massachusetts. From about 1915 Forbes would write a summary of the day's activities in the 1″ × 2″ section for each date.

[21] Forbes to Alys Lucas, 12 Dec. 1916 (first quotation); and Alexander Forbes, "Keith Lucas," *Science,* 1916, *44:*808–810.

[22] Forbes to William Cameron Forbes, 25 Jan. 1912, Forbes papers; and Forbes pocket diary for 1912, 27 April.

[23] Walter Bradford Cannon to A. V. Hill, 8 Aug. 1912, Hill papers.

[24] Forbes to Paul Dudley White, 17 July 1959, plus attached rough notes, Forbes papers.

chine immediately thereafter. By November 1912 he had decided to spend £50 of his own money—equivalent at the time to about one quarter of the average American's annual income—to have Cambridge Scientific build him his own version of the Lucas capillary analyzing machine. In addition, he acquired over the next year the usual panoply of minor items needed to set up shop in electrophysiology: platinum and silver-coated wires, resistance boxes and condensers, a slab of stone for mounting, and special Zeiss apochromats for the optical systems.[25]

Forbes also found himself a kindred soul and technical consultant: Horatio B. Williams. Williams (M.D. Syracuse, 1905) was a young associate professor of physiology at the College of Physicians and Surgeons, Columbia University, whose strong undergraduate background in calculus and physics led him around 1909 to collaborate with a clinician who had acquired one of the very early Edelmann string galvanometers. Together they published the first electrocardiograms made in the United States.[26] Fascinated more by the instrumental than the physiological aspects of the work, Williams spent the summer of 1911 with Einthoven in Leiden and then borrowed Einthoven's plans in order to duplicate the entire massive apparatus at Columbia Presbyterian Hospital. By about 1914 Williams was sufficiently dissatisfied with the imprecision of the Edelmann unit and the unwieldiness of the original Einthoven model that he redesigned the instrument from first principles, producing the Williams-Hindle galvanometer, which became the standard instrument in American electrocardiography for more than two decades.[27] He also made himself expert in drawing and silvering quartz strings of theretofore unsurpassed thinness and flexibility, as well as in electronic techniques for protecting the highly delicate strings from current surges that would burn off their conducting coats.[28] By May 1913 Forbes had put himself in contact with Williams, and in June and again in December he spent a week and a half in Williams's laboratory learning more about string galvanometers. Thereafter Williams became Forbes's supplier of quartz strings—ultra-high-quality ones with diameters of 2.5 microns and deflection times of 2.5 milliseconds—and there was a steady exchange of brief letters on technical subjects between Boston and New York.[29]

[25] Forbes letters to and from suppliers, boxes 58–59, Forbes papers.

[26] Horatio B. Williams, "Contraction of a Muscle during Voluntary Innervation," *Proceedings of the Society for Experimental Biology and Medicine,* 1910, *7:*126–127; Walter B. James and Horatio B. Williams, "The Electrocardiogram in Clinical Medicine, I: The String Galvanometer and the Electrocardiogram in Health," "II: The Electrocardiogram in Some Familiar Diseases of the Heart," *American Journal of the Medical Sciences,* 1910, *140:*408–421, 644–669; and undergraduate transcript of Horatio B. Williams, Office of the Registrar, Syracuse University. On Williams see also *National Cyclopedia of American Biography,* Vol. XLIII, pp. 242–243; Walter S. Root, Ralph H. Kruse, and Kenneth S. Cole, "H. B. Williams, Physician and Physiologist," *Science,* 1956, *124:*527; *New York Times,* 2 Nov. 1955, p. 35; *Who Was Who in America,* Vol. III, p. 921; *Journal of the American Medical Association,* 1955, *159:*1672; and *Journal of the Optical Society of America,* 1956, *46:*366–367.

[27] Alfred E. Cohn, with a letter of Horatio B. Williams, "Recollections Concerning Early Electrocardiography in the United States," *Bull. Hist. Med.,* 1955, *29:*469–474; and Cohn to G. E. Burch, 26 June 1953, Bakken Library, Minneapolis. The original Williams-Hindle string galvanometer is now in the Smithsonian Institution.

[28] Horatio B. Williams, "On the Silvering of Quartz Fibers by the Cathode Spray," *Physical Review,* 1914, *4:*517–521; and Horatio B. Williams, "Note on the Protection of String Galvanometer Circuits against External Electric Disturbances," *American Journal of Physiology,* 1916, *40:*230–237.

[29] Forbes pocket diary for 1913, 1 May, 20–29 June, 17–26 December; and Williams to Forbes, 22 July 1913, Forbes papers (which contain 115 letters between Williams and Forbes, most written before 1925).

By early December 1913 Forbes was ready to start on the series of experiments that were to establish his reputation: using the string galvanometer to record in mammals (cats) the action currents passing in nerves from the spinal cord to effector muscles, that is, in the outgoing (efferent) portion of a reflex arc, after stimulation of the incoming (afferent) segment of the arc.[30] He could also record from the attached muscle and compare the electrical signal there with the action current in the nerve that gave rise to it. And finally, he could compare the hump recorded from a natural reflex response to the sharper spike recorded when the same nerve was stimulated directly. Forbes worked steadily on these experiments from December 1913 to March 1915, having as an occasional helpmate a young Harvard medical student, Alan Gregg, who later went on to become president of the Rockefeller Foundation—one of the primary supporters of medical, and especially neurophysiological, research in the United States from the mid 1920s to the early 1950s.[31]

In retrospect, Forbes's records are not impressive. The reflex action current in nerve caused a string galvanometer excursion on the film only some 3–7 millimeters high and 7–9 millimeters long. But even with such records he could perform the essential task of cutting the muscle out of the study of the reflex arc, thereby eliminating a half a century's obstacle to understanding what was really happening in the nerve. Forbes could time the reflex arc with a precision unattainable before. He could show that the electrical signal and mechanical contraction of the attached muscle were not always to be taken as faithful representations of nerve messages. He could describe more precisely what was meant by reflex fatigue, separating the processes in the muscle from those truly inherent in the reflex arc itself. And he could show that the message in the nerve was longer and broader than the signal caused by direct stimulation of the nerve. If, as Forbes believed, the individual nerve fibers in the trunk obeyed the all-or-none law, then this meant that some axons carried their impulses in advance of others. The impulses in the many neurons that made up the reflex, rather than being fired simultaneously "in a volley," were discharged slightly out of phase, in "platoon fire."[32] Although Forbes wrote to Sherrington in 1916 that it "was very interesting to me developing the ideas," he had had little reaction among fellow physiologists to indicate "that it proved to be more than waste paper." The papers were long, and for their time formidably technical, but the Nobelist John C. Eccles later judged them to be "landmarks, because for the first time there was the skilled application of electrical recording to central reflex phenomena."[33]

At this juncture real changes came to Forbes's life from outside physiology. He had become increasingly involved in support for the Allied cause, publishing in *Atlantic Monthly* letters from Sherrington, Lucas, and Hill as examples of scientists deeply involved in the British war effort.[34] With his background and interests Forbes

[30] Alexander Forbes and Alan Gregg, "Electrical Studies in Mammalian Reflexes, I: The Flexion Reflex," "II. The Correlation between Strength of Stimuli and the Direct and Reflex Nerve Response," *Amer. J. Physiol.*, 1915, *37:*118–176; 1916, *39:*172–235.

[31] Wilder Penfield, *The Difficult Art of Giving: The Epic of Alan Gregg* (Boston: Little, Brown, 1967).

[32] Forbes and Gregg, "Flexion Reflex" (cit. n. 30), pp. 148–149.

[33] John C. Eccles, "Alexander Forbes and His Achievement in Electrophysiology," *Perspectives in Biology and Medicine,* 1969/70, *13:*388–404, quoting from p. 389.

[34] Forbes to Hill, 30 May, 14 Aug. 1916, Hill papers 3/18; and Alexander Forbes, "The Laboratory Reacts," *Atlantic Monthly,* Oct. 1916, *118:*544–551. See also the strong pro-British sentiments in Forbes to Hill, 5 June 1915, Hill papers.

reckoned that he would be most useful as a volunteer in electronics, and so in February 1916 he started working two to three half-days a week at the Croft Laboratory in Cambridge, the radio laboratory of the Harvard physics department. The laboratory was developing a light wireless apparatus for airplanes, and Forbes was put to work on a problem of inductance that utilized his knowledge of stimulating coils. "It's queer the way things connect up," he wrote to Hill later that month, "here is this wireless job establishing a link between Nernst's theory of excitation as modified by A. V. Hill, and military aeroplanes." In the meanwhile, Forbes wrote, "I have three books on Wireless and some on aeroplanes which I study in my leisure moments (which at present don't exist)."[35]

It was during this volunteer research that he first encountered at first-hand the primitive triode vacuum tube—the *audion*, as it was called at the time. Forbes was therefore well placed to understand when Williams wrote from New York in early April 1916:

> I am planning to introduce some wireless stuff into physiological laboratory practice before long. Some of our physical men here have been working on the characteristics of the vacuum tube amplifiers ("Audions") and from what they tell me I think it may be possible to amplify the vagus current and some of the other feeble currents without distortion either of amplitude or phase.

Williams touched on the same subject a week later in another letter:

> I am going down to Washington tomorrow to the Physical Society meeting and expect to find out some things I want to know about the vacuum amplifier scheme I wrote you of. I do not mean the ordinary audion, but a tube pumped to an extremely high vacuum like the Coolidge X-ray tube. The G.E. is making them under the name of "Pleiotron," but I believe they are not available commercially as yet.[36]

Although the Englishman John Ambrose Fleming had described a thermionic two-electrode (diode) "valve" in 1905, it was fit solely for rectification of high frequency oscillations, and hence useful mostly as a detector in primitive wireless telegraphy. Lee de Forest, beginning in 1907, added the grid for control of electron flow and thereby started development of the three-electrode (triode) suitable for more-powerful wireless telegraphy. From 1912 on, Harold Arnold, first working in Robert Millikan's laboratory at the University of Chicago and later leading a team at the Bell Telephone Laboratories, developed the high-vacuum triode for use in amplifying circuits for government radios and in long-distance telephony.[37] It was for this reason that Williams and Forbes only encountered these more sophisticated tubes in military applications.

Forbes came to know these "bulbs" intimately over the next three years. In June

[35] Forbes pocket diary for 1916; and Forbes to Hill, 27 Feb. 1916, Forbes papers (quotations). Forbes's pocket diary for 1916 lists such books as Vaughan Cornish, *Wave Action;* W. S. Franklin, *Electric Waves; McGraw-Hill's Standard Handbook for Electrical Engineers;* Zenneck T. Seelig, *Wireless Telegraphy;* J. J. Thompson, *Conduction of Electricity through Gases;* and J. A. Mauborgne, *Practical Uses of the Wave Meter in Wireless Telegraphy.*

[36] Williams to Forbes, 11, 19 April 1916, Forbes papers.

[37] The most detailed survey of the early development of the vacuum tube, although also maddening in its disorganization and tube-collector mentality, is Gerald F. J. Tyne, "The Saga of the Vacuum Tube" (22 parts), *Radio News,* March 1943–April 1946.

1916 he started working part-time on radios at the Boston navy yard. He enrolled as a specialist in radio electronics for a training cruise in August and September 1916. By March 1917 he was traveling almost weekly to Newport to work on boats, and when the United States entered the war in April, he started going on periodic patrol cruises. Between his laboratory work on electronics at the Croft, his field tests in Newport, and his cruises, physiology dropped out almost entirely for the rest of 1917. This pattern continued through the first half of 1918, with occasional duty at the Washington naval yard, where he began to specialize in "radio compasses"—radio direction finders used both for navigation and to home in on radio transmissions from German submarines. To install and troubleshoot radio compasses on American destroyers, he was sent to Britain from June through December 1918.[38]

The posting to England was a sad reminder to Forbes of how much the war had changed the life of his English friends. Lucas was dead, killed on 5 October 1916 in a midair collision over Salisbury Plain as he was carrying out instrumental flight tests. Adrian, after training at Queen Square Hospital, London, in neurology, had been posted to the army hospital at Aldershot to treat soldiers with nerve injuries. It was thus with great somberness that Forbes went to visit Hill at Cambridge on a weekend in late October 1918. They got the key to Lucas's laboratory and

> we went in there and payed silent homage to the memory of the greatest physiologist, in my opinion of our day. There was all the familiar apparatus I had so enjoyed working with in 1912, set up exactly as it had been in the old laboratory. Now it was lying idle and had not been touched for many a day. At the end of the room was a fine portrait of Lucas. It was very melancholy to me to see it all.[39]

But the war had also given Forbes the expertise to carry neurophysiological research in a new direction: electronic amplification. On 10 February 1919, just a few days after being officially demobilized, Forbes went to the Western Electric Company in New York, the manufacturing arm of Bell Telephone. There "I consulted Dr. Arnold about the problem of using Audion tubes for amplifying action currents in the nervous system in recording them with the string galvanometer, a job I planned to take up as soon as I got back to the laboratory. He advised me which type of tube to use and helped me with the theory." A week later Forbes wrote to Edwin H. Colpitts, Western Electric's chief engineer, to order two each of the D-tube and the L-tube. Colpitts responded by sending the four items, but on loan, since these advanced models tubes were, strictly speaking, not commercially available.[40] Such a decision was no doubt made easier both by Forbes's navy experience and by his long-standing family connection to the Bell Telephone Company.

The next ten months were filled with the exhilarating and also exacting work of development, most of it carried out with a recently hired research assistant, Catherine Thacher. From the very beginning Forbes focused on using a single tube to carry out the amplification. Diary notations record both the advances and the setbacks of designing circuits, learning how to protect the galvanometer so as not to blow the

[38] Forbes pocket diaries for 1916 and 1917, *passim;* and Forbes, "War Journal," 1918–1919, Forbes papers; for a short printed version see Alexander Forbes, "A Radio Compass Officer in Time of War," *The Open Road,* May 1922, pp. 17–22, 62.

[39] Forbes, "War Journal," 27 Oct. 1918.

[40] *Ibid.,* 10 Feb. 1919 (quotation); and Forbes to Colpitts, 18 Feb. 1919, Colpitts to Forbes, 25 Feb. 1919, Forbes papers.

string with current surges, and eventually carrying out experiments that succeeded. Forbes scribbled:"C. T. Bulbs came"(1 March); "Fixed Audion switch board" (6 March); "Got first pictures of audion ampl. c̄ C. T." (5 April); "recorded a c [action current] of cat's n [nerve] c̄ bulb. enlarged pictures" (10 April); "Worked on carpentry & wiring of permanent audion hookup. C. T. came in P.M. & helped in final wiring. Worked well" (21 October); "Did real expt c̄ Audion amplifying" (23 October); "New string broke" (29 October); "Did crucial expt on audion" (12 November); "Studied curves & calculated values at home. In town before noon. Big experiment on audion in P.M. compound method developed film" (20 November).[41] In the midst of these investigations Forbes was always looking for better equipment and new information. In the fall he bought a new Williams-Hindle string galvanometer, smaller and yet more powerful because it had an air gap of only 1.5 millimeters between the poles. He even took a brief trip to Chicago at the end of November to attend a session on the electron tube at the annual meeting of the American Physical Society, and he may have reported briefly on his own results.[42]

Forbes certainly let his fellow physiologists know about his work. In May 1919 he wrote to Williams, who was still in the Army Engineers, describing the promising preliminary results and the circuitry, and invited the older physiologist to share in the work. Williams, who was not demobilized until September, finally wrote back that he did not have "any mortgage on the notion of applying amplifiers to physiological work," and he was delighted that Forbes had "the opportunity and energy to get after it." Forbes had already told Adrian about amplification in March, and the Englishman responded: "The valve idea for magnifying the electric response sounds an excellent idea; if you don't make it work we shall have to breed a new kind of frog with a large electric response." Adrian wrote again in July: "How is your wireless valve string galvanometer? I should be extremely interested to hear how it is getting on. We must fire one up here if it is really going to amplify enough"; he then asked Forbes about the condenser-coupling method he had adopted.[43] In late December 1919 Forbes went to Cincinnati to attend the annual meeting of the American Physiological Society, and made a brief preliminary report. It was concerned largely to explain the principles and circuitry, but did report that with appropriate tissue resistance of 50,000 ohms, "about 25-fold amplification may be obtained."[44] Forbes wrote up the long paper in a concentrated flurry of effort in February and March 1920, and it appeared three months later.

The article, "Amplification of Action Currents with the Electron Tube in Recording with the String Galvanometer," was typical for those announcing new technical breakthroughs; three-quarters of the sixty-two pages were devoted to description of the circuits, analysis of possible methods—condenser, transformer, resistance bridge—of coupling to the string galvanometer, protecting against external interference, and the experimental and theoretical search for maximum

[41] Forbes pocket diaries for 1919 and 1920. See also "Audion Experiments, 1919–1920," box 62, folder 1853, Forbes papers.

[42] See the letters to and from Hindle, Forbes papers; and *Phys. Rev.*, 1920, *15:*127. A diary entry for 28 Nov. 1919 reads: "Wrote on Audion paper. Arr Chicago 1.15 AM."

[43] Forbes to Williams, 29 May 1919; Williams to Forbes, 1 Oct. 1919; Adrian to Forbes, 21 March 1919; and Adrian to Forbes, 16 July 1919; Forbes papers.

[44] Forbes pocket diary for 1919, 28–31 Dec.; and Alexander Forbes and Catharine Thacher, "Electron Tube Amplification with the String Galvanometer," *Amer. J. Physiol.*, 1920, *51:*177–178.

amplification. The physiological part consisted of some records taken from artificially stimulated frog and cat sciatic nerves. Since the amplification obtained depended on the resistivity of the tissue, frog nerves produced better records, with forty- to forty-five-fold amplification. Whereas without amplification an excursion would be a barely perceivable 1 millimeter, with less than maximum gain it was a clear spike of 25 millimeters or more. Whereas the normal human myogram was an irregular waveform of about 5 millimeters from trough to peak, with amplification the excursion would frequently go off the 35-millimeter film—less magnification than in the case of frog sciatic nerve, but still a remarkable increase in resolution.[45]

The pattern of Forbes's work over the next three years was, remarkably enough, only moderately influenced by his new technical breakthrough. He was inordinately busy and productive, with at least seven separate projects on reflexes and excitation going at one time, with almost as many collaborators. All of these used the string galvanometer, but none really needed amplification—or the kinds of technical headaches that came with its use. Only in two studies published in 1923 and 1924, additions to his prewar papers on the electrical study of mammalian reflexes, did Forbes use amplification extensively, and there only to see the fine structure of an effect already established without it.[46] Forbes had a well-established research program within which his wartime interest had produced a technical breakthrough, but that same program did not contain a set of problems that the innovation could solve uniquely. Hence, in the press of numerous other investigations, the innovation went relatively little used for the moment.

A slightly younger and much less busy physiologist, Herbert Spencer Gasser at Washington University in St. Louis, was ready to take over where Forbes had left off. Following Forbes's preliminary report to the American Physiological Society in December 1919, Gasser began to explore the idea of applying the triode to electrophysiology. In this he may well have been encouraged by his chairman. In the decade before World War I, Joseph Erlanger had worked on electrical conduction in the heart, and he had some experience with electrophysiology—which his protégé at that time did not. Gasser got help in launching the project from Harry Sidney Newcomer, a classmate at Johns Hopkins Medical School with a strong background in mathematics and physics, who had charge of a laboratory in the Department of Mental and Nervous Diseases at the Pennsylvania Hospital in Philadelphia. Newcomer in turn discussed the idea with a former roommate of his, Irving B. Crandall, an electrical engineer working for Western Electric in New York City. Crandall drew up a design utilizing Western Electric's newest V-type tubes and sent the diagrams to Newcomer, who assembled the amplifier in Philadelphia. Newcomer wheedled a two-week vacation from his new employer, and drove to St. Louis with the amplifier to do the experiments.[47]

The amplifier's key feature was its three-stage construction, the output of the first

[45] Alexander Forbes and Catharine Thacher, "Amplification of Action Currents with the Electron Tube in Recording with the String Galvanometer," (received 6 April 1920), *Amer. J. Physiol.*, 1920, *52*:409–471; the physiological part occupies pp. 456–464.

[46] Alexander Forbes, Stanley Cobb and Helen Cattell, "Electrical Studies in Mammalian Reflexes, III: Immediate Changes in the Flexion Reflex after Spinal Transection," and Forbes and McKeen Cattell, "Electrical Studies in Mammalian Reflexes, IV: The Crossed Extension Reflex," *Amer. J. Physiol.*, 1923, *65*:30–44; 1924, *70*:140–173.

[47] Interview with Harry Sidney Newcomer in Menlo Park, California, 9 March 1978. Gasser wrote a third-person autobiography that was published posthumously, with an excellent bibliography, as

tube becoming the input of the second, and so with the second and the third. Switches allowed the output from one, two, or three tubes to be sent to the string galvanometer, giving amplifications of approximately 25-, 500-, and over 5,000-fold on nerve preparations with resistances of around 100,000 ohms. So potentially sensitive an instrument had to be shielded and placed far away from other galvanometer apparatus, in order to protect it from currents of the lantern, galvanometer magnet, and camera motor. The camera motor and the tuning fork used as a time marker had also to be separately shielded. Altogether, it was a highly powerful instrument—so much so that two stages were all that Gasser and Newcomer could use effectively. When they switched in the third stage, even with the galvanometer shunted, the apparatus became so prone to adventitious current surges that they burned out one string after another.[48]

As a source of action currents they chose the phrenic nerve of the dog, down which a two-second cascade of nerve impulses would flow in order to move the diaphragm, so that the animal could breathe. Over 1910–1912 Rudolf Dittler and Siegfried Garten at Leipzig had used an unamplified string galvanometer to record from the phrenic nerves of rabbits, cats, and dogs, but the excursions were so miniscule as to be barely visible. Gasser and Newcomer, with one tube engaged, could get an irregular wave of 4-millimeter amplitude, and with two stages, a rapid oscillation of 35–40 millimeters. They could even hook up a second, unamplified, galvanometer to record muscle action currents in the diaphragm, project both instruments onto the same stretch of film, and find a satisfying correspondence between the rates in nerve and muscle, each in the range of 70–110 spikes per second. Moreover, one could see that the currents varied in amplitude and in spacing, suggesting that there was not some relatively simple and constant rate of firing in all the fibers.[49] When Gasser reported on this work to the Washington University Medical Society in May 1921, he explained his results thus:

> The waves vary in size, in distance apart, and in shape. If we accept the all-or-none law the differences in size must be due to differences in the number of fibres involved. The variation in the intervals is irregular. The differences in shape are most likely due to failure of the impulse to start in all fibres simultaneously, the impulse being therefore somewhat out of phase in the several fibers.

Erlanger, who was clearly proud of the good investigation being done in his department, pointed out to the audience that although several laboratories had demonstrated the effectiveness of amplification, "the records heretofore published have been almost if not quite illegible." Gasser and Newcomer had not only "amplified the currents, but [had] gotten legible records" that showed "that natural action currents of nerve and muscle are not timed or spaced in the nerve or muscle but in the

---

"Herbert Spencer Gasser, 1888–1963," *Experimental Neurology,* Suppl. 1, 1964, pp. v-vii, 1–38. For a recent overview of the work of Gasser, Erlanger, and their associates see Louise H. Marshall, "The Fecundity of Aggregates: The Axonologists at Washington University, 1922–1942, " *Perspect. Biol. Med.,* 1983, *26:*613–636.

[48] H. S. Gasser and H. S. Newcomer, "Physiological Action Currents in the Phrenic Nerve: An Application of the Thermionic Vacuum Tube to Nerve Physiology," *Amer. J. Physiol.,* 1921, *57:*1–26.

[49] *Ibid.,* pp. 14–25.

central nervous system from which they come in volleys"—Erlanger here used exactly the word Forbes had rejected when describing the same phenomenon.[50]

Gasser's amplifier was in fact too powerful for his graphic recording device, but that was a problem already in the process of being solved by the cathode-ray oscilloscope. In December 1920, a year after Forbes and Thacher had made their first report, the American Physiological Society happened to meet in Chicago at the same time as the Physical Society. Gasser met Williams by chance in the hallway there, and the older physiologist-cum-engineer suggested that he might wish to hear a paper by J. B. Johnson of Western Electric, on a new form of low-voltage, hot-cathode oscilloscope.[51] Whereas the old Braun tubes in use since the 1890s needed 10,000–50,000 volts of steady potential to deflect the electron beam, making them expensive, nonportable, and dangerous, Johnson's new version could operate on 500 volts—well within the range of potential produced by a three-stage amplifier. Almost immediately Gasser and Erlanger set to work obtaining one, and by a year later, in December 1921, could report preliminary results.[52]

But the cathode-ray oscilloscope, as Gasser and Erlanger came to use it from mid 1921 on, had its own strange constraints. The actinic value of the phosphors available before the mid 1930s was low, and so a photographic plate put up against the screen could not record a single excursion. Gasser and Erlanger had to link the stimulating circuit to the sweep circuit so that multiple identical events would be traced on the film. They therefore had to use artificial stimulation, and such old physiological warhorses as the frog sciatic nerve as their experimental materials. In breaking through to a new program of research, they had to relinquish the ability to record natural events, in favor of recording repeated events that could be analyzed to 0.10 millisecond of precision.

IV. ADRIAN, SINGLE-UNIT RECORDING, AND THE VINDICATION OF THE ALL-OR-NONE PRINCIPLE

Adrian came back to Cambridge in the spring of 1919, took over Lucas's laboratory, and tried to pick up the threads of his research. He wrote to Forbes in August 1920: "I have got the capillary electrometer going again during the last two months & it seems to me that an amplified electrometer would be about the last word in recording nerve impulses." By later in the year the two had hatched a plan for Forbes to spend part of 1921 in Cambridge, and Adrian, for his part, felt: "I have so much to learn from you that I shall look forward very eagerly to your visit." Specifically, he hoped to learn more from Forbes about amplification and string galvanometers. Adrian wrote to Boston in March 1921:

> Do please bring over some valves; I believe they can be obtained over here but at any rate you know the habits of those you have used.

[50] H. S. Gasser and H. S. Newcomer, "The Application of the Thermionic Vacuum Tube to the Study of Nerve Physiology," *Journal of the Missouri State Medical Association,* 1921, *18:*461–462 (includes Erlanger's comments).

[51] "Herbert Spencer Gasser, 1888–1963" (cit. n. 47), p. 9.

[52] J. B. Johnson, "A Low Voltage Cathode Ray Oscillograph," *Phys. Rev.,* 1921, *17:*420–421; Johnson, "A Low Voltage Cathode Ray Oscillograph," *Journal of the Optical Society and Review of Scientific Instruments,* 1922, *6:*701–712; and H. S. Gasser and J. Erlanger, "The Cathode Ray Oscillograph as a Means of Recording Nerve Action Currents and Induction Shocks," *Amer. J. Physiol.,* 1922, *59:*473–474.

> There is a man at the engineering labs here named Turner: he has just written a book on wireless etc & knows where to lay hands on most of the apparatus. We are just getting a new C.S.I. string galvanometer & the lab. should be particularly good for valve amplification as our 100 D.C. mains are supplied by accumulators.

Forbes obliged by arranging for the European chief engineer of Western Electric, in London, to deliver two V-type tubes to Cambridge for their work.[53]

To judge from the entries in his pocket diary, Forbes's visit in England, from mid May to late October 1921, was a busy mutual exchange of techniques. Forbes set up a one-stage amplifier and got the string galvanometer going. He also took the lead in preparing mammalian nerves and reflexes for experiments, since most of Adrian's previous work had been on frogs. Adrian taught Forbes techniques used for decrement measurements in narcotized chambers and worked with him mastering the capillary electrometer. Together they applied techniques of narcotization, string galvanometry, and reflex measurement to the study of the all-or-none response of sensory fibers, especially in the cat saphenous nerve. They could demonstrate via narcotizing experiments that mammalian motor nerves showed clear all-or-none responses. Sensory nerves were more equivocal; certain aspects showed indirect evidence of all-or-none behavior, yet two different methods failed to show a decrement.[54]

The equivocal nature of these results was to be a foretaste of a season of dissatisfaction for Adrian. Over the next three years he and a student, Sybil Cooper, tried to use more complicated mammalian preparations to support the dual ideas of all-or-none and decrement, but the results were not clear-cut. By about 1924, despite being elected to the Royal Society the year before, Adrian was not happy about the way his research was going. Many years later he wrote that he had taken up string galvanometer recording from the central nervous system "in the hope of being able to find out exactly what was coming down the nerve fibres when the muscle contracted." Were these impulses trains of steady or variable frequency? How were they controlled? He felt he had learned a great deal from Forbes about string galvanometers and mammalian preparations, "but the experiments I had started became more and more unprofitable. You know the sort of thing that happens—they become more and more complicated and the evidence more indirect, and after a time it was quite clear that I was getting nowhere at all."[55] Neither the analytical methods of Lucas nor the recording methods of Forbes seemed to be leading anywhere.

Nor was Adrian's discomfiture helped by his good friend Forbes, who started attacking the problem of the decrement. One day in late October 1923, as Forbes was dozing on a train, it occurred to him that there might be another explanation for decrement–one testable by his apparatus. "N.Y. Thought of explanation of 'Decrement.,'" he wrote in his diary. He and a younger colleague, Hallowell Davis, then did experiments on decrement almost weekly, until early March 1924, after which Forbes turned most of the ongoing research over to Davis.[56] Davis carried it out with

[53] Adrian to Forbes, 20 Aug. 1920, 1 March 1921; and Western Electric to Forbes, 16 May, 6 June 1921; Forbes papers.

[54] E. D. Adrian and Alexander Forbes, "The All-or-Nothing Response of Sensory Nerve Fibres," *J. Physiol.,* 1922, *56:*301–330.

[55] E. D. Adrian, "Memorable Experiences in Research," *Diabetes,* 1954, *3:*26–27.

[56] Forbes pocket diary for 1923, 20 Oct., and Forbes pocket diaries for 1923 and 1924, *passim.*

a student and an assistant, reporting preliminary results to the American Physiological Society meeting in December 1924, and finishing the full papers in the autumn of 1925.[57]

The protocol was elegantly simple. Forbes and Davis ran the peroneal nerve of a cat through a narcotizing chamber, placing three recording electrodes equidistant within the area to be narcotized. A fourth recording electrode was placed distally, outside the narcotizing chamber, to test the state of conduction after the nerve impulse emerged from the narcotic. The results were equally straightforward—and not in Adrian's favor. At each point in time after the alcohol vapor began to flow, the three electrodes recording from within the chamber showed approximately equal action currents "at all points in the narcotized region; in short, there is no progressive decrement." Previous evidence that had been interpreted to support a decrement was merely experimental error due to diffusion of the narcotic near the edge of the narcotizing chamber, or to spread of the stimulating current. As partial comfort to Adrian, Davis and Forbes noted that since the action current was reduced uniformly at all points to a degree depending on the depth of narcosis, and therefore "the size of the impulse depends only on the condition of the nerve at each point and not on its previous history," it followed "that the energy of the impulse comes not from the stimulus but from the fiber, and the all-or-none law is more firmly established than ever."[58] Forbes had already, in the course of 1924, communicated the main lines of their findings by correspondence—if such forewarnings were any consolation.[59]

Adrian was being attacked not only by friends, but by distant strangers as well. Unbeknownst to the English or the Americans, the German-trained Japanese physiologist Genichi Kato had for years been duplicating Adrian's prewar narcotizing experiments, but using the much larger and longer sciatic nerves of the giant Japanese bullfrog. These enabled him to prove what Forbes and Davis had independently discovered with more rigorous experiments, that the appearance of decrement was due to the leakage of narcotic vapor and the spread of stimulating current. Kato embodied his results in a book written in rather curious English and published in Japan in 1924: *The Theory of Decrementless Conduction in Narcotized Region of Nerve.* He sent a copy to Adrian in the fall of 1924, but received no reply.[60]

Everyone agreed that decrement, Adrian, and perhaps the all-or-none principle—in that order—were in trouble. Forbes and Davis read Kato's book in late 1924 and saw that it only confirmed, with rather looser methods, what the Harvard laboratory had already found concerning decrement, and they wrote Adrian about it.[61] As Davis remarked in a letter to an American protégé of Sherrington, John Fulton at Yale, it seemed as though several lines led them all "to Adrian's door, armed with brick-

[57] Hallowell Davis, Alexander Forbes, David Brunswick and Anne McHenry Hopkins, "Conduction without Progressive Decrement in Nerve under Alcohol Narcosis," *Amer. J. Physiol.,* 1925, *72:*177–178; Davis and Brunswick, "Studies of the Nerve Impulse, I: A Quantitative Method of Electrical Recording," *ibid.,* 1926, *75:*497–531; and Davis, Forbes, Brunswick, and Hopkins, "Studies of the Nerve Impulse, II: The Question of Decrement," *ibid., 76:*448–471.

[58] Davis, Forbes, Brunswick, and Hopkins, "Conduction without Progressive Decrement."

[59] See esp. nine letters exchanged between Adrian and Forbes, Nov. 1923–May 1925, Forbes papers.

[60] Genichi Kato to Forbes, 24 Jan. 1925, Forbes papers; and Kato, *The Theory of Decrementless Conduction in Narcotized Region of Nerve* (Tokyo: Nankodo, 1924).

[61] Forbes to Frederick W. Ellis, 6 Nov. 1924, Forbes papers; and Hallowell Davis to Sybil Cooper, 3 Jan. 1925. Unless otherwise noted, letters to and from **Hallowell Davis** are in his possession.

bats, so to speak." Adrian wrote back that he was reconciled to giving up decrement, provided that the all-or-none principle survived, which he believed it might. But as Fulton remarked in his letters to Davis, Adrian's position was unenviable. Moreover, Sherrington had noted privately that Adrian's ideas for interpreting the central nervous system, based upon decrement and the all-or-none principle, could safely be ignored. It seemed, Fulton said, that Adrian had hardly a leg to stand on.[62]

It was at this point that Adrian turned to Gasser. The St. Louis physiologist had been in England and on the continent since mid 1923, learning new physiological techniques courtesy of a Rockefeller grant. He had worked with Hill in London and had become friends with Adrian and his young wife, coming out to Cambridge for occasional weekends. He had even joined the Adrians for a skiing trip in Switzerland in January 1925, in the period when Adrian was feeling most under attack. Thus the way was paved for Adrian to write to Gasser in London for details on the St. Louis amplifier, which he then had made up, with some modifications, by a local Cambridge electronics firm.[63]

It was probably about early spring of 1925 when Adrian tried out his new three-stage amplifier. It had a amplification of about two-thousandfold and was "terribly microphonic"; he confessed later that "I knew very little about it and was rather afraid of all the complications."[64] He decided to test it using Lucas's old capillary electrometer, rather than the string galvanometer. This was probably financial prudence as much as anything else. If a current surge overloaded your recording device, the worse that could happen would be a little spilled mercury; whereas if you overloaded the filament of a string galvanometer, it fried the conductive coating and you were out £5.

To get a baseline Adrian put a pair of standard nonpolarizable electrodes on the nerve of a standard frog nerve-muscle preparation, set up in a shielded chamber. He was distressed, but not greatly surprised, to find that the baseline was not steady. When he closed the circuit he got a constant rapid oscillation, which he naturally suspected was some kind of artifact, and which he hoped did not presage weeks of aimless fiddling. As he began to readjust the preparation and apparatus, he found that the baseline sometimes oscillated, and sometimes was steady. Then he found that the oscillation was present only if the muscle was hanging down freely; if it was supported on a glass plate, the baseline was steady.

> The explanation suddenly dawned on me, and that was a time when I was very pleased indeed. A stretched muscle, a muscle hanging under its own weight, ought, if you come to think of it, to be sending sensory impulses up the nerves coming from the muscle spindles, signalling the stretch on the muscle. When you relax the stretched muscle, when you support it, those impulses ought to cease.[65]

Within an hour Adrian could show that this was indeed the case and could make the first photographic records of naturally occuring nerve impulses.

These were in fact the proprioceptive impulses that Sherrington's work had

[62] Davis to John F. Fulton, 1 Jan. 1925; Cooper to Davis, 25 Jan. 1925; and Fulton to Davis, 1 Feb., 11 May 1925.

[63] Cooper to Davis, 25 Jan. 1925; and Adrian "Memorable Experiences" (cit. n. 55), pp. 126–127.

[64] Adrian, "Memorable Experiences," pp. 126–127.

[65] *Ibid.*

predicted as long ago as the 1890s, and that Forbes and Williams, in one of their few digressions using the one-stage amplifier and string galvanometer, had recorded in 1923.[66] But Forbes had generated the impulses in the afferent nerve by snapping the muscle, whereas Adrian, because of the much greater amplification with which he worked, was getting them with just the weight of the muscle itself. More important was that these were naturally occurring. Returning to the telephone analogy, Adrian's technique was like tapping into a trunk cable and recording just a dozen or so natural conversations, simultaneously. He was recording unique, natural events, not repetitive, stimulated events. Moreover, he could analyze these raw capillary electrometer records using Lucas's old analyzing machine and show that the naturally occurring action currents resembled those propagated by artificial electrical stimulation.

For what Adrian wanted to do, the arrangement of instrumentation and biological materials fitted together perfectly and had distinct advantages over any setup used in other European or American laboratories. As he pointed out in his first paper on this research, the string galvanometer used in this way would have had distortions that could not be corrected, and the actinic value of the cathode-ray oscillograph was too low to allow photographs of single excursions.[67] The capillary electrometer, although out of favor with electrophysiologists because it was relatively insensitive and its records had to be corrected, was perfect. With amplification its lack of sensitivity was not a problem, and records could be analyzed in a few minutes using Lucas's machine. The internal resistance of the capillary electrometer was practically infinite, so that it drew very little current. If a large accidental voltage spike occurred in the system, the worst that could happen was some loss of mercury from the tube, or some electrolysis and bubbles of gas that were easily cleaned out.

In early July 1925 Adrian could report a host of results to a meeting of the Physiological Society at Oxford.[68] He had done numerous controls to show that these signals were not artifacts. Hanging a 50-gram weight from the gastrocnemius muscle caused a burst of action currents in the sciatic nerve that would diminish in frequency for two or three minutes, although occasional impulses were still being sent out ten minutes later. If Adrian pinched the skin he could also get a quick succession of action currents, which then decreased rapidly. In all, it was a remarkable start, and Gasser, who attended the meeting, must have been pleased that his amplifier was being put to such good use. Adrian was doubtless the happiest of all. The next month he wrote to Forbes from vacation in the French Alps:

> I have been having a fine time with an amplifier & the capillary electrometer—in fact I could hardly bring myself to go off on a holiday. The chief thing it has shown so far is the electric response in sensory nerves—cutaneous as well as proprioceptive but I am extremely puzzled by the fact that they seem to be so much in phase with one another. I shall have to get a statistician on to the job.[69]

When Adrian returned to Cambridge from his vacation in October 1925, he had not only a fascinating new problem, but also a new assistant. The young Swedish

[66] Alexander Forbes, Clarence J. Campbell, and Horatio B. Williams, "Electrical Records of Afferent Nerve Impulses from Muscular Receptors," *Amer. J. Physiol.*, 1924, *69:*283–303.

[67] E. D. Adrian, "The Impulses Produced by Sensory Nerve Endings," *J. Physiol.*, 1926, *61:*49–72, esp. pp. 49–50.

[68] E. D. Adrian and Sybil Cooper, "Action Currents in Sensory Nerve Fibres," *J. Physiol.*, 1925, *60:*xlii–xliii.

[69] Adrian to Forbes, 30 Aug. 1925, Forbes papers.

physiologist Yngve Zotterman had worked in the Cambridge Laboratory briefly in the summer of 1920 and was now back for more advanced work, courtesy of a Rockefeller traveling fellowship. The problem to be addressed was one of complexity; there were simply too many sensory organs generating messages. Attempting to overcome this, Adrian reached into Lucas's bag of tricks. He tried recording from the nerve leading to the cutaneus dorsi, since it had so few nerve fibers. Unfortunately, he and Zotterman could not detect any sensory action currents when the muscle was stretched. But Adrian recalled that Lucas had also mentioned another muscle, the sternocutaneus, which had only one muscle spindle.[70] They tried mounting it, but discovered that it too had a complex pattern of discharge, indicating more than one sensory ending. But no matter, they could still use it for overall studies of relations between stimulus and discharge rates.

But then one Tuesday afternoon, 3 November 1925, as they were finishing up with such a preparation of the sternocutaneus, Zotterman, whose job it was to dissect out the muscle and attached nerve, suggested that they try reducing the number of end-organs that were firing by cutting away strips of muscle. As he did so, the pattern became clearer and clearer, until finally they realized that they were recording the activity in a single axon from a single end-organ—even though that active axon lay inside a mass of inactive ones. They loaded the tiny muscle with gradually decreasing weights and could see the difference in frequency of initial discharge burst: one gram gave 33 per second, half a gram gave 27 per second, and a quarter gram yielded 21 per second. Or they could load a constant weight and analyze the variation in frequency over time: first there would be a burst of impulses, which decreased in frequency as the end-organ accommodated. Perhaps just as important as having a single fiber to record from, they could use Lucas's machine to analyze the distorted excursions recorded by the capillary electrometer into their true forms: the spikes then all had the same shape and height, varying only in frequency.

It is clear that from the very beginning Adrian and Zotterman viewed their findings as a direct pictorial vindication of the all-or-none principle, without any need for recourse to the indirect evidence invoking decrement. It was this realization, certainly on Adrian's part, that drove them on. Zotterman later recalled of the scene of their first recording from a single fiber: "Under strong emotional stress we hurried on, recording the response to different degrees of stimulation. Adrian ran in and out, controlling the recording apparatus in the darkroom and developing the photographic plates. We were excited, both of us quite aware that what we now saw had never been observed before." Zotterman wrote a few days later to a fellow physiologist about a "beautiful plate showing impulses with exactly the same spike height, at constant intervals, at constant stimulation. It is only the frequency that varies with the tension." The emotional surge had settled somewhat when Adrian wrote to Forbes in the middle of November that they were "getting on quite friendly terms with the sensory endings in muscles & we are very proud of having got records of the impulses set up by a single nerve ending. They are quite regular, like a beating

[70] E. D. Adrian and Yngve Zotterman, "The Impulses Produced by Sensory Nerve Endings, Part 2: The Response of a Single End-Organ," *J. Physiol.,* 1926, *61:*151–171, esp. p. 151. Adrian says that Lucas mentions the sternocutaneus "in a footnote," but gives no specific reference; a search of Lucas's published papers reveals no such footnote, so it is likely that Lucas simply mentioned the fact in conversation. For Zotterman see his *Touch, Tickle and Pain,* 2 pts. (Oxford: Pergamon Press, 1969, 1971).

heart . . . & I am glad to say that they are all the same size whatever the strength of the stimulus."[71]

The following winter and spring Adrian and Zotterman systematically explored other kinds of tactile sense receptors, according to the pattern set by those first experiments in November and December 1925: record from a sensory system under different kinds of stimulus, and then try to manipulate the materials or the stimulus so that as few units as possible are active. They recorded, for example, from the internal plantar nerve leading to the cat's paw. Then, by using a single pinprick, they hoped to be able to activate only one of the sensory axons in the nerve trunk. Unfortunately, the cat's paw turned out to be rather densely innervated with pressure receptors, and hence the spike trains they recorded were complex rather than simple. Other tactile receptors, such as Pacinian corpuscles in the mesentery or skin pain receptors, proved equally difficult to record from. The results they got were acceptable and certainly conformed to the emerging patterns, but they were nowhere as striking and brilliant as the single-unit recordings from the sternocutaneus.[72]

Meanwhile, publications describing their results started to appear. The first, in February 1926, by Adrian alone, described the apparatus and the preliminary findings. The second, by Adrian and Zotterman, on recording from single fibers, appeared in April. It raised the most notice among the educated public and was recapitulated in newspapers in London and New York. The third, on the cat's paw work by Adrian and Zotterman, came out in August, and the last, on pain receptors by Adrian alone, in October.[73]

A touch of drama was added because the International Physiological Congress was held at Stockholm in August 1926. Adrian and Zotterman both gave papers in the session on neurophysiology, with Adrian describing the apparatus, both authors explaining their joint work on afferent impulses from muscles, and Adrian showing the records. The Stockholm newspapers gave the discoveries much play, emphasizing the role that a Swede had played in them, and also the sensitivity of the apparatus.[74] There was a final sad irony to the occasion in that Kato and his assistants had also come to the Congress. They had made the long trek from Japan on the Trans-Siberian Railway, bringing with them a collection of giant toads, just to show the experiments that disproved decrement. They must have been disappointed to find that Adrian had moved on to new, highly visual ways of proving the all-or-none principle. In addition, all their toads died on the trip.[75]

The logical next step was carried out by Adrian in 1928, in collaboration with yet

[71] Zotterman, *Touch, Tickle and Pain,* Pt. I, p. 220; Yngve Zotterman to Göran Liljestrand, 9 Nov. 1925; Liljestrand papers, Royal Academy of Sciences, Stockholm; and Adrian to Forbes, 19 Nov. 1925, Forbes papers.

[72] E. D. Adrian and Yngve Zotterman, "The Impulses Produced by Sensory Nerve Endings, Part 3: Impulses Set Up by Touch and Pressure," *J. Physiol.,* 1926, *61:*466–483.

[73] Adrian, "Impulses" (cit. n. 67); Adrian and Zotterman, Impulses, Part 2" (cit. n. 70); "Listening-in to the Nerves," *The Observer* (London), 28 March 1926; Forbes to Adrian, 21 May 1926, Forbes papers; Adrian and Zotterman, "Impulses, Part 3" (cit. n. 72); and E. D. Adrian, "The Impulses Produced by Sensory Nerve-Endings, Part 4: Impulses from Pain Receptors," *J. Physiol.,* 1926/27, *62:*33–51.

[74] See, e.g., *Svenska Dagbladet,* 5 Aug. 1926, pp. 3, 18; and *Dagens Nyheter,* 5 Aug. 1926, pp. 1, 8. Abstracts of Adrian and Zotterman's presentations appear in "Proceedings of the XIIth International Physiological Congress Held at Stockholm," *Skandinavisches Archiv für Physiologie,* 1926, *49:*79–80.

[75] "Proceedings of the XIIth International Physiological Congress," pp. 158–160.

another visiting American, the physicist-turned-physiologist Detlev Bronk. Together they attacked the problem of whether the motor system coded messages in the same way. As their biological preparation in these experiments they used the mammalian phrenic nerve, in which Gasser and Newcomer's earliest amplified records had demonstrated that the central nervous system sent trains of impulses down the phrenic nerve at regular intervals to initiate the rhythmic movement of the diaphragm. Since Adrian and Bronk could not control the source of stimulation, they had to take a different approach. They split up the nerve and gradually cut down the interior fibers until they were recording from a single active axon. They even added a slightly different fillip to the instrumental side. For preliminary assessments of the preparation, without having to make photographs each time they made a cut, Adrian and Bronk simply plugged their amplifier into a loudspeaker. When they got down to recording from a single axon, they could recognize the spike train as a monotone hum in their ears, rather than a cacophony. Their expectations were exactly realized: The phrenic nerve showed the same regular sequence of identical impulses as in recording from single sensory units, especially once the records had been analyzed with the Lucas machine.[76]

V. FROM IDEAS TO RECORDS, VIA INSTRUMENTS AND BIOLOGICAL TECHNIQUES

By the time Hill came to write his nominating letter in 1930, the conception of nervous function had been transformed. What had started as a principle more like a scholastic proposition, ended as a set of pictures. Logic yielded to images created by instruments.

There has been a recent tendency to interpret instruments as objects that reify the conceptions that motivated their construction. While this view is a useful corrective to a naive belief in the "objectivity" of instrumentally derived results, it can in itself be historically naive. The danger in experimental biology is not so much that concepts drive instrumentation as that instrumentation drives concepts. Despite Forbes's role as the first to bring amplification into neurophysiology, the central instrument in his program remained the string galvanometer. During the 1920s, as before the war, this instrument, in amplified or unamplified form, defined his studies of reflexes as those concerned with time, amplitude, and phase relations between input stimulus and output reflex response. Gasser and Erlanger very quickly dropped the string galvanometer as a graphic display device and concentrated on the cathode-ray oscilloscope, because it promised the most distortion-free recording. But throughout the 1920s its low luminosity, in conjunction with the relatively slow photographic emulsions then available, meant that the Washington University physiologists had to confine their attention to artificially stimulated events, repeated dozens of times to make a trace on the plate.

This story illustrates another aspect of physiological apparatus: the tendency of instruments to be coupled in sequence. Indeed, the multistage amplifier, in which the output of one tube becomes the input of the next, is an allegory for the nature of these instrumental chains. Matteuci's galvanometer, Marey's capillary electrometer,

[76] E. D. Adrian and D. W. Bronk, "The Discharge of Impulses in Motor Nerve Fibres, Part I: Impulses in Single Fibres of the Phrenic Nerve," *J. Physiol.*, 1928, *66:*81–101; and Adrian and Bronk, "The Discharge of Impulses in Motor Nerve Fibres, Part II: The Frequency of Discharge in Reflex and Voluntary Contractions," *J. Physiol.*, 1929, *67:*119–151.

and even Einthoven's string galvanometer were typical of nineteenth-century instruments in that each contained within itself all elements of its performance. To improve the output, you attempted to improve the instrument per se. The intercalation of the vacuum-tube amplifier between experiment and display device introduced a second element that could be varied. An experimenter had to be skilled not just in the idiosyncrasies of each piece of apparatus but in the greater challenge of making both work together well. Under such circumstances it is understandable that Forbes and Adrian, when putting their vacuum-tube circuits into operation, each chose the display devices with which they had the most experience. Such chaining of instruments was to grow rapidly after the 1930s. By the early 1950s Adrian's prime protégés Alan Hodgkin and Andrew Huxley, Nobel laureate themselves in 1963, worked with four or five elements in sequence: intracellular microelectrode plus cathode follower (to inhibit distortion) plus multistage amplifier plus voltage clamp feedback circuits plus cathode-ray oscilloscope. The investigator could skirt total insanity only because as the number and variety of devices in the sequence multiplied, some of them became standardized and perhaps even commercially available. Little wonder that mid-twentieth-century scientists came to use *instrumentation* as a mass noun that expressed the totality of linked devices.

How did physiologists come to possess, or have access to, the technological skills they needed for instrumental innovation? An older pattern of relationship between investigator and expertise is exemplified in Lucas: an unusual biologist gifted with extraordinary physical or engineering talents who carried out the work of developing his instruments essentially himself. Such a relationship was difficult to maintain into the twentieth century, as the physical and engineering basis for instruments became vastly more complicated. Forbes developed his apparatus from the ground up, but only by virtue of a three-year wartime apprenticeship in using the quirky audions. Even then he had need of constant technical consultation—on tube properties from the engineers at Western Electric, on circuit design from Harvard physicists across the river in Cambridge, and on string galvanometer protection from Williams in New York. Gasser had a closer technical collaboration with Newcomer, who in turn built the amplifier from designs of his former roommate, Crandall, by then one of those engineers at Western Electric. Adrian thereafter took over the design directly from Gasser, along lines of colleaguely contact, and had his amplifier built as a special order by a commercial supplier. By the early 1930s it was neither necessary nor usually expected that a neurophysiologist would have mastered all the electronics in his instrumentation, although the very best of them—Hodgkin for example—could still build a complete setup from scratch on a breadboard.

The cases of Lucas and Adrian demonstrate, however, that era-defining science does not proceed from instrumental innovation alone; ingenious experimental technique and appropriate biological materials are critical. In contrast to the more pragmatic approaches taken in Boston and St. Louis, the two Cambridge physiologists much more clearly designed their experiments to test concepts and interpretations. In Lucas's experiments of 1904 on the all-or-none direct response of muscle fibers, in the extension of that principle in 1908 by stimulation through the motor nerve, and in Adrian's narcotizing chamber experiments of 1912–1914 arguing the applicability of the all-or-none principle to nerve itself, the protocols derived from an incisive logical analysis of the possibilities and were designed to give results that spoke unequivocally. None was in itself arduous or lengthy. The real work—the thinking—was done before the experiments were even begun.

Ingenious techniques linked to more powerful instrumentation also gave varying results according to the biological materials used. The standard materials of late nineteenth- and early twentieth-century electrophysiology were frog sciatic and cat peroneal nerves. They were long and relatively easy to isolate, could be kept viable long enough to carry out experiments, and their properties were well enough known that results could be widely accepted and reproduced. Although Lucas and Adrian frequently used such standard materials, their real brilliance as investigators lay in their ability to seek out an anatomical or physiological anomaly that constituted a special window into phenomena. Hence the fruitfulness of the cutaneous dorsi and the sternocutaneous as nerve-muscle systems. Such unusual biological materials simplified the phenomena, disaggregating the activity of one element from that of tens or hundreds more in the same microscopic region. Without such simplification even the most sophisticated instrumentation can detect only hopelessly confused, conflated, and compound events. Adrian's work began the pattern of neurophysiologists searching for better and better "single-unit recording," made possible through the judicious combination of instruments, experimental techniques, and unusual biological materials.

In modern experimental biology all three—instrumentation, biological materials, and technique—are bound together, if not inextricably, at least with such strong and numerous mutual interrelationships that the ensemble acquires an identity of its own. The whole constitutes the investigator's *system.* Scientists use the word in phrases such as "What system is she using?" or "There are some problems with his system." Investigators prize the unique properties of their systems, because it is the possession of such a system, as well as the results that it produces, that constitutes a claim to attention within the scientific community. In this sense the construction of such a system is also a fame-seeking strategy. Why, after all, should any laboratory duplicate the system of another, when the only outcome would thereby be to duplicate another's results?

The tacit recognition of this strategy, along with investment of thousands of dollars, years of effort, and a lifetime of special expertise, gives such systems a powerful inertia that is not based merely on the presence of an instrument itself. It is instructive to recall that by the mid 1920s Forbes's laboratory at Harvard had a multistage amplifier, a duplicate of Lucas's capillary electrometer setup, and even a Lucas analyzing machine—but Forbes did not take his program in the direction Adrian took his. Forbes was happy with his own instruments, his own biological preparations, his own techniques. It was only Adrian's sense of frustration that led him to change his instrumentation, and thereby to change the system within which it was embedded. Neurophysiology's own change of prospects from 1904 to 1932 reminds us that although it is important to understand the engineering features of instruments, it is ultimately this embeddedness in a scientific program that constitutes their historical significance.

Bibliography

The following bibliography is a selection of the more general sources cited in the footnotes plus additional citations suggested by the contributors. It is not comprehensive. Further, it does not address the enormous bibliography of material on scientific instruments as artifacts or the equally large bibliography of material on the history of experiment, both of which intersect with the contextual approach to instruments presented in this volume. *The Uses of Experiment: Studies in the Natural Sciences,* edited by David Gooding, Trevor Pinch, and Simon Schaffer (listed below), contains a good bibliography on the history of experiment. The Scientific Instruments Commission of the International Union of the History and Philosophy of Science publishes an annual bibliography in its newsletter, which can be obtained from the secretary of the commission, G. L'E. Turner, History of Science Group, Sherfield Building, Imperial College, London SW7 2AZ.

Achinstein, Peter. *Concepts of Science: A Philosophical Analysis.* Baltimore: Johns Hopkins University Press, 1968.

Altick, Richard D. *The Shows of London.* Cambridge, Mass.: Belknap Press, 1978.

Bachelard, Gaston. *L'activité rationaliste de la physique contemporaine.* Paris: Presses Universitaires de France, 1951.

Barnes, John. *Catalogue of the Barnes Museum of Cinematography.* 5 vols. (St. Ives, Cornwall: Barnes Museum of Cinematography, 1967–).

———. "The Projected Image: A Short History of Magic Lantern Slides." *The New Magic Lantern Journal,* 1985, *3*(3):2–7.

Bedini, Silvio A. "The Role of Automata in the History of Technology." *Technology and Culture,* 1964, *5:*24–42.

Beekman, G. E. W. "New Glory for the Ancient Beijing Observatory." *New Scientist* (London), 20 September 1984, p. 54.

Bennett, J. A. *The Divided Circle: A History of Instruments for Astronomy, Navigation, and Surveying.* Oxford: Phaidon, Christie's, 1987.

———. "The Mechanic's Philosophy and the Mechanical Philosophy." *History of Science,* 1986, *24:*1–24.

———. "A Viol of Water or a Wedge of Glass." In *The Uses of Experiment,* ed. Gooding, Pinch, and Schaffer (q.v.), pp. 105–114.

Brahe, Tycho. *Tycho Brahe's Description of His Instruments and Scientific Work as Given in* Astronomiae instauratae mechanica *(Wandesburgi, 1589).* Translated and edited by Hans Raeder, Elis Strömgren, and Bengt Strömgren. Copenhagen: Munksgaard, 1946.

[Brander]. *G. F. Brander, 1713–1783: Wissenschaftliche Instrumente aus seiner Werkstatt.* Alto Brachner (Projektleitung) [et al.]. Munich: Deutsches Museum, 1983.

Brenni, P.; W. D. Hackmann. "Gli strumenti scientifici." In *L'eridità scientifica di Leopoldo Nobili* (Reggio Emilia, 1984), pp. 29–45.

Brown, Harold I. "Galileo on the Telescope and the Eye." *Journal of the History of Ideas,* 1985, *46:*487–501.

Brown, Laurie M.; Lillian Hoddeson, Editors. *The Birth of Particle Physics.* Cambridge: Cambridge University Press, 1983.

Burnett, John. "The Use of New Materials in the Manufacture of Scientific Instruments, c. 1880–c. 1920." In *The History and Preservation of Chemical Instrumentation,* ed. John T. Stock and Mary V. Orna (Dordrecht/Boston: Reidel, 1986), pp. 217–238.

Chapuis, Alfred; Edouard Gélis. *Le monde des automates: Étude historique et technique.* 2 vols. Neuchâtel: Chapuis, 1928.

Chouillet-Roche, Anne-Marie. "Le clavecin oculaire du P. Castel." *Dix-huitième siècle,* 1976, *8:*141–166.

Coates, Vary; Bernard Finn. *A Retrospective Technology Assessment: Submarine Telegraphy—The Transatlantic Cable of 1866.* San Francisco: San Francisco Press, 1979.

Cohen, John. *Human Robots in Myth and Science.* London: George Allen & Unwin, 1966.
Constant, E. W. "Scientific Theory and Technological Testability: Science, Dynamometers, and Water Turbines in the Nineteenth Century." *Technology and Culture,* 1983, *24:* 183–198.
Daumas, Maurice. "Les appareils d'expérimentation de Lavoisier." *Chymia,* 1950, *3:*45–62.
———. "Precision of Measurement and Physical and Chemical Research in the Eighteenth Century." In *Scientific Change: Historical Studies,* ed. Alistair C. Crombie (London: Heinemann, 1963), pp. 418–430.
———. *Scientific Instruments of the Seventeenth and Eighteenth Centuries and Their Makers.* London: Portman, 1989.
Dear, Peter. "Jesuit Mathematical Science and the Reconstruction of Experience in the Early Seventeenth Century." *Studies in the History and Philosophy of Science,* 1987, *18:* 133–175.
Dennis, Michael Aaron. "Graphic Understanding: Instruments and Interpretation in Robert Hooke's *Micrographia.*" *Science in Context,* 1989, *3:*309–364.
Derrida, Jacques. *Margins of Philosophy.* Translated and edited by Alan Bass. Chicago: University of Chicago Press, 1982. See chapters "Tympan," "Differánce," and "Signature Event Context."
DeVorkin, David H. "Electronics in Astronomy: Early Applications of the Photoelectric Cell and Photomultiplier for Studies of Point-Source Celestial Phenomena." *Proceedings of the IEEE,* 1985, *73:*1205–1220.
Dorling, Jon. "Demonstrative Induction: Its Significant Role in the History of Physics." *Philosophy of Science,* 1973, *40:*360–372.
Doyon, André; Lucien Liaigre. *Jacques Vaucanson: Mécanicien de génie.* Paris: Presses Universitaires de France, 1966.
Drake, Stillman. *Galileo at Work: His Scientific Biography.* Chicago: University of Chicago Press, 1978.
———. "Renaissance Music and Experimental Science." *Journal of the History of Ideas,* 1970, *31:*483–500.
Eamon, William. "Technology and Magic in the Late Middle Ages and in the Renaissance." *Janus,* 1983, *70:*171–212.
Erhardt-Siebold, Erika von. "Some Inventions of the Pre-Romantic Period and Their Influence upon Literature." *Englische Studien,* 1931, *66:*347–363.
Feldman, Theodore S. "Late Enlightenment Meteorology." In *The Quantifying Spirit in the Eighteenth Century,* ed. Tore Frängsmyr, J. L. Heilbron, and Robin E. Rider (Berkeley/Los Angeles: University of California Press, 1990), pp. 143–177.
Field, J. V. "What Is Scientific about a Scientific Instrument?" *Nuncius,* 1988, *3*(2):3–26.
Fleck, Ludwik. *Genesis and Development of a Scientific Fact.* Edited by Thaddeus J. Trenn and Robert K. Merton; translated by Fred Bradley and Thaddeus J. Trenn. Chicago: University of Chicago Press, 1979.
Frank, Robert G. "The Telltale Heart: Physiological Instruments, Graphic Methods, and Clinical Hopes, 1854–1914." In *The Investigative Enterprise: Experimental Physiology in Nineteenth-Century Medicine,* ed. William Coleman and Frederic L. Holmes (Berkeley/Los Angeles: University of California Press, 1988), pp. 211–290.
Franklin, Allan. "Discovery, Pursuit, and Justification." *Perspectives on Science,* 1993, *1*(2):252–284.
Franssen, Maarten. "The Ocular Harpsichord of Louis-Bertrand Castel: The Science and Aesthetics of an Eighteenth-Century *Cause célèbre.*" *Tractrix,* 1991, *3:*15–77.
Galison, Peter. "Bubble Chambers and the Experimental Workplace." In *Observation, Experiment, and Hypothesis in Modern Physical Science,* ed. Peter Achinstein and Owen Hannaway (Cambridge, Mass.: MIT Press, 1985), pp. 309–373.
———. "History, Philosophy, and the Central Metaphor." *Science in Context,* 1988, *2:*197–212.
———. *How Experiments End.* Chicago: University of Chicago Press, 1987.
———. "Theoretical Predispositions in Experimental Physics: Einstein and the Gyromagnetic Experiments, 1915–1925." *Historical Studies in the Physical Sciences,* 1982, *12:* 285–323.

Galison, Peter; Bruce Hevly; Rebecca Lowen. "Controlling the Monster." In *Big Science: The Growth of Large-Scale Research,* ed. Galison and Hevly (Stanford, Calif.: Stanford University Press, 1992), pp. 46–77.

Golinski, Jan. "The Theory of Practice and the Practice of Theory: Sociological Approaches in the History of Science." *Isis,* 1990, *81:*492–505.

Gooding, David C. "How Do Scientists Reach Agreement about Novel Observations?" *Studies in the History and Philosophy of Science,* 1986, *17:*205–230.

Gooding, David; Trevor Pinch; Simon Schaffer, Editors. *The Uses of Experiment: Studies in the Natural Sciences.* Cambridge: Cambridge University Press, 1989.

Goody, Graeme. "Precision Measurement and the Genesis of Physics Teaching Laboratories in Victorian Britain." *British Journal for the History of Science,* 1990, *23:*25–51.

Guerlac, Henry. "Chemistry as a Branch of Physics: Laplace's Collaboration with Lavoisier." *Historical Studies in the Physical Sciences,* 1976, *7:*193–276.

Hacking, Ian. "Do We See through a Microscope?" *Pacific Philosophical Quarterly,* 1981, *62:*305–322.

———. *Representing and Intervening: Introductory Topics in the Philosophy of Natural Science.* Cambridge: Cambridge University Press, 1983.

Hackmann, W. D. *Electricity from Glass: The History of the Frictional Electrical Machine, 1600–1850.* Alphen aan den Rijn, Neth.: Sijthoff & Noordhoff, 1978.

———. "Instrumentation in the Theory and Practice of Science: Scientific Instruments as Evidence and as an Aid to Discovery." *Annali dell'Istituto e Museo di Storia della Scienza de Firenze,* 1985, *10*(2):87–115.

———. "The Relationship between Concept and Instrument Design in Eighteenth-Century Experimental Science." *Annals of Science,* 1979, *26:*205–224.

———. "Scientific Instruments: Models of Brass and Aids to Discovery." In *The Uses of Experiment,* ed. Gooding, Pinch, and Schaffer (q.v.), pp. 31–65.

Hahn, Roger. "New Considerations on the Physical Sciences of the Enlightenment Era." *Studies on Voltaire and the Eighteenth Century,* 1989, *264:*789–796.

Hanson, Norwood Russell. "Galileo's Discoveries in Dynamics." *Science,* 1965, *147:*471.

Harré, Rom. *Great Scientific Experiments: Twenty Experiments that Changed Our View of the World.* Oxford: Phaidon, 1981.

Heilbron, J. L. *Electricity in the Seventeenth and Eighteenth Centuries: A Study of Early Modern Physics.* Berkeley/Los Angeles: University of California Press, 1979.

———. "Introductory Essay." In *The Quantifying Spirit in the Eighteenth Century,* ed. Tore Frängsmyr, J. L. Heilbron, and Robin E. Rider (Berkeley/Los Angeles: University of California Press, 1990), pp. 1–23.

Henry, John. "Occult Qualities and the Experimental Philosophy: Active Principles in Pre-Newtonian Matter Theory." *History of Science,* 1988, *24:*335–381.

Hesse, Mary. *Models and Analogies in Science.* Notre Dame, Ind.: University of Notre Dame, 1966.

Hoff, H. E.; L. A. Geddes. "The Rheotome and Its Prehistory: A Study in the Historical Interrelation of Electrophysiology and Electromechanics." *Bulletin of the History of Medicine,* 1957, *31:*212–234, 327–347.

Hoke, Donald. *Ingenious Yankees: The Rise of the American System of Manufactures in the Private Sector.* New York: Columbia University Press, 1990.

Holmes, Frederic Lawrence. *Eighteenth-Century Chemistry as an Investigative Enterprise.* Berkeley: Office for History of Science and Technology, University of California, 1989.

Hounshell, David. *From the American System to Mass Production, 1800–1932: The Development of Manufacturing Technology in the United States.* Baltimore: Johns Hopkins University Press, 1984.

Hughes, Thomas Parke. *Networks of Power: Electrification in Western Society, 1880–1930.* Baltimore: Johns Hopkins University Press, 1983.

Hughes, Thomas Parke; Trevor J. Pinch, Editors. *The Social Construction of Technological Systems: New Directions in the Sociology and History of Technology.* Cambridge, Mass.: MIT Press, 1987.

Hunt, Bruce J. "Michael Faraday, Cable Telegraphy and the Rise of Field Theory." *History of Technology,* 1991, *13:*1–19.

Hutchison, Keith. "What Happened to Occult Qualities in the Scientific Revolution?" *Isis,* 1982, *73:*233–252.

Kassler, Jamie Croy. "Man—A Musical Instrument: Models of the Brain and Mental Functioning before the Computer." *History of Science,* 1984, *22:*59–92.

King, Henry C. With the assistance of John R. Milburn. *Geared to the Stars: The Evolution of Planetariums, Orreries, and Astronomical Clocks.* Toronto University Press, 1978. See Richard Steele, "The Englishman."

Koppes, Clayton. *JPL and the American Space Program: A History of the Jet Propulsion Laboratory.* New Haven: Yale University Press, 1982.

Koyré, Alexandre. "An Experiment in Measurement." *Proceedings of the American Philosophical Society,* 1953, *97:*222–237. Reprinted in Koyré, *Metaphysics and Measurement: Essays in Scientific Revolution,* ed. M. A. Hoskin (London: Chapman & Hall; Cambridge, Mass.: Harvard University Press, 1968), pp. 89–117.

———. "Galileo and the Scientific Revolution of the Seventeenth Century." *The Philosophical Review,* 1943, *52:*333–348. Reprinted in Koyré, *Metaphysics and Measurement,* ed. Hoskin, pp. 1–15.

———. "Traduttore-Traditore: A propos de Copernic et de Galilée." *Isis,* 1943, *34:*209–210.

Kuhn, Thomas S. "The Function of Measurement in Modern Physical Science." *Isis,* 1961, *52:*161–190. Reprinted in Kuhn, *The Essential Tension: Selected Studies in Scientific Tradition and Change* (Chicago: University of Chicago Press, 1977), pp. 178–224.

———. "Mathematical Versus Experimental Traditions in the Development of Physical Science." *Journal of Interdisciplinary History,* 1976, *7:*1–31. Reprinted in Kuhn, *Essential Tension,* pp. 31–65.

Kutschmann, Werner. "Scientific Instruments and the Senses: Towards an Anthropological Historiography of the Natural Sciences." *International Studies in the Philosophy of Science,* 1986, *1:*106–123.

Latour, Bruno. "Give Me a Laboratory and I Will Raise the World." In *Science Observed: Perspectives on the Social Study of Science,* ed. Karen Knorr-Cetina and Michael Mulkay (London/Beverly Hills, Calif: Sage, 1983), pp. 141–170.

———. *Science in Action: How to Follow Scientists and Engineers through Society.* Cambridge, Mass.: Harvard University Press, 1987.

Latour, Bruno; Steve Woolgar. *Laboratory Life: The Social Construction of Scientific Facts.* London/Beverly Hills, Calif.: Sage, 1979. Revised edition, Princeton: Princeton University Press, 1986.

Leatherdale, W. H. *The Role of Analogy, Models, and Metaphor in Science.* Amsterdam/Oxford: North Holland; New York: American Elsevier, 1974.

Lenoir, Timothy. "Models and Instruments in the Development of Electrophysiology." *Historical Studies in the Physical and Biological Sciences,* 1986, *17:*1–54.

———. "Practice, Reason, Context: The Dialogue between Theory and Experiment." *Science in Context,* 1988, *2:*3–22.

Levere, Trevor H. "Lavoisier: Language, Instruments, and the Chemical Revolution." In *Nature, Experiment, and the Sciences,* ed. Levere and W. R. Shea (Dordrecht: Kluwer Academic Publishers, 1990), pp. 207–233.

Lodwig, T. H.; W. A. Smeaton. "The Ice Calorimeter of Lavoisier and Laplace and Some of Its Critics." *Annals of Science,* 1974, *31:*1–18.

Lynch, A. C. "History of the Electrical Units and Early Standards." *Proceedings of the IEEE,* 1985, *132A:*564–573.

MacKenzie, Donald. *Inventing Accuracy: A Historical Sociology of Nuclear Missile Guidance.* Cambridge, Mass.: MIT Press, 1990.

MacLachlan, James. "A Test of an 'Imaginary' Experiment of Galileo." *Isis,* 1973, *64:*374–379.

Maddison, Francis R. "Early Astronomical and Mathematical Instruments: Brief Survey of Sources and Modern Studies." *History of Science,* 1963, *2:*17–50.

———. *Medieval Scientific Instruments and the Development of Navigational Instruments in the Fifteenth and Sixteenth Centuries,* III. Coimbra: Agrupamento de Estutos de Cartografia Antiga, 1969.

Mayr, Otto. *Authority, Liberty, and Automatic Machinery in Early Modern Europe.* Baltimore: Johns Hopkins University Press, 1986.

McConnell, Anita. *Geophysics and Geomagnetism: Catalogue of the Science Museum Collection.* London: The Science Museum, 1986.

Millburn, John R. *Benjamin Martin: Author, Instrument-Maker and "Country Showman."* Leiden: Noordhof, 1976.

Misa, Thomas J. "Military Needs, Commercial Realities, and the Development of the Transistor, 1948–1958." In *Military Enterprise and Technological Change,* ed. Merritt Roe Smith (Cambridge, Mass.: MIT Press, 1985), pp. 253–287.

Naylor, R. H. "Galileo: Real Experiment and Didactic Demonstration." *Isis,* 1976, *67:* 398–419.

Needham, Joseph. *Science and Civilisation in China.* Cambridge: Cambridge University Press, 1954–. Vol. III: *Mathematics and the Sciences of the Heavens and the Earth.* See "The Development of Astronomical Instruments," pp. 284–390, and "The Time of the Jesuits," pp. 437–458.

Needham, Joseph, Lu Gwei-djen, John Combridge, and John Major. *The Hall of Heavenly Records: Korean Astronomical Instruments and Clocks.* Cambridge: Cambridge University Press, 1986.

North, J. D. "Science and Analogy." In *On Scientific Discovery,* ed. Mirko D. Grmek, Robert S. Cohen, and Guido Cimino (Dordrecht/Boston: Reidel, 1981), pp. 115–140.

Pearce, Susan, Editor. *Objects of Knowledge.* London: Athlone Press, 1990.

Pipping, Gunnar. *The Chamber of Physics: Instruments in the History of Sciences Collections of the Royal Swedish Academy of Sciences, Stockholm.* Stockholm: Almqvist & Wiksell, 1977.

Price, Derek J. de Solla. "Automata and the Origins of Mechanism and the Mechanical Philosophy." *Technology and Culture,* 1964, *5:*9–23.

———. "Of Sealing Wax and String." *Natural History,* 1984, *93:*48–56.

———. "Philosophical Mechanisms and Mechanical Philosophy: Some Notes towards a Philosophy of Scientific Instruments." *Annali dell'Istituto e Museo di Storia della Scienza de Firenze,* 1980, *5*(1):75–85.

Pumfrey, Stephen. "Mechanizing Magnetism in Restoration England: The Decline of Magnetic Philosophy." *Annals of Science,* 1987, *44:*1–21.

Roberts, Lissa. "A Word and the World: The Significance of Naming the Calorimeter." *Isis,* 1991, *82:*198–222.

Rosen, Edward. *The Naming of the Telescope.* New York: Henry Schuman, 1947.

Rotman, Brian. *Ad Infinitum—The Ghost in Turing's Machine: Taking God Out of Mathematics and Putting the Body Back In.* Stanford, Calif.: Stanford University Press, 1993.

Schaffer, Simon. "Glass Works: Newton's Prisms and the Uses of Experiment." In *The Uses of Experiment,* ed. Gooding, Pinch, and Schaffer (q.v.), pp. 67–104.

———. "Measuring Virtue." In *The Medical Enlightenment of the Eighteenth Century,* ed. Andrew Cunningham and Roger French (Cambridge: Cambridge University Press, 1990), pp. 281–318.

———. "Natural Philosophy and Public Spectacle in the Eighteenth Century." *History of Science,* 1983, *21:*1–43.

Schweber, Silvan S. "The Empiricist Temper Regnant: Theoretical Physics in the United States, 1920–1950." *Historical Studies in the Physical and Biological Sciences,* 1986, *17:*55–98.

Settle, Thomas B. "An Experiment in the History of Science." *Science,* 1961, *133:*19–23.

Shapin, Steven. "Pump and Circumstance: Robert Boyle's Literary Technology." *Social Studies of Science,* 1984, *14:*481–520.

Shapin, Steven; Simon Schaffer. *Leviathan and the Air-Pump: Hobbes, Boyle, and the Experimental Life.* Princeton: Princeton University Press, 1985.

Sherwood, Merriam. "Magic and Mechanics in Medieval Fiction." *Studies in Philology,* 1947, *44:*567–592.

Shigeru, Nakayama. *A History of Japanese Astronomy: Chinese Background and Western Impact.* Cambridge, Mass.: Harvard University Press, 1969.

Sivin, Nathan. *Cosmos and Computation in Early Chinese Mathematical Astronomy.* Leiden: E. J. Brill, 1969.

Smith, Crosbie; M. Norton Wise. *Energy and Empire: A Biographical Study of Lord Kelvin.* Cambridge: Cambridge University Press, 1990.

Smith, Robert. *The Space Telescope: A Study of NASA, Science, Technology, and Politics.* New York: Cambridge University Press, 1989.

Smith, Robert W.; Joseph N. Tatarewicz. "Replacing Technology: The Large Space Telescope and CCDs." *Proceedings of the IEEE,* 1985, *73:*1221–1235.

Stewart, Larry. "Public Lectures and Private Patronage in Newtonian England." *Isis,* 1986, *77:*47–58.

Stocking, George W., Editor. "Objects and Others: Essays on Museums and Material Culture." Madison: University of Wisconsin Press, 1985.

Stuewer, Roger H. *The Compton Effect: Turning Point in Physics.* New York: Science History Publications, 1975.

Swenson, Loyd. *The Ethereal Affair: A History of the Michelson-Morley-Miller Ether Drift Experiments, 1880–1930.* Austin: University of Texas Press, 1972.

Tatarewicz, Joseph N. *Space Technology and Planetary Astronomy.* Bloomington: Indiana University Press, 1990.

Taylor, E. G. R. *The Mathematical Practitioners of Tudor and Stuart England.* Cambridge: Cambridge University Press, for the Institute of Navigation, 1954.

———. *The Mathematical Practitioners of Hanoverian England, 1714–1840.* Cambridge: Cambridge University Press, 1966.

Trigg, George L. *Landmark Experiments in Twentieth-Century Physics.* New York: Crane, Russak, 1975.

Turner, A. J. "Mathematical Instruments and the Education of Gentlemen." *Annals of Science,* 1973, *30:*51–88.

Turner, Gerard L'E. "Animadversions on the Origin of the Microscope." In *The Light of Nature: Essays in the History and Philosophy of Science presented to A. C. Crombie,* ed. J. D. North and J. J. Roche (Dordrecht: Martinus Nijhoff, 1985), pp. 193–207.

———. "Mathematical Instrument-Making in London in the Sixteenth Century." In *English Map-Making, 1500–1650,* ed. Sarah Tyacke (London: British Library, 1983), pp. 93–106.

———. "The Microscope as a Technical Frontier in Science." In *Historical Aspects of Microscopy,* ed. Savile Bradbury and Turner (Cambridge: W. Heffer, 1967), pp. 175–199.

———. "Scientific Instruments." In *Information Sources in the History of Science and Medicine,* ed. Pietro Corsi and Paul Weindling (London: Butterworth Scientific, 1983), pp. 243–258.

Van Helden, Albert. "The 'Astronomical Telescope,' 1611–1650." *Annali dell'Istituto e Museo di Storia della Scienza de Firenze,* 1974, *1:*13–36.

———. "The Birth of the Modern Scientific Instrument, 1550–1700." In *The Uses of Science in the Age of Newton,* ed. John G. Burke (William Andrews Clark Memorial Library) (Berkeley/Los Angeles: University of California Press, 1983), pp. 49–84.

———. "The Invention of the Telescope." *Transactions of the American Philosophical Society,* 1977, *67*(4).

———. "The Telescope in the Seventeenth Century." *Isis,* 1974, *65:*38–58.

Vincenti, Walter. *What Engineers Know and How They Know It: Analytical Studies from Aeronautical History.* Baltimore: Johns Hopkins University Press, 1990.

Warner, Deborah Jean. "What Is a Scientific Instrument, When Did It Become One, and Why?" *British Journal for the History of Science,* 1990, *23:*83–93.

Wellek, Albert. "Farbenharmonie und Farbenklavier: Ihre Entstehungsgeschichte im 18. Jahrhundert." *Archiv für die Gesämte Psychologie,* 1935, *94:*347–375.

Wheatland, David. *The Apparatus of Science at Harvard, 1765–1800.* Cambridge, Mass.: Harvard University Press, 1968.

Wilson, Catherine. "Visual Surface and Visual Symbol: The Microscope and the Occult in Early Modern Science." *Journal of the History of Ideas,* 1988, *49:*85–108.

Winkler, Mary G.; Albert Van Helden. "Representing the Heavens: Galileo and Visual Astronomy." *Isis,* 1992, *83:*195–217.

Wise, M. Norton; Crosbie Smith. "Measurement, Work, and Energy in Lord Kelvin's Britain." *Historical Studies in the Physical and Biological Sciences,* 1986, *17:*146–173.

Worrall, J. "The Pressure of Light: The Strange Case of the Vacillating 'Crucial Experiment.'" *Studies in the History and Philosophy of Science,* 1982, *12:*133–171.

Wylie, Alexander. "The Mongol Astronomical Instruments in Peking." In Wylie, *Chinese Researches* (Shanghai, 1897; Taipei: Chengwen, 1966), pp. 1–27.

Notes on Contributors

Thatcher E. Deane, an unaffiliated scholar of Chinese history, bureaucracy, and science, received his Ph.D. from the University of Washington. He is currently employed as a management systems analyst for the Seattle Department of Parks and Recreation.

Robert G. Frank, Jr., teaches medical history at the School of Medicine, and history of biology in the Department of History, of the University of California at Los Angeles. He has written on the history of the laboratory medical sciences in nineteenth- and twentieth-century Europe and America, and on physiology, medicine, practitioners, and diseases in early modern England.

Jan Golinski is Assistant Professor in the Department of History and the Humanities Program at the University of New Hampshire. He is the author of *Science as Public Culture: Chemistry and Enlightenment in Britain, 1760–1820* (Cambridge, 1992).

Bruce Hevly is Assistant Professor of History at the University of Washington. He carried out the research for the article in this volume while a postdoctoral scholar in the Program of History of Science at Stanford University, with support from the Stanford Linear Accelerator Center.

Thomas L. Hankins is Professor of History at the University of Washington. He is the author of *Sir William Rowan Hamilton* (Johns Hopkins, 1980) and of *Science and the Enlightenment* (Cambridge, 1985). He is currently studying the history of scientific instruments.

Bruce J. Hunt is Associate Professor of History at the University of Texas at Austin. He recently published *The Maxwellians* (Cornell, 1991) and is now working on a study of the interaction between cable telegraphy and electrical science in the mid-nineteenth century.

Timothy Lenoir is Professor of History and Cochair of the Program in History and Philosophy of Science at Stanford University. His books include *The Strategy of Life* (Reidel, 1982), *Politik im Tempel der Wissenschaft* (Campus Verlag, 1992), *Instituting Science* (Stanford, 1994), and *Reforming Vision: Optics, Aesthetics, and Ideology in Germany, 1845–1890* (Chicago, forthcoming).

Simon Schaffer is Reader in History and Philosophy of Science at the University of Cambridge. He is a coauthor (with Steven Shapin) of *Leviathan and the Air-Pump: Hobbes, Boyle, and the Experimental Life* (Princeton, 1985) and a coeditor (with David Gooding and Trevor Pinch) of *The Uses of Experiment: Studies in the Natural Sciences* (Cambridge, 1989).

Robert W. Smith is a historian at the Smithsonian Institution's National Air and Space Museum. His publications include *The Space Telescope: A Study of NASA, Science, Technology, and Politics* (Cambridge, 1989; 1993). He also holds an adjunct position at the Johns Hopkins University.

Joseph N. Tatarewicz is the author of *Space Technology and Planetary Astronomy* (Indiana, 1990) and an independent scholar. He is now at work on a book sponsored by NASA, *Exploring the Planets: The Planetary Geosciences since Galileo;* he also teaches history at the University of Maryland, Baltimore County.

Albert Van Helden, Lynette S. Autrey Professor of History at Rice University, specializes in the history of telescopic astronomy. His works include *The Invention of the Telescope* (APS, 1977), *Measuring the Universe* (Chicago, 1985), and an edition and translation of Galileo's *Sidereus Nuncius* (Chicago, 1989). He is at present working on a book on Galileo and the telescope.

Deborah Warner is a curator at the National Museum of American History and the editor of *Rittenhouse,* a quarterly journal devoted to the enterprise of instrument making in America.

Index